Statistical Tables for the Design of Clinical Trials

DAVID MACHIN & MICHAEL J CAMPBELL

Medical Statistics and Computing, University of Southampton

BLACKWELL SCIENTIFIC PUBLICATIONS

OXFORD LONDON EDINBURGH

BOSTON PALO ALTO MELBOURNE

© 1987 by Blackwell Scientific Publications
Editorial offices:
Osney Mead, Oxford, OX2 0EL
8 John Street, London, WC1N 2ES
9 Forrest Road, Edinburgh, EH1 2QH
52 Beacon Street, Boston, Massachusetts 02108, USA
744 Cowper Street, Palo Alto, California 94301, USA
107 Barry Street, Carlton, Victoria 3053, Australia

First published 1987

Set by Text Processing Ltd, Clonmel, Eire and printed and
bound in Great Britain by Billing & Sons Ltd,
Worcester

DISTRIBUTORS

USA
　Blackwell Mosby Book Distributors
　11830 Westline Industrial Drive
　St Louis, Missouri 63141

Canada
　Blackwell Mosby Book Distributors
　120 Melford Drive, Scarborough, Ontario M1B 2X4

Australia
　Blackwell Scientific Book Distributors
　31 Advantage Road, Highett, Victoria 3190

British Library
　Cataloguing in Publication Data

Machin, David
　Statistical tables for the design of clinical trials.
　1. Medical Statistics
　I. Title　II. Campbell, Michael J.
　312　RA407

ISBN 0-632-01275-7

Contents

Preface, vii

Statistical and Mathematical Symbols, ix

1 Basic Design Considerations, 1

2 The Normal Distribution, 5
Table 2.1 The normal distribution function, 8
Table 2.2 Percentage points of the normal distribution, z_γ, 9
Table 2.3 Values of $\theta(1 - \alpha, 1 - \beta) = (z_{1-\alpha} + z_{1-\beta})^2$, 9

3 Comparing Two Binomial Proportions, 10
Table 3.1 Sample sizes for comparison of proportions, 18
Table 3.2 Multiplying factor for comparing proportions when using a chi-squared test with continuity correction, 34

4 Testing Equivalence of Two Binomial Proportions, 35
Table 4.1 Sample sizes for testing equivalence of proportions, 38

5 Confidence Limits for a Binomial Proportion, 54
Table 5.1 Sample sizes for a given confidence interval, 58
Table 5.2 Multiplying factors for finite populations, 59

6 Post Marketing Surveillance, 60
Table 6.1 Sample sizes required to produce adverse reactions with given probability and incidence, 65
Table 6.2 Sample sizes for detection of an adverse reaction: background incidence known, 66
Table 6.3 Sample sizes for detection of a specified adverse reaction : background incidence unknown, 68
Table 6.4 Sample sizes for detection of 50 specified adverse reactions: background incidence unknown, 71
Table 6.5 Sample sizes for detection of 100 specified adverse reactions: background incidence unknown, 73
Table 6.6 Sample sizes for detection of adverse reactions: background incidence known, 75
Table 6.7 Sample sizes for detection of adverse reactions: background incidence known and allowing for examination of 100 types of reaction, 77

7 Comparing Two Means, 79
Table 7.1 Sample sizes for one sample t – test, 83
Table 7.2 Sample sizes for two sample t – test, 86

8 Significance of the Correlation Coefficient, 89
Table 8.1 Sample sizes for detecting a statistically significant correlation coefficient, 92

9 Comparing Two Survival Curves (Logrank Comparisons), 94
Table 9.1 Number of critical events for comparison of survival rates (logrank test), 98
Table 9.2 Number of subjects for comparison of survival rates (logrank test), 115
Table 9.3 Survival rate corresponding to a given ratio of median survival times, 131

10 Comparing Two Survival Curves (Exponential Survival), 132

    Table 10.1 Number of critical events for comparison of median survival times, 136
    Table 10.2 Duration of study required to compare two survival distributions, 139

11 Phase II Trials (Gehan's Method), 169

    Table 11.1 Probability of a given number of successive treatment failures, 172
    Table 11.2 Number of patients, n_1, required for the first stage of a Phase II trial, 173
    Table 11.3 Additional number of patients required for the second stage of a Phase II trial, 174

12 Phase II Trials (Fleming's Single Stage Procedure), 178

    Table 12.1 Sample sizes required for Fleming's single stage Phase II procedure, 181

13 Randomization, 197

    Table 13.1 10,000 Random digits, 200

14 Bibliography of other Statistical Tables and Texts, 203

Author Index, 208

Subject Index, 000

Preface

A good clinical trial design should be one that can answer the questions posed by an investigator using the minimum number of experimental units. The simplest design is essentially one in which two treatments or procedures are allocated to the subjects at random and an evaluation of the efficacy of that treatment assessed for each individual patient. The basic question that is then asked is: How many patients must be recruited to this particular clinical study?

It is well known that there is no simple answer to this question. It depends not only on the size of the effects ultimately observed in the two patient groups but also on the statistical requirements of the Type I and Type II errors (related to test size and power respectively). The subject numbers also depend on the particular type of measurement, for example observing a response as either success or failure is one type of measurement whereas the time between critical events is another. There has been much recent work on the particular problems associated with the latter types of measurement particularly with respect to survival studies.

Many sets of tables have been published to enable the experimenter to decide on appropriate patient numbers, although many such tables, especially those produced in journals, are often limited in scope because of space restrictions. In recent years there have been several new tables associated with survival studies in particular. The purpose of this volume is to bring together old and new tables in a format that will be of everyday use to clinicians and statisticians. We have also expanded some previously published tables to encompass a more useful range of values.

It can be argued that all clinical studies should be designed with the help of a medical statistician. We recognise, however, that this is neither practical nor entirely necessary in many situations. This book of tables, therefore, does not merely reproduce the tables that have been published elsewhere but describes each in such a way that a clinician can utilize the tables at the planning stage of his study. Our intention is to create a bench book which will find a place in many laboratories.

We have concentrated mainly on the problems of design in which there is a simple comparison between the two groups. We recognize that this does not cover every case for which sample size calculations need to be considered and a bibliography of tables for the comparison of more than two groups is included. We have made a personal judgement of which tables are in fact most commonly required and acknowledge that some research workers may be disappointed with the omissions. We have deliberately omitted those tables which are of particular use in analysis, for example chi-squared tables, tables of Student's t statistic etc. because they are readily available in many standard books of tables as well as in most medical statistics text books.

The introductory chapter is provided for completeness so that the tables can be used without reference to other texts. For example, if a clinician does only occasional trials, he may not wish to be constantly referred to other texts to discover what the power of a test is. In each chapter there is a section giving the formulae on which the table was based, so that the investigator can obtain more precise estimates for his own trial if he so wishes. One could ask, with computing equipment readily available, why not just supply the formulae and omit the tables? In justification for the tables we argue that a formula simply provides a single number and that in most practical situations an experimenter wants a range of numbers, which are more easily obtained from a table. Often he has an idea of the size of the trial, or in fact is forced by time and financial restrictions to conduct a trial of a certain size. With a table he can let his eye wander and discover what size differences he is likely to discover, and at what significance level and power. In a number of the tables we have rounded the numbers upwards to the nearest ten, hundred or even thousand to avoid spurious or unnecessary accuracy and to make them simpler to use.

The writing of this book has been stimulated by medical colleagues seeking our assistance in the design of their particular studies. Not only have they illustrated the need for such a collection but have provided us with the necessary stimulus and a wide range of interesting problems.

We would be grateful for any criticism and suggestions for improvements.

In preparing this text we are particularly grateful to Mrs Wendy M. Couper for typing the manuscript.

Southampton

David Machin
Michael J Campbell

Statistical and Mathematical Symbols

The appropriate chapters in which symbols are first defined are given in square brackets. Some symbols are used for more than one purpose

α	Test size, significance level or probability of Type I error [1]
α'	Modified test size to take into account multiple testing [6]
β	Probability of Type II error, $1 - \beta$ corresponds to the power [1]
γ	Area under the standardised normal probability distribution [2]. Similarly γ_1, γ_2 [2]
δ	A pre-specified difference between proportions [1]
$\triangle$	Shift of location between two alternative treatments [7]
	Ratio of median survival times of pre-specified interest [10]
ε	Pre-specified, usually small difference, between two proportions [4]
	Pre-specified width of $100(1 - \alpha)\%$ confidence interval [5]
	Precision [11]
$\theta(\gamma_1, \gamma_2)$	$(z_{\gamma_1} + z_{\gamma_2})^2$ [2]
λ	Expected incidence of adverse reactions [6]
	Appropriate multiples for power calculations for one or two sided t-test [7]
	Hazard function corresponding to an exponential survival distribution [10]
λ_0	Background incidence of adverse reaction [6]
λ_1	Additional incidence of adverse reaction caused by use of drug under study [6]
	Hazard rate corresponding to an exponential survival distribution in group 1 [10]. Similarly λ_2 [10]
$\overline{\lambda}$	Weighted average of background and additional incidence of an adverse reaction [6]
μ_0	Postulated mean of subjects from a particular group [7]. Similarly μ_1 [7]
v	Appropriate multiples for power calculations for one or two sided t-test [7]
π	Ratio of the circumference to the diameter of a circle, $\pi = 3.14159\ldots$ [2]
	Proportion of individuals having a particular attribute in a population. [11]
	True prevalence [5]
π_0	Largest response proportion in a Phase II trial which indicates that treatment does not warrant fuller investigation [12]
π_a	True proportion of successes in the acupuncture group [1]
	Smallest response proportion in a Phase II trial which indicates treatment warrants fuller investigation [12]
π_c	True proportion of successes in the placebo group [1]
π_L	Lower $100(1 - \alpha)\%$ confidence limit of a prevalence [5]
π_U	Upper $100(1 - \alpha)\%$ confidence limit of a prevalence [5]
π_1	Proportion of successes in group 1 [3]. Similarly π_2 [3]
$\overline{\pi}$	Weighted average of π_1 and π_2 [3]
π_1'	Nearest tabulated proportion of successes in group 1 [3] Similarly π_1'', π_2', π_2'' [3]
ρ	Population correlation coefficient. [8]
σ	Population standard deviation. [7]
τ	Median survival time [9]

ix

τ_1	Median survival time of subjects in group 1 [9] Similarly τ_2 [9]
$\phi(t)$	Probability density function of the standardised normal distribution [2]
$\Phi(z_\gamma)$	Area under the standardised normal curve to the left of z_γ [2]
χ^2	Chi-squared [3]
ω_1	Suspected ratio, under the null hypothesis, of the proportion of matched pairs with a success on treatment 1 and a failure on treatment 2 to the total number of pairs with only one success [3]
ω_2	As ω_1 but under the alternative hypothesis [3]
a	Number of successes in a particular treatment group [3]
	Number of subjects experiencing adverse reactions [6]
	Patient entry rate per unit time to a clinical trial [10]
b	Number of failures in a particular treatment group [3]
c	Number of successes in a particular treatment group [3]
d	The observed difference between proportions [1]
	Number of failures in a particular treatment group [3]
d_t	Postulated standardised difference between two means [7]
e	The base of natural logarithms e = 2.71828 ... [2]
	Number of subjects with successes under both treatments [3]
	Number of events to be observed [9]
f	Number of subjects with success with treatment 2 and failure with treatment 1 [3]
	Divisor required for non parametric sample size calculations [7]
	Expected ratio of between to within groups standard deviations [14]
F	The variance ratio [14]
$F_1(x)$	Probability of a response in group 1 being less than x [7]. Similarly $F_2(x)$ [7]
g	Number of subjects with failure with treatment 2 but success with treatment 1 [3]
h	Number of subjects with failure with both treatments [3]
	The relative hazard of two treatment groups [9]
k	The number of contradictory pairs [3]
	Number of subject groups [14]
log	Logarithms to the base e [8]
m	Number of subjects recruited to the smaller of two treatment groups [3]
	Median survival time [10]
m'	Nearest tabulated value of sample size for a particular test size and power [3]. Similarly m'' [3]
	Number of subjects recruited to the smaller of two treatment groups allowing for withdrawals [9]
m_c	Number of subjects to be recruited to the smaller of the treatment groups when Yates' correction to be used [3]
n	Number of subjects recruited to the larger of two treatment groups [3]
	Number of subjects recruited to a prevalence study [5]
	Number of subjects recruited to each treatment group [10]
n_1	Number of patients to be recruited to the first stage of a Phase II trial [11]
n_2	Number of subjects to be recruited to the second stage of a Phase II trial [11]
n'	Number of subjects recruited to a prevalence study in a small population [5]
	Number of subjects recruited to each treatment group allowing for withdrawals [10]
N	Total number of subjects recruited to a study [3]
	Total population size [5]
	Number of subjects to be recruited to a Phase II trial [11]

p	Observed prevalence [5]
	Estimate of the efficacy of a drug after a Phase II trial [11]
p_1	Observed proportion of successes in first stage of a Phase II trial [11]
p_a	Observed proportion of successes in the acupuncture group [1]
p_c	Observed proportion of successes in the control group [1]
P	Probability [1]
r	Ratio of the number of subjects in two treatment groups [3]
	Number of cases found in a prevalence study [5]
	Number of successive patient failures on a Phase II drug [11]
R	Odds ratio [3]
s	Number of adverse reactions being simultaneously monitored [6]
	Standard deviation [7]
$S(t)$	Survivorship function [10]
t	Student's t [7]
	Survival time [10]
T	Time period of patient entry to a study [10]
T_L	Lower limit for estimate of required time period of patient entry to a study [10]
u_ρ	Standardised normal random variable corresponding to a particular value of the correlation coefficient ρ [8]
u'_ρ	Initial estimate of u_ρ [8]
x	Particular value of a response [7]
	Anticipated withdrawal rate from a particular study [9]
y	Particular value of a response [8]
$z_{1-\alpha}$	Value of a standardised normal deviate corresponding to an area of $1-\alpha$ [2]. Similarly $z_{1-\beta}$, z_γ, z_{γ_1}, z_{γ_2}, [2]

Chapter 1
Basic Design Considerations

1.1 INTRODUCTION

This chapter reviews some of the basic considerations for planning any clinical investigation. Concepts such as test size, one or two-tailed tests, power, the null hypothesis and statistical significance are discussed. The need for randomized allocation of subjects to treatments is emphasised. For purposes of illustration it is assumed that a prospective, randomized, two arm clinical trial is planned. It should be emphasized, however, that the tables included in this book are just as appropriate for the design of epidemiological and laboratory based studies.

1.2 DESIGN

In designing any clinical trial it is necessary to have firm objectives in view. Thus, it is preferable to have clearly different and well defined alternative treatments, clear patient entry criteria and unambiguous definitions of end points. All subjects entered into the trial and randomized should be followed up in the same manner, irrespective of whether or not the treatment continues for that individual subject. Since treatment advantages may be small, one should choose treatments to be as different as possible at the planning stage. If real but small differences are to be demonstrated then large numbers of patients need to be recruited. Thus, trials comparing two treatments are often preferable to trials involving three or more treatment comparisons.

Subjects should be allocated at random to the alternative treatments and blind assessment of their progress should, if possible, be made. The double blind procedure in which both the doctor and patient are ignorant of the actual treatment is to be desired.

A clear knowledge and understanding of the natural history of the particular disease under study is essential before conducting any trial. It is usually preferable to deal with a single disease entity or diagnosis.

1.3 END POINTS

As indicated above, specific end points for a trial must be defined at the planning stage. Multiple end points should be avoided if possible. Definitions of end points can be difficult, particularly for example, in trials to assess pain relief. It is beyond the scope of this text to discuss appropriate assessment systems and end point definitions. Nevertheless, these are worthy of detailed consideration. Whatever definitions are used they must be framed in such a way that they can be unambiguously evaluated for each subject. It is important to realise that the efficacy of a particular treatment may inhibit the calculation of the end points. For example, if a particular treatment is not effective then a patient may not return for the visit at which the intended evaluation was to take place. Methods of incorporating such patients into the evaluation should be included.

1.4 SAMPLE SIZE

The number of patients necessary to recruit to a particular trial depends on several factors. These factors are the anticipated clinical differences between the alternative treatments, the level of statistical significance the investigator considers appropriate and the chance of detecting that difference. For purposes of illustration we will consider a trial of a placebo (control) against acupuncture in the relief of pain in a particular diagnosis (Lewith & Machin 1983). The patients are randomized to receive either placebo or acupuncture (we omit details of how a placebo acupuncture can be arrived at but this is clearly an important consideration). In addition we assume that pain relief is assessed at a fixed time after randomization and is defined in such a way as to be unambiguously evaluable for each patient at that time. In our randomized control trial we assume the aim is to estimate the true difference δ between the true success rate π_a of the real acupuncture under study and the true success rate π_c of the control. The acupuncture group of patients yield a treatment success rate p_a which is an estimate of π_a and the control group give a success rate p_c which is an estimate of π_c. Thus, $d = p_a - p_c$, the observed difference is an estimate of the true difference δ.

1.5 THE NULL HYPOTHESIS

The Null Hypothesis that acupuncture and placebo are equally effective implies that $\pi_a = \pi_c$. Even when that is true observed differences $d = p_a - p_c$ other than zero will occur. The probability, P, of obtaining the observed difference d or a more extreme one given that $\pi_a = \pi_c$ can be calculated. if under the null hypothesis the difference actually observed, d or a more extreme one, has a very small probability of occurring by chance, then we would reject this null hypothesis and conclude the two treatments do indeed differ in efficacy. We have to decide, however, at what level of probability we would reject this null hypothesis and hence conclude the two treatments do indeed differ in efficacy.

1.6 TEST SIZE, SIGNIFICANCE LEVEL OR TYPE I ERROR

The critical value we take for the probability P is arbitrary, and we denote it by α. If $P \leq \alpha$ one rejects the null hypothesis, conversely if $P > \alpha$ one does not reject the null hypothesis. For example $\alpha = 0.05$ could be specified as indicating statistically significant differences between treatments. However, even when the null hypothesis is in fact true there is a risk of rejecting it. Clearly the probability of rejecting the null hypothesis when it is true is α. The quantity α can be referred to as either the test size, significance level or probability of a Type I error.

1.7 TYPE II ERROR

The clinical trial could yield an observed difference d that would lead to $P > \alpha$ even though the null hypothesis is really not true (i.e. π_a is indeed not equal to π_c). We then make the error of not rejecting the null hypothesis when it is in fact false. The false negative error is called a Type II error and its probability is denoted by β.

Power

The probability of a Type I error is calculated on the assumption that the null hypothesis is true, that is $\delta = \pi_a - \pi_c = 0$. The probability of a Type II error is based on the assumption that the null hypothesis is not true, that is $\delta = \pi_a - \pi_c \neq 0$. There are clearly many such values of δ in this instance and each would give a difference value for β. The power is defined as one minus the probability of a Type II error. Therefore, if the probability of a Type II error is 0.2, the power $1 - \beta = 0.8$ or 80 per cent.

1.8 ONE SIDED AND TWO SIDED TESTS

It is plausible to assume in the acupuncture trial that the placebo is in some sense 'inactive' and that any 'active' treatment will have to perform better than the 'inactive' treatment if it is to be adopted into clinical practice. Thus, the alternative hypothesis may be that the acupuncture has an improved success rate i.e. $\pi_a > \pi_c$. This leads to a one-sided or one-tailed test. On the other hand if we cannot make any assumption about the new treatment, then the alternative hypothesis is that π_a and π_c differ, i.e. $\pi_a \neq \pi_c$. In general, a one-sided test is more powerful than the corresponding two-sided test. However, a decision to use a one-sided test should never be made after looking at the data and observing the direction of the departure. Such decisions should be made at the design stage and one should use a one-sided test only if it is quite certain that departures in one particular direction will always be ascribed to chance, and therefore regarded as non-significant however large they are. Armitage (1971) argues that this situation rarely arises in practice.

1.9 TESTING FOR EQUIVALENCE

The acupuncture trial compared an active treatment versus a placebo, and one anticipates (if anything) an improvement of pain relief associated with active acupuncture treatment. In certain circumstances, however, it could be important to demonstrate the equivalence of an alternative treatment. Suppose a multicompound chemotherapy adjuvant treatment for breast cancer is very toxic and that the toxicity can be identified with one drug of the combination. Then if one could obtain the same or similar efficacy with a chemotherapy combination without this drug this would clearly be desirable even though life expectancy under this regimen may not in fact improve. Makuch & Simon (1979) propose that testing for equivalence is appropriate in such a situation.

1.10 SIGNIFICANCE TESTS AND CONFIDENCE INTERVALS

Many medical statisticians, for example Altman & Gore (1982), have suggested that there is an over emphasis on tests of significance in the reporting of trial results and they argue that wherever possible confidence intervals should be quoted. A P value alone gives the reader who wishes to make use of the published results of a particular study little information, whereas an estimate of the effect together with the corresponding confidence limits enables him to judge better the efficacy or otherwise of the alternative treatments. For the purposes of these tables, and in the planning stages of a trial, discussion is easier in terms of statistical significance but nevertheless it should be emphasized that confidence limits should be quoted in the final report. An alternative approach is described by McHugh and Le (1984).

1.11 ETHICAL CONSIDERATIONS

Altman & Gore (1982) have discussed statistical and ethical problems in the conduct of medical research and they argue that the misuse of statistics is unethical, emphasize ethical aspects in study design and discuss sample size requirements. They point out that it may be unethical to conduct a trial of inadequate power.

1.12 RANDOMIZATION

Of fundamental importance to the design of any clinical trial is the random allocation of subjects to the alternative treatments. Such allocation safeguards in particular against selection bias and is the necessary basis for the subsequent statistical tests. Discussion of this important topic is given by Altman & Gore (1982) and by Pocock (1983).

1.13 USE OF THE TABLES

It is hoped that these tables will prove useful in a number of ways, in particular, in those three listed below.

1. Number of Subjects

Before conducting the acupuncture clinical trial a researcher believes that the placebo group will yield a positive response rate of 30%. How many subjects need he test, if he wants to show a positive response rate for acupuncture of 70% at a given significance level and power?

2. Power of Test

A common situation is one where the number of patients is governed by other forces such as time, money, manpower, disease prevalence and disease incidence rather than purely scientific criteria. The researcher may then wish to know what probability he has of detecting a certain difference in treatment efficacy with a trial of his intended size.

3. Confidence Intervals

In this case the experiment has been completed and a non-significant result observed. The question then is what range of response rates for placebo and acupuncture are consistent with this result? The tables can be used to determine appropriate confidence intervals for differences between treatments. This is particularly relevant, for example, when surveying the literature of different clinical trials of a new drug.

In addition there is a chapter on randomization and a bibliography of other statistical tables and texts.

1.14 GENERAL STRUCTURE OF SUBSEQUENT CHAPTERS

Introduction
Theory and Formulae
Bibliography
Description of Tables
Use of Tables
References
Tables

1.15 REFERENCES

Altman D.G. & Gore S.M. (1982) *Statistics in Practice*. British Medical Association, London.
Armitage P. (1971) *Statistical Methods in Medical Research*. Blackwell Scientific Publications, Oxford.
Lewith G.T. & Machin D. (1983) On the evaluation of the clinical effects of acupuncture. *Pain*, **16**, 111–127.
Makuch R. & Simon R. (1978) Sample size requirements for evaluating a conservative therapy. *Cancer Treatment Reports*, **62**, 1037–1040.
McHugh R.B. & Le C.T. (1984) Confidence estimation and the size of a clinical trial. *Controlled Clinical Trials*, **5**, 157-163.
Pocock S.J. (1983) *Clinical Trials: A Practical Approach*. Wiley, Chichester.

Chapter 2
The Normal Distribution

2.1 INTRODUCTION

The normal distribution plays a central role in statistical theory and frequency distributions resembling the normal distribution are often observed in practice.

2.2 THEORY AND FORMULAE

Of particular importance is the normal distribution with mean 0 and standard deviation 1. The probability density function $\phi(t)$ of such a normally distributed random variable t is given by

$$\phi(t) = \frac{1}{\sqrt{(2\pi)}} \exp(-\tfrac{1}{2}t^2). \tag{2.1}$$

The curve (2.1) is shown in Fig 2.1

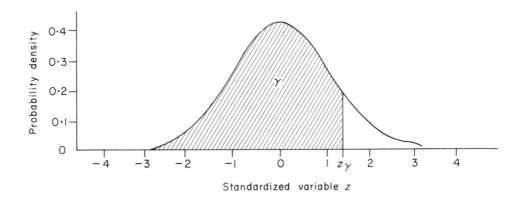

Fig 2.1 The probability density function of a standardized normal distribution

For many purposes in this book we shall need to calculate the area under any part of this normal distribution. To do this use is made of the symmetrical nature of the distribution and the fact that the total area under a probability density function is unity.

We can find any particular area if we tabulate values of z_γ corresponding to the shaded areas in Fig 2.1 which has area γ ($\gamma \geqslant 0.5$). For $\gamma < 0.5$ we can clearly use the symmetry of the distribution and calculate the unshaded area.

More formally we have to solve the following equation for z_γ

$$\gamma = \int_{-\infty}^{z_\gamma} \phi(t)dt \tag{2.2}$$

the right hand side of which is often denoted $\Phi(z_\gamma)$.

For example if $\gamma = 0.5$, then one can see from Fig 2.1 that $z_\gamma = z_{0.5} = 0$.

It is also useful to be able to find the value of γ for a given value of z_γ, for example if $z_\gamma = 1.96$ it turns out that $\gamma = 0.975$.

Of particular use for many calculations of sample sizes for clinical trials is the tabulation of $\theta(\gamma_1, \gamma_2)$ where

$$\theta(\gamma_1, \gamma_2) = (z_{\gamma_1} + z_{\gamma_2})^2 \qquad (2.3)$$

and γ_1 and γ_2 are two relevant areas under the standardized normal curve. In the situations discussed in this book $\gamma_1 = 1 - \alpha$ and $\gamma_2 = 1 - \beta$, where α and $1 - \beta$ are the appropriate test size and power.

2.3 BIBLIOGRAPHY

Tables corresponding to equation (2.2) are usually reproduced in standard statistical texts such as Armitage (1971) and Bailey (1981) who also discuss the importance of the normal distribution in detail. Specialized tables are also available, for example Fisher & Yates (1963) and Lindley & Scott (1984). Tables corresponding to equation (2.3) are given by Pocock (1983).

2.4 DESCRIPTION OF THE TABLES

Table 2.1

The function tabulated is $\gamma = \phi(z_\gamma)$ which is the probability that a normally distributed random variable, with zero mean and standard deviation 1, will be less than z_γ.

Table 2.2

The table gives the percentage points z_γ corresponding to the area under the curve being γ ($\gamma \geq 0.5$).

Table 2.3

The table gives the values of $\theta(1 - \alpha, 1 - \beta) = (z_{1-\alpha} + z_{1-\beta})^2$ for values of α and β often occurring in practice, $\theta(1 - \alpha, 1 - \beta)$ is tabulated for both one and two sided values of α.

2.5 USE OF THE TABLES

Table 2.1

Example 2.1

In calculating the power of a trial comparing two treatments, an investigator had used equation (3.6) to obtain $z_{1-\beta} = 1.05$, and he would like to know the corresponding power $1 - \beta$.

In the terminology of Table 2.1, the investigator needs to find γ for $z_\gamma = 1.05$. Direct entry into the table with $z_\gamma = 1.05$ gives the corresponding $\gamma = 0.8531$. Thus the power of the test would be approximately $1 - \beta = 0.85$ or 85%.

Table 2.2

Example 2.2

At the planning stage of a randomized trial an investigator is considering using a test size α (two-tailed) of 0.05 and a power $1 - \beta = 0.8$. What are the values of $z_{1-\alpha}$ and $z_{1-\beta}$ that he needs for his calculations? At a later

2.3–2.5 6

stage in the planning he is led to believe that a one-sided test would be more appropriate, how does this effect his calculations?

For a two-tailed test at test size α one requires a probability of $\alpha/2$ in each tail of the corresponding standardized normal distribution. The investigator thus requires to find $z_\gamma = z_{1-\alpha/2}$ for $\alpha = 0.05$. He therefore enters Table 2.2 with $\gamma = 1 - 0.025 = 0.975$, the corresponding value of $z_{0.975} = 1.9600$.

To find $z_{1-\beta}$ for $1 - \beta = 0.80$ he enters the table directly to find $z_{0.80} = 0.8416$.

For a one-tailed test the investigator would require $z_{0.95} = 1.6449$.

Table 2.3

Example 2.3

What values of $\theta(1 - \alpha, 1 - \beta)$ would the investigator described in Example 2.2 require?

Two-sided test: $\alpha = 0.975$, $1 - \beta = 0.80$, direct entry into Table 2.3 gives $\theta(0.975, 0.80) = 7.849$.

One-sided test: $\alpha = 0.05$, $1 - \beta = 0.80$ and $\theta(0.95, 0.80) = 6.183$.

2.6 REFERENCES

Armitage, P. (1971) *Statistical Methods in Medical Research*, Blackwell Scientific Publications, Oxford.

Bailey N.T.J. (1981) *Statistical Methods in Biology*, 2nd ed., Hodder & Stoughton, London.

Fisher R.A. & Yates F. (1963). *Statistical Tables for Biological, Agricultural and Medical Research*, 6th ed. Oliver & Boyd, Edinburgh.

Lindley D.V. & Scott W.F. (1984) *New Cambridge Elementary Statistical Tables*. Cambridge University Press, Cambridge.

Pocock S.J. (1983) *Clinical Trials: A Practical Approach*. Wiley, Chichester.

Table 2.1 The normal distribution function. Probability that a normally distributed variable is less than z.

z	0.00	0.01	0.02	0.03	0.04	0.05	0.06	0.07	0.08	0.09
0.0	0.50000	0.50399	0.50798	0.51197	0.51595	0.51994	0.52392	0.52790	0.53188	0.53586
0.1	0.53983	0.54380	0.54776	0.55172	0.55567	0.55962	0.56356	0.56749	0.57142	0.57535
0.2	0.57926	0.58317	0.58706	0.59095	0.59483	0.59871	0.60257	0.60642	0.61026	0.61409
0.3	0.61791	0.62172	0.62552	0.62930	0.63307	0.63683	0.64058	0.64431	0.64803	0.65173
0.4	0.65542	0.65910	0.66276	0.66640	0.67003	0.67374	0.67724	0.68082	0.68439	0.68793
0.5	0.69146	0.69497	0.69847	0.70194	0.70540	0.70884	0.71226	0.71566	0.71904	0.72240
0.6	0.72575	0.72907	0.73237	0.73565	0.73891	0.74215	0.74537	0.74857	0.75175	0.75490
0.7	0.75804	0.76115	0.76424	0.76730	0.77035	0.77337	0.77637	0.77935	0.78230	0.78524
0.8	0.78814	0.79103	0.79389	0.79673	0.79955	0.80234	0.80511	0.80785	0.81057	0.81327
0.9	0.81594	0.81859	0.82121	0.82381	0.82639	0.82894	0.83147	0.83398	0.83646	0.83891
1.0	0.84134	0.84375	0.84614	0.84849	0.85083	0.85314	0.85543	0.85769	0.85993	0.86214
1.1	0.86433	0.86650	0.86864	0.87076	0.87286	0.87493	0.87698	0.87900	0.88100	0.88298
1.2	0.88493	0.88686	0.88877	0.89065	0.89251	0.89435	0.89617	0.89796	0.89973	0.90147
1.3	0.90320	0.90490	0.90658	0.90824	0.90988	0.91149	0.91309	0.91466	0.91621	0.91774
1.4	0.91924	0.92073	0.92220	0.92364	0.92507	0.92647	0.92785	0.92922	0.93056	0.93189
1.5	0.93319	0.93448	0.93574	0.93699	0.93822	0.93943	0.94062	0.94179	0.94295	0.94408
1.6	0.94520	0.94630	0.94738	0.94845	0.94950	0.95053	0.95154	0.95254	0.95352	0.95449
1.7	0.95543	0.95637	0.95728	0.95818	0.95907	0.95994	0.96080	0.96164	0.96246	0.96327
1.8	0.96407	0.96485	0.96562	0.96638	0.96712	0.96784	0.96856	0.96926	0.96995	0.97062
1.9	0.97128	0.97193	0.97257	0.97320	0.97381	0.97441	0.97500	0.97558	0.97615	0.97670
2.0	0.97725	0.97778	0.97831	0.97882	0.97932	0.97982	0.98030	0.98077	0.98124	0.98169
2.1	0.98214	0.98257	0.98300	0.98341	0.98382	0.98422	0.98461	0.98500	0.98537	0.98574
2.2	0.98610	0.98645	0.98679	0.98713	0.98745	0.98778	0.98809	0.98840	0.98870	0.98899
2.3	0.98928	0.98956	0.98983	0.99010	0.99036	0.99061	0.99086	0.99111	0.99134	0.99158
2.4	0.99180	0.99202	0.99224	0.99245	0.99266	0.99286	0.99305	0.99324	0.99343	0.99361
2.5	0.99379	0.99396	0.99413	0.99430	0.99446	0.99461	0.99477	0.99492	0.99506	0.99520
2.6	0.99534	0.99547	0.99560	0.99573	0.99585	0.99598	0.99609	0.99621	0.99632	0.99643
2.7	0.99653	0.99664	0.99674	0.99683	0.99693	0.99702	0.99711	0.99720	0.99728	0.99736
2.8	0.99744	0.99752	0.99760	0.99767	0.99774	0.99781	0.99788	0.99795	0.99801	0.99807
2.9	0.99813	0.99819	0.99825	0.99831	0.99836	0.99841	0.99846	0.99851	0.99856	0.99861
3.0	0.99865	0.99869	0.99874	0.99878	0.99882	0.99886	0.99889	0.99893	0.99896	0.99900
3.1	0.99903	0.99906	0.99910	0.99913	0.99916	0.99918	0.99921	0.99924	0.99926	0.99929
3.2	0.99931	0.99934	0.99936	0.99938	0.99940	0.99942	0.99944	0.99946	0.99948	0.99950
3.3	0.99952	0.99953	0.99955	0.99957	0.99958	0.99960	0.99961	0.99962	0.99964	0.99965
3.4	0.99966	0.99968	0.99969	0.99970	0.99971	0.99972	0.99973	0.99974	0.99975	0.99976
3.5	0.99977	0.99978	0.99978	0.99979	0.99980	0.99981	0.99982	0.99983	0.99983	
3.6	0.99984	0.99985	0.99985	0.99986	0.99986	0.99987	0.99988	0.99988	0.99989	
3.7	0.99989	0.99990	0.99990	0.99990	0.99991	0.99991	0.99992	0.99992	0.99992	0.99992
3.8	0.99993	0.99993	0.99993	0.99994	0.99994	0.99994	0.99994	0.99995	0.99995	0.99995
3.9	0.99995	0.99995	0.99996	0.99996	0.99996	0.99996	0.99996	0.99996	0.99997	0.99997

Table 2.1   8

Table 2.2 Percentage points of the normal distribution.

α		z
2-sided	1-sided	
0.0010	0.0005	3.2905
0.0050	0.0025	2.8070
0.0100	0.0050	2.5758
0.0200	0.0100	2.3263
0.0250	0.0125	2.2414
0.0500	0.0250	1.9600
0.1000	0.0500	1.6449
0.2000	0.1000	1.2816
0.3000	0.1500	1.0364
0.4000	0.2000	0.8416
0.5000	0.2500	0.6745
0.6000	0.3000	0.5244
0.7000	0.3500	0.3853
0.8000	0.4000	0.2533

Table 2.3 Values of $0(1 - \alpha, 1 - \beta)$.

α		Power $1-\beta$										
2-sided	1-sided	0.30	0.40	0.50	0.65	0.70	0.75	0.80	0.85	0.90	0.95	0.99
0.0010	0.0005	7.651	9.224	10.828	13.512	14.554	15.721	17.075	18.723	20.904	24.358	31.549
0.0050	0.0025	5.210	6.521	7.879	10.191	11.098	12.121	13.313	14.772	16.717	19.819	26.352
0.0100	0.0050	4.208	5.394	6.635	8.768	9.611	10.565	11.679	13.048	14.879	17.814	24.031
0.0200	0.0100	3.247	4.297	5.412	7.353	8.127	9.005	10.036	11.308	13.017	15.770	21.648
0.0250	0.0125	2.948	3.952	5.024	6.900	7.650	8.502	9.505	10.744	12.411	15.103	20.864
0.0500	0.0250	2.061	2.913	3.841	5.500	6.172	6.940	7.849	8.978	10.507	12.995	18.372
0.1000	0.0500	1.255	1.936	2.706	4.122	4.706	5.379	6.183	7.189	8.564	10.822	15.770
0.2000	0.1000	0.573	1.057	1.642	2.778	3.261	3.826	4.508	5.373	6.569	8.564	13.017

Table 2.2–Table 2.3

Chapter 3
Comparing Two Binomial Proportions

3.1 INTRODUCTION

A common method of summarizing data from clinical trials is to examine events that have occurred during a fixed period after treatment. For example, to compare two methods of treating cancer, the proportion of subjects surviving in each group for 5 years is often quoted. Discussion of the appropriate length of follow-up is not within the scope of this book, but clearly it should not be so short that only a few deaths have occurred, nor so long that practically all the subjects have died.

Data from this type of experiment can be summarised in a 2×2 table as shown in Fig. 3.1 in which N patients are assigned at random to one of the treatments: m to treatment 1 and n to treatment 2. Any analysis of the data compares the proportion of successes (or failures) in the two treatment groups.

Fig. 3.1: Notation for 2×2 contingency table (unpaired case)

Treatment	'Success'	'Failure'	Total	Estimated proportion of sucess
1	a	b	m	a/m
2	c	d	n = rm	c/n
Total	a + c	b + d	N	

The traditional tests for comparing proportions are Fisher's exact test, and the χ^2 test with Yates' continuity correction (Armitage 1971, p131). Berkson (1978) and Upton (1982) have claimed that these tests are less powerful, and so have less chance of detecting a given difference in proportions, than the χ^2 test without the continuity correction. The derivation of the exact test assumes that subtotals m, n, $a+c$, $b+d$ of Fig. 3.1 are fixed and in a clinical trial, before it is carried out, the number of successes, a+c is a random variable. This led Pocock (1982) to argue that the χ^2 test without continuity correction was appropriate. However, Yates (1984) has demonstrated that the procedure of fixing the number of subjects in advance, and then randomly assigning treatments to subjects leads to the exact test, or the continuity corrected χ^2 test. The choice of the appropriate test is not merely an academic exercise, since in some cases the sample size required to detect a small difference in proportions with a given significance level and power can be considerably larger if the χ^2 test with continuity correction is employed in the analysis than if the χ^2 test without continuity correction is used. Clearly one should use the same test for the planning as for the analysis.

In cross-over trials, usually for the treatment of chronic conditions, the subjects have both treatments. In this case the results from one treatment are unlikely to be independent of the results of the other on the same subject. In other trials the subjects on the two treatments may be closely matched, e.g. for age, disease severity, blood group etc. The usual test of significance in these cases is McNemar's and we rewrite Fig. 3.1 as Fig. 3.2.

Fig. 3.2: Notation for 2×2 contingency table (matched or paired case)

		Treatment 1		
		'Success'	'Failure'	Total
Treatment 2	'Success'	e	f	c
	'Failure'	g	h	d
	Total	a	b	m

The null hypothesis is that the number of contradictory pairs, i.e. success on one treatment but failure on the other is the same for each treatment. Thus the test is whether f to g or equivalently $f/(f + g)$ to 0.5.

3.2 THEORY AND FORMULAE

Sample size

At the design stage we have the option either to randomize equally to the two alternative treatments or to randomize with unequal allocation. In the former case $n = m$, in the latter we take $n = rm$ $(r > 1)$ and we have to choose the allocation ratio r. Suppose we wish to detect a difference in proportions $\delta = \pi_2 - \pi_1$, where $\pi_2 > \pi_1$, with significance level α, power $1 - \beta$ and denote the weighted average $(\pi_1 + r\pi_2)/(r + 1)$ by $\bar{\pi}$.

Without correction for continuity

The required sample size m, without correction for continuity, is given by

$$m = \frac{[z_{1-\alpha}\sqrt{\{(r+1)\bar{\pi}(1 - \bar{\pi})\}} + z_{1-\beta}\sqrt{\{r\pi_1(1 - \pi_1) + \pi_2(1 - \pi_2)\}}]^2}{r\delta^2}.$$ (3.1)

For the special case of equal allocation $r = 1$ and (3.1) becomes

$$m = \frac{[z_{1-\alpha}\sqrt{\{(2\bar{\pi}(1 - \bar{\pi}))\}} + z_{1-\beta}\sqrt{\{\pi_1(1 - \pi_1) + \pi_2(1 - \pi_2)\}}]^2}{\delta^2}.$$ (3.2)

It can be shown that (3.2) can be approximated by

$$m = \frac{(z_{1-\alpha} + z_{1-\beta})^2\{\pi_1(1 - \pi_1) + \pi_2(1 - \pi_2)\}}{\delta^2},$$ (3.3)

the accuracy of which is unaffected by the values of π_1 and π_2. The advantage of equation (3.3) is that it can be easily calculated using Table 2.3 which gives $(z_{1-\alpha} + z_{1-\beta})^2$ for different values of α and β.

With continuity correction

When using the χ^2 test with continuity correction the required sample size is:

$$m_c = \frac{m}{4}\left[1 + \sqrt{\{1 + \frac{2(r + 1)}{rm\delta}\}}\right]^2$$ (3.4)

where m is given by equation (3.1).

Power

In contrast if one were asked for the power of a trial for a given π_1, π_2, δ, m and r, then if a correction for continuity is not involved

$$z_{1-\beta} = \frac{\delta\sqrt{m} - z_{1-\alpha}\sqrt{\{(r + 1)\pi(1 - \pi)\}}}{\sqrt{\{r\pi_1) + \pi_2(1 - \pi_2)\}}}.$$ (3.5)

If a correction for continuity is involved, then

$$z_{1-\beta} = \frac{\sqrt{[\{m_c - v + v^2/(4m_c)\} r \delta^2]} - z_{1-\alpha}\sqrt{\{(r+1)\pi(1-\pi)\}}}{\{r\pi_1(1-\pi_1) + \pi_2(1-\pi_2)\}} \qquad (3.60)$$

where $v = (r+1)/\delta r$. Table 2.2 can then be used to obtain the power $1 - \beta$ from $z_{1-\beta}$.

Allocation of subjects to treatments

Most trials allocate equal numbers of subjects to the two competing treatments. Equation (3.5) enables us to investigate the power of tests which provide unequal allocation for a given total sample size. When $N = m + n$ is very large it can be shown that choosing $r = [\pi_2(1-\pi_2)/\{\pi_1(1-\pi_1)\}]^{\frac{1}{2}}$ in fact maximizes the possible power. This is because the two estimated proportions have variances proportional to $\pi_1(1-\pi_1)$ and $\pi_2(1-\pi_2)$ respectively and one would weight the subjects inversely in proportion to their variances. That is, one would choose more subjects from the treatment whose expected success rate is nearer 0.5 as that treatment would have the higher variance. That this is not true for all N has been demonstrated by Ury & Fleiss (1980). For $1 < r \leqslant 2$ the increased power obtained by unequal allocation is very small, and is obtained at a considerably increased complexity at the randomization stage of the trial. In most practical situations π_1 and π_2 are not going to differ so much as to make unequal allocation worthwhile.

Interpolation

For values of π_1 and π_2 not given by the tables a quick approximation to the appropriate sample size is calculated as follows. Let m' be the number of subjects required for a significance level α and power $1 - \beta$ to test π_1 against π_2, and π_1' and π_2' and the nearest tabulated proportions. Then the required number of subjects in each arm for comparison of π_1, π_2 is given approximately by

$$m = \frac{m'(\pi_1' - \pi_2')^2}{(\pi_1 - \pi_2)^2} \qquad (3.7)$$

for a test without correction for continuity.

Suppose π_1 is close to a tabulated value π_1', but that π_2 falls midway between two tabulated values, π_2' and π_2''. Two possible estimates of the required number of subjects can thus be obtained from equation (3.7) yielding m' and m'' subjects respectively. A better estimate may be obtained by the average of these two.

One proportion known

In some situations one may in fact know, with a fair degree of certainty, the proportion of successes under one of the treatments. For example, a large number of very similar clinical trials may have been conducted with a particular drug, showing that the success rate is about 20%. In a clinical trial to test a new product under identical conditions, it may not seem worthwhile treating patients with the old drug. Suppose the new drug is expected to produce a success rate of π_2. Then the required number of subjects, for a significance level α and power $1 - \beta$ without correction for continuity, comparing π_2 with the established success rate π_1 is?

$$m = \frac{[z_{1-\alpha}\sqrt{\{\pi_1(1-\pi_1)\}} + z_{1-\beta}\sqrt{\{\pi_2(1-\pi_2)\}}]^2}{\delta^2}. \qquad (3.8)$$

When π_1 and π_2 are close, equation (3.8) yields a required number of subjects which is roughly half that given by equation (3.3). Now since we do not have to allocate any subjects to the established treatment this implies that we have reduced the total number of subjects to 25% of that required for a trial in which both proportions are to be estimated. However, unlike (3.3) equation (3.8) it is not symmetrical in π_1 and π_2 unless $z_{1-\alpha} = z_{1-\beta}$. In most situations we can halve the entries in the tables corresponding to (3.3) and ignore the lack of symmetry.

It shold be noted that equation (3.8) arises in the context of Phase II clinical trials as equation (12.1) which is tabulated in Table 12.1. Thus use of Table 12.1 would also give sample sizes corresponding to (3.8) when $\pi_2 > \pi_1$.

Odds ratio

If the probability of success under treatment 1 is π_1 the odds associated with it are $\pi_1 / (1 - \pi_1)$. Similarly, the odds associated with success under treatment 2 are $\pi_2 / (1 - \pi_2)$. In some situations it may be difficult to propose a value for δ, the increased proportion of successes under treatment 2 that a trial is designed to detect. Given the value of π_1, however, the odds ratio $R = \pi_2 (1 - \pi_1) / \{ \pi_1 (1 - \pi_2) \}$ can often be proposed. The value of π_2 is then obtained from

$$\pi_2 = \frac{R \pi_1}{R \pi_1 + (1 - \pi_1)} .$$
(3.9)

Thus, rather than pre-specifying π_1 and π_2, an investigator may pre-specify π_1 and R, and use (3.9) to obtain π_2 for entry into the tables.

Sample Size : Matched or Paired Case

In this case we need to specify not the proportions of successes under treatments 1 and 2, but the proportion of pairs with a success on treatment 1 and failure on treatment 2 to the number of pairs with only one success overall.

Under the null hypothesis, $f / (f + g)$ is expected to be 0.5 , and we label this ω_1. It may be difficult to prescribe a value for $f/(f + g)$ under an alternative hypothesis. As a rough guide one can obtain a result by assuming that the matching factors are independent of the probability of success under either treatment. The alternative hypothesis is that π_1 and π_2 are the probabilities of success under treatments 1 and 2 respectively and ω_2 is the expected value of $f/(f + g)$ under this hypothesis. Then

$$\omega_2 = \frac{\pi_1 (1 + \pi_2)}{\pi_1 (1 - \pi_2) + \pi_2 (1 - \pi_1)} .$$
(3.10)

We know $\omega_1 = 0.5$, so we can use ω_1 and ω_2 in place of π_1 and π_2 in equation (3.8) to obtain the numbers of contradictory pairs required k. Again, making the assumption that matching does not affect the probability of success, we can obtain an overall estimate of the number of subjects required as

$$m = \frac{k}{\pi_1 (1 - \pi_2) + \pi_2 (1 - \pi_1)} .$$
(3.11)

In practice one might use equation (3.10) to obtain a reasonable value for ω_2, and thus obtain a value for k. For a cross-over design, provided that the time each patient is in the trial is not too long, one would conduct the trial until the number of contradictory pairs, k, was achieved.

3.3 BIBLIOGRAPHY

The most widely used set of tables for calculating the number of subjects required for comparing proportions with a given power and significance are those of Cochran & Cox (1957, p24–25). They require an equal number of subjects to be allocated to each group. Similarly, more extensive tables have been given by Cohen (1977, p186-197). Casagrande *et al.* (1978) derived an approximate formula which gives numbers very close to those required by Fisher's exact test. Fleiss *et al.* (1980) and Fleiss *et al* (1982) generalized this formula to the situation where there are an unequal number of subjects in each group. This formula was improved by Diegert & Diegert (1981). Further discussion of unequal allocation has been given by Campbell (1982) and Fleiss *et al* (1982). The equation for the number of subjects, based on the uncorrected formula was given by Fleiss (1981) while Pocock (1982) gave the simpler version of the formula given in equation (3.3). Mantel (1983) discussed interpolation and the use of the tables for McNemar's test.

3.4 DESCRIPTION OF THE TABLES

Table 3.1

Table 3.1 gives the sample sizes required for the comparison of two binomial proportions for a range of two-sided test sizes of α from 0.01 to 0.20, one sided values from 0.005 to 0.1 and power $1 - \beta$ ranging from 0.50 to 0.99. The smaller binomial proportion is taken as π_1 and varies from 0.05 (0.05) 0.50 whilst the corresponding π_2 increases in steps of 0.05 from $\pi_1 + 0.05$. The entries in the tables are calculated using equation (3.1) and are rounded upwards to the nearest whole number. If $\pi_1 > 0.5$ then rather than defining π_1 as the proportion of successes, calculate the proportion of failures, $1 - \pi_2$ and $1 - \pi_1$ and enter the table with these in place of π_1, π_1.

Table 3.2

If sample sizes are estimated making use of the continuity correction then the subject numbers in Table 3.1 should be multiplied by the corresponding entry for $\pi_2 - \pi_1$ and m in Table 3.2. This product gives sample sizes derived from equation (3.4). This method may give a slightly larger sample size than that obtained directly from equation (3.6). This is because the subject numbers in Table 3.1 have been rounded up to the nearest integer.

3.5 USE OF THE TABLES

Table 3.1

Example 3.1

A trial of transcutaneous electrical stimulation (TNS) for the relief of pain in patients with osteoarthritis of the knee is planned on the basis of the preliminary results published by Smith, Lewith & Machin (1983) who obtained a 25% placebo and a 65% TNS response after four weeks of treatment. A response was defined as a pre-specified reduction in pain experienced by the patient following treatment. How many patients need to be recruited to such a trial with $\alpha = 0.05$, (one-sided) and power $1 - \beta = 0.90$?

Here $\pi_1 = 0.25$, $\pi_2 = 0.65$ and the corresponding entry in the table is $m = 25$. Thus one would plan to recruit 50 patients to such a trial who would be randomized equally to the alternative therapies.

Had a two-sided test appeared more appropriate, again with $\alpha = 0.05$ and power $1 - \beta = 0.90$, then the corresponding $m = 31$.

Example 3.2

Familiari *et al* (1981) compared two drugs in the treatment of peptic ulcer. Forty days after treatment they found 23/30 (77%) of subjects were healed using pirenzepine and 18/31 (58%) using trithiozine. They stated that there was no significant difference between the two drugs at the 5% level, on a two sided test.

If, in fact, $\pi_1 = 0.58$ and $\pi_2 = 0.77$, what is the probability of detecting this difference at the 5% level?

The numbers of patients recruited to each treatment are approximately equal to 30, further $\pi_1 \simeq 0.60$ and $\pi_2 \simeq 0.75$. Now since π_1 is greater than 0.5, we calculate $1 - \pi_2 = 0.25$, $1 - \pi_1 = 0.40$ and use the table with these values for π_1 and π_2. In the row for $\alpha = 0.05$ the smallest value of m is 75, which corresponds to a power of 0.50. The trial, however, only recruited 30 patients to each arm and so had a power less than 0.50.

The exact power can be calculated by means of equation (3.5) with $m = 30, r = 31/30 = 1.033, \pi_1 = 0.23$, $\pi_2 = 0.58$ giving $\bar{\pi} = 0.8265$, and $z_{1-\beta} = 0.6564$. Table 2.1 then gives power $1 - \beta = 0.25$. Thus, there is only a 1 in 4 chance of declaring such an anticipated difference as significant with a trial of that size.

Example 3.3

Elwood & Sweetnam (1979) conducted a clinical trial of aspirin for the treatment of patients who had a myocardial infarction. They were hoping to obtain a 25% reduction in mortality, based on a mortality of about 15% in the year following infarction. In this case $\pi_2 = 0.15$ and $\pi_1 = 0.75 \times 0.15 = 0.1125$ thus $\delta = 0.0375$. The closest values in the table are $\pi_1 = 0.10$ and $\pi_2 = 0.15$ which for 80% power and a two-sided significance level of 0.05 yields 686 patients in each group.

From equation (3.7) a corrected number of subjects is $686 \times (0.15 - 0.10)^2/(0.15 - 0.1125)^2 = 1220$. Direct application of equation (3.3) suggests 1270 subjects per treatment group. One can see that when π_1 and π_2 are close, comparatively small changes in either can result in very large changes in the required number of subjects.

Example 3.4

In fact Elwood and Sweetnam (1979) recruited about 850 patients to each treatment group. What is the probability of their detecting a significant result at the 5% level, given that the underlying proportions are $\pi_1 = 0.10$ and $\pi_2 = 0.15$?

From Table 3.1 it can be seen that the power, for two-sided $\alpha = 0.05$ lies between 0.85 and 0.90. If in fact, we chose $\pi_1 = 0.1125$, then from (3.5) we obtain $z_{1-\beta} = 0.33$, and from Table 2.1 power $1 - \beta = 0.63$. Thus there is quite a high chance that the trial would have missed detection of a 25% reduction in mortality.

Example 3.5

The rate of wound infection over one year in an operating theatre was 10%. This figure has been confirmed from several other operating theatres with the same scrub-up preparation. An investigator wishes to test the efficacy of a new scrub-up preparation. How many operations does he need to examine in order to be 90% confident that the new procedure only produces a 5% infection rate?

In this case Table 3.1, with $\pi_1 = 0.05$, $\pi_2 = 0.10$, and assuming one-sided $\alpha = 0.05$, $1 - \beta = 0.90$, gives

474 operations. Since π_2 is known we can halve this to 237 operations. Direct application of equation (3.8) yields 239 operations, indicating that the approximation holds up well in this case.

Example 3.6

In a clinical trial for the treatment of pre-menstrual syndrome, suitable women are allocated at random to one ot the two treatments for one menstrual cycle, and then, after a further cycle to reduce any carry-over effects, the other treatment for the thrid cycle. The outcome is measured as either 'obtained relief' or 'did not obtain relief'. The expected proportion of women obtaining relief for treatments 1 and 2 are 0.2 and 0.3 respectively. How many women should be included in the trial, for a two-sided significance level 0.05 and power 0.90?

In this case, the proportion of women obtaining relief from treatment 1 but not from treatment 2 is expected to be half of the women obtaining relief from only one treatment. From equation (3.10) with $\pi_1 = 0.2$ and $\pi_2 = 0.3$, the alternative hypothesis yields a proportion $\omega_2 = 0.37$. Using Table 3.1, with $\pi_1 = 0.40$ and $\pi_2 = 0.50$ we obtain 519 but with $\pi_1 = 0.35$ and $\pi_2 = 0.50$ we obtain 227. To obtain the number required for $\pi_1 = 0.37$ and $\pi_2 = 0.5$ we can use equation (3.7) with either $\pi_1 = 0.4$, $\pi_2 = 0.5$, and $m' = 519$ or $\pi_1 = 0.35$, $\pi_2 = 0.5$ and $m' = 227$. In both cases we obtain $m \simeq 300$. We require half this number in the trial, i.e. about 150 women who only obtained relief from one treatment. Equation (3.11) then implies that about 400 women will be required overall.

Note that we can also use Table 12.1 in this example with two-sided $\alpha = 0.05$, power $1 - \beta = 0.90$, $\pi_0 = 0.50$, $\pi_a = 0.65$, and we would require 113 subjects. More exactly with $\pi_a = 0.63$ $(= 1 - 0.37)$ interpolation in formula (3.7), gives 151 subjects almost exactly the same as in the previous paragraph.

Example 3.7

In the trial referred to in Example 3.1, Smith, Lewith & Machin (1983) obtain a 40% difference in the percentage success rate between placebo and TNS. If another investigator believes that with his patients the placebo response is likely to be 65%, he will be unable to observe a 40% difference in response due to TNS, since this would require a 105% response in his treatment group! However, the odds ratio in favour of TNS has been found to be $\{0.65/(1 - 0.65)\}/\{0.25/(1 - 0.25)\} = 5.57$. How many subjects would the investigator require to detect an odds ratio of 5.57, with $\pi_1 = 0.65$, significance level $\alpha = 0.05$ (one-sided) and power $1 - \beta = 0.90$?

From equation (3.9), to obtain an odds ratio of 5.57, with $\pi_1 = 0.65$, he should put $\pi_2 = 0.91$. Since both π_1 and π_2 are above 0.5, we calculate $\pi_1 = 1 - \pi_2 \simeq 0.01$, $\pi_2 = 1 - \pi_1 = 0.35$ and with one-sided $\alpha = 0.05$ and $1 - \beta = 0.90$, Table 3.1 then gives $m = 46$. This is close to the 50 subjects required when $\pi_1 = 0.25$, $\pi_2 = 0.65$ in Example 3.1.

Table 3.2

Example 3.8

It is anticipated that the results of the trial planned in Example 3.1 will be analysed using Yates' correction for continuity. What influence does this have on the number of patients to be recruited?

For $\pi_1 = 0.25$, $\pi_2 = 0.65$ we have $\pi_2 - \pi_1 = 0.40$ and $m = 25$. Table 3.2 gives the multiplying factor as lying between 1.237 and 1.161, say about 1.2. Thus we would need $25 \times 1.2 = 30$ subjects per treatment group. For a two-sided test the corresponding $m = 31$, so the multiplying factor is about 1.161, and the required number of subjects is 36 per group.

Example 3.9

In Example 3.3, what is the effect on the power if the investigators planned to use the continuity correction.

From (3.6), with $\pi_1 = 0.1125$, $\pi_2 = 0.15$ and $v = 53.33$, $z_{1-\beta} = 0.266$, and power $1 - \beta = 0.60$. Thus, use of the continuity correction results in a test of lower power.

3.6 REFERENCES

Armitage P. (1971) *Statistical Methods in Medical Research,* Blackwell Scientific Publications, Oxford.
Berkson J. (1978) In dispraise of the exact test. *J. Statist. Plan. Inf,* **2** 27–42.
Campbell M. J. (1982) The choice of relative group sizes for the comparison of independent proportions. *Biometrics,* **38,** 1093.
Casagrande J. T., Pike M. C. & Smith P. G. (1978) An improved approximate formula for comparing two binomial distributions. *Biometrics,* **34,** 483–486.
Cochran W. G. & Cox G. M. (1957) *Experimental Designs,* 2nd edn. Wiley, New York.
Cohen J. (1977) *Statistical Power Analysis for the Behavioral Sciences,* revised edn. Academic Press, New York.
Diegert C. & Diegert K. V. (1981) Note on inversion of Casagrande-Pike-Smith approximate sample-size formula for Fisher-Irwin test on 2 × 2 tables. *Biometrics,* **37,** 595.
Elwood P. C. & Sweetnam P. M. (1979) Aspirin and secondary mortality after myocardial infarction. *Lancet,* **ii,** 1313–1315.
Familiari L., Postorino S., Turiano S. & Luzza G. (1981) Comparison of pirenzepine and trithiozine with placebo in treatment of peptic ulcer. *Clinical Trial J,* **18,** 363–368.
Fleiss J. L. (1981) *Statistical Methods for Rates and Proportions,* 2nd edn. Wiley, New York.
Fleiss J. L., Tytun A. & Ury S. H. K. (1980) A simple approximation for calculating sample sizes for comparing independent proportions. *Biometrics,* **36,** 343–346.
Fleiss J. L., Tytun A. & Ury S. H. K. (1982) Response to 'The choice of relative group sizes for the comparison of independent proportions'. *Biometrics,* **38,** 1093–1094.
Mantel N. (1983) Extended use of binomial sample-size tables. *Biometrics,* **39,** 777–779.
Pocock S. J. (1982) Statistical aspects of clinical trial design. *The Statistician,* **31,** 1–18.
Smith C. R., Lewith G. T. & Machin D. (1983) A preliminary study to establish a controlled method of assessing transcutaneous nerve stimulation (TNS) as a treatment for the pain caused by osteoarthritis (OA) of the knee. *Physiotherapy,* **69,** 266–268.
Upton G. J. G. (1982) A comparison of alternative tests for the 2 × 2 comparative trial. *J.Roy.Statist.Soc(A),* **145,** 86–105.
Ury H. K. & Fleiss J. L. (1980) On approximate sample sizes for comparing two independent proportions with the use of Yates' correction, *Biometrics,* **36,** 347–351.
Yates F. (1984) Tests of significance for 2 × 2 contingency tables (with discussion) *J.Roy.Statist.Soc.(A),* **147,** 426 463.

Table 3.1 Sample sizes for comparison of proportions.

π_2	α 2-sided	1-sided	0.50	0.65	0.70	0.75	Power $1-\beta$ 0.80	0.85	0.90	0.95	0.99
						$\pi_1 = 0.05$					
0.10	0.01	0.005	369	487	533	586	647	723	824	986	1329
0.10	0.02	0.010	301	408	451	499	556	626	721	872	1197
0.10	0.05	0.025	214	305	342	385	435	497	582	719	1015
0.10	0.10	0.050	151	229	261	298	343	398	474	598	871
0.10	0.20	0.100	92	154	181	212	250	298	363	473	719
0.15	0.01	0.005	120	158	173	190	209	233	266	318	427
0.15	0.02	0.010	98	132	146	162	180	202	232	281	385
0.15	0.05	0.025	70	99	111	125	141	161	188	231	326
0.15	0.10	0.050	49	74	85	97	111	129	153	193	280
0.15	0.20	0.100	30	50	59	69	81	96	117	152	231
0.20	0.01	0.005	65	85	93	102	113	125	143	170	228
0.20	0.02	0.010	53	71	79	87	97	109	125	151	206
0.20	0.05	0.025	38	54	60	67	76	86	101	124	174
0.20	0.10	0.050	27	40	46	52	60	69	82	103	149
0.20	0.20	0.100	16	27	32	37	43	52	63	81	123
0.25	0.01	0.005	43	56	61	67	73	82	93	111	148
0.25	0.02	0.010	35	47	52	57	63	71	81	98	133
0.25	0.05	0.025	25	35	39	44	49	56	65	80	113
0.25	0.10	0.050	18	26	30	34	39	45	53	67	96
0.25	0.20	0.100	11	18	21	24	28	34	41	53	79
0.30	0.01	0.005	31	40	44	48	53	59	67	79	106
0.30	0.02	0.010	26	34	37	41	46	51	58	70	95
0.30	0.05	0.025	18	25	28	32	36	40	47	58	80
0.30	0.10	0.050	13	19	22	25	28	32	38	48	69
0.30	0.20	0.100	8	13	15	18	20	24	29	38	56
0.35	0.01	0.005	24	31	34	37	41	45	51	60	80
0.35	0.02	0.010	20	26	29	31	35	39	44	53	72
0.35	0.05	0.025	14	20	22	24	27	31	36	44	61
0.35	0.10	0.050	10	15	17	19	21	25	29	36	52
0.35	0.20	0.100	6	10	12	13	16	18	22	29	43
0.40	0.01	0.005	19	25	27	29	32	36	40	48	63
0.40	0.02	0.010	16	21	23	25	28	31	35	42	57
0.40	0.05	0.025	11	16	17	19	22	24	28	34	48
0.40	0.10	0.050	8	12	13	15	17	20	23	29	41
0.40	0.20	0.100	5	8	9	11	12	15	18	22	33
0.45	0.01	0.005	16	20	22	24	26	29	33	39	51
0.45	0.02	0.010	13	17	19	21	23	25	29	34	46
0.45	0.05	0.025	10	13	14	16	18	20	23	28	38
0.45	0.10	0.050	7	10	11	12	14	16	19	23	33
0.45	0.20	0.100	4	7	8	9	10	12	14	18	27
0.50	0.01	0.005	14	17	19	20	22	24	27	32	42
0.50	0.02	0.010	11	14	16	17	19	21	24	28	38
0.50	0.05	0.025	8	11	12	13	15	17	19	23	32
0.50	0.10	0.050	6	8	9	10	12	13	15	19	27
0.50	0.20	0.100	4	6	6	7	8	10	12	15	22
0.55	0.01	0.005	12	15	16	17	19	20	23	27	35
0.55	0.02	0.010	10	12	13	15	16	18	20	24	31
0.55	0.05	0.025	7	9	10	11	12	14	16	19	26
0.55	0.10	0.050	5	7	8	9	10	11	13	16	22
0.55	0.20	0.100	3	5	5	6	7	8	10	12	18
0.60	0.01	0.005	10	13	14	15	16	17	19	23	29
0.60	0.02	0.010	8	11	11	12	14	15	17	20	26
0.60	0.05	0.025	6	8	9	10	11	12	14	16	22
0.60	0.10	0.050	4	6	7	7	8	9	11	13	19
0.60	0.20	0.100	3	4	5	5	6	7	8	10	15

Table 3.1

18

Table 3.1 Continued.

π_2	α 2-sided	1-sided	0.50	0.65	0.70	0.75	Power $1-\beta$ 0.80	0.85	0.90	0.95	0.99
						$\pi_1 = 0.05$					
0.65	0.01	0.005	9	11	12	13	14	15	17	19	25
0.65	0.02	0.010	7	9	10	11	12	13	14	17	22
0.65	0.05	0.025	5	7	8	8	9	10	12	14	18
0.65	0.10	0.050	4	5	6	6	7	8	9	11	16
0.65	0.20	0.100	3	4	4	5	5	6	7	9	13
0.70	0.01	0.005	8	10	10	11	12	13	14	16	21
0.70	0.02	0.010	7	8	9	9	10	11	12	14	19
0.70	0.05	0.025	5	6	7	7	8	9	10	12	16
0.70	0.10	0.050	4	5	5	6	6	7	8	10	13
0.70	0.20	0.100	2	3	4	4	5	5	6	7	11
0.75	0.01	0.005	7	8	9	10	10	11	12	14	18
0.75	0.02	0.010	6	7	8	8	9	10	11	12	16
0.75	0.05	0.025	4	5	6	6	7	8	8	10	13
0.75	0.10	0.050	3	4	4	5	5	6	7	8	11
0.75	0.20	0.100	2	3	3	4	4	4	5	6	9
0.80	0.01	0.005	6	7	8	8	9	10	11	12	15
0.80	0.02	0.010	5	6	7	7	8	8	9	11	13
0.80	0.05	0.025	4	5	5	6	6	7	7	8	11
0.80	0.10	0.050	3	4	4	4	5	5	6	7	9
0.80	0.20	0.100	2	3	3	3	3	4	4	5	7
0.85	0.01	0.005	6	7	7	7	8	8	9	10	13
0.85	0.02	0.010	5	6	6	6	7	7	8	9	11
0.85	0.05	0.025	3	4	4	5	5	6	6	7	9
0.85	0.10	0.050	3	3	3	4	4	4	5	6	8
0.85	0.20	0.100	2	2	2	3	3	3	4	4	6
0.90	0.01	0.005	5	6	6	6	7	7	8	9	10
0.90	0.02	0.010	4	5	5	5	6	6	7	8	9
0.90	0.05	0.025	3	4	4	4	4	5	5	6	7
0.90	0.10	0.050	2	3	3	3	4	4	4	5	6
0.90	0.20	0.100	2	2	2	2	3	3	3	4	5
0.95	0.01	0.005	5	5	5	6	6	6	7	7	8
0.95	0.02	0.010	4	4	5	5	5	5	6	6	7
0.95	0.05	0.025	3	3	3	4	4	4	4	5	6
0.95	0.10	0.050	2	3	3	3	3	3	3	4	5
0.95	0.20	0.100	2	2	2	2	2	2	3	3	4

Table 3.1

Table 3.1 Continued.

π_2	α 2-sided	1-sided	0.50	0.65	0.70	0.75	Power $1-\beta$ 0.80	0.85	0.90	0.95	0.99
						$\pi_1=0.10$					
0.15	0.01	0.005	581	767	841	924	1021	1140	1300	1556	2098
0.15	0.02	0.010	474	643	711	787	877	988	1137	1377	1889
0.15	0.05	0.025	337	481	540	607	686	785	918	1135	1603
0.15	0.10	0.050	237	361	412	470	540	628	748	945	1376
0.15	0.20	0.100	144	243	285	335	394	469	574	747	1135
0.20	0.01	0.005	170	224	245	269	297	331	377	451	608
0.20	0.02	0.010	139	187	207	229	255	287	330	399	547
0.20	0.05	0.025	98	140	157	177	199	228	266	329	464
0.20	0.10	0.050	69	105	120	137	157	182	217	274	398
0.20	0.20	0.100	42	71	83	97	115	136	166	216	328
0.25	0.01	0.005	86	112	123	135	149	166	189	226	303
0.25	0.02	0.010	70	94	104	115	128	144	165	200	273
0.25	0.05	0.025	50	71	79	89	100	114	133	164	231
0.25	0.10	0.050	35	53	60	69	79	91	109	137	198
0.25	0.20	0.100	22	36	42	49	57	68	83	108	163
0.30	0.01	0.005	54	70	77	84	92	103	117	140	187
0.30	0.02	0.010	44	59	65	72	79	89	102	123	168
0.30	0.05	0.025	31	44	49	55	62	71	82	101	142
0.30	0.10	0.050	22	33	38	43	49	57	67	84	122
0.30	0.20	0.100	14	22	26	30	36	42	51	67	100
0.35	0.01	0.005	38	49	53	58	64	71	81	96	129
0.35	0.02	0.010	31	41	45	50	55	62	71	85	116
0.35	0.05	0.025	22	31	34	38	43	49	57	70	98
0.35	0.10	0.050	16	23	26	30	34	39	46	58	84
0.35	0.20	0.100	10	16	18	21	25	29	35	46	69
0.40	0.01	0.005	28	36	40	43	48	53	60	71	95
0.40	0.02	0.010	23	31	34	37	41	46	52	63	85
0.40	0.05	0.025	17	23	26	29	32	36	42	52	72
0.40	0.10	0.050	12	17	20	22	25	29	34	43	62
0.40	0.20	0.100	7	12	14	16	18	22	26	34	50
0.45	0.01	0.005	22	28	31	34	37	41	46	55	73
0.45	0.02	0.010	18	24	26	29	32	36	40	48	65
0.45	0.05	0.025	13	18	20	22	25	28	33	40	55
0.45	0.10	0.050	9	14	15	17	20	22	26	33	47
0.45	0.20	0.100	6	9	11	12	14	17	20	26	39
0.50	0.01	0.005	18	23	25	27	30	33	37	44	58
0.50	0.02	0.010	15	19	21	23	25	28	32	39	52
0.50	0.05	0.025	11	14	16	18	20	22	26	32	44
0.50	0.10	0.050	8	11	12	14	16	18	21	26	37
0.50	0.20	0.100	5	7	9	10	11	13	16	21	30
0.55	0.01	0.005	15	19	20	22	24	27	30	35	47
0.55	0.02	0.010	12	16	17	19	21	23	26	31	42
0.55	0.05	0.025	9	12	13	15	16	18	21	26	35
0.55	0.10	0.050	6	9	10	11	13	15	17	21	30
0.55	0.20	0.100	4	6	7	8	9	11	13	17	24
0.60	0.01	0.005	13	16	17	19	20	22	25	29	38
0.60	0.02	0.010	10	13	14	16	17	19	22	26	34
0.60	0.05	0.025	7	10	11	12	14	15	17	21	29
0.60	0.10	0.050	5	8	8	9	11	12	14	17	24
0.60	0.20	0.100	3	5	6	7	8	9	11	14	20
0.65	0.01	0.005	11	13	15	16	17	19	21	24	32
0.65	0.02	0.010	9	11	12	13	15	16	18	21	28
0.65	0.05	0.025	6	9	9	10	11	13	15	18	24
0.65	0.10	0.050	5	6	7	8	9	10	12	14	20
0.65	0.20	0.100	3	4	5	6	7	8	9	11	16

Table 3.1

Table 3.1 Continued.

	α		Power $1-\beta$								
π_2	2-sided	1-sided	0.50	0.65	0.70	0.75	0.80	0.85	0.90	0.95	0.99
						$\pi_1 = 0.10$					
0.70	0.01	0.005	9	12	12	13	15	16	18	21	26
0.70	0.02	0.010	8	10	11	11	12	14	15	18	24
0.70	0.05	0.025	6	7	8	9	10	11	12	15	20
0.70	0.10	0.050	4	6	6	7	8	9	10	12	17
0.70	0.20	0.100	3	4	4	5	6	6	8	9	13
0.75	0.01	0.005	8	10	11	12	12	14	15	17	22
0.75	0.02	0.010	7	8	9	10	11	12	13	15	20
0.75	0.05	0.025	5	6	7	8	8	9	10	12	16
0.75	0.10	0.050	4	5	5	6	7	7	8	10	14
0.75	0.20	0.100	2	3	4	4	5	5	6	8	11
0.80	0.01	0.005	7	9	9	10	11	12	13	15	19
0.80	0.02	0.010	6	7	8	8	9	10	11	13	16
0.80	0.05	0.025	4	6	6	7	7	8	9	10	14
0.80	0.10	0.050	3	4	5	5	6	6	7	8	11
0.80	0.20	0.100	2	3	3	4	4	5	5	7	9
0.85	0.01	0.005	6	8	8	9	9	10	11	12	15
0.85	0.02	0.010	5	6	7	7	8	9	9	11	14
0.85	0.05	0.025	4	5	5	6	6	7	7	9	11
0.85	0.10	0.050	3	4	4	4	5	5	6	7	9
0.85	0.20	0.100	2	3	3	3	3	4	5	5	8
0.90	0.01	0.005	6	7	7	7	8	8	9	10	13
0.90	0.02	0.010	5	6	6	6	7	7	8	9	11
0.90	0.05	0.025	4	4	5	5	5	6	6	7	9
0.90	0.10	0.050	3	3	3	4	4	5	5	6	8
0.90	0.20	0.100	2	2	2	3	3	3	4	5	6
0.95	0.01	0.005	5	6	6	6	7	7	8	9	10
0.95	0.02	0.010	4	5	5	5	6	6	7	8	9
0.95	0.05	0.025	3	4	4	4	4	5	5	6	7
0.95	0.10	0.050	2	3	3	3	4	4	4	5	6
0.95	0.20	0.100	2	2	2	2	3	3	3	4	5

Table 3.1

Table 3.1 Continued.

π_2	α 2-sided	1-sided	0.50	0.65	0.70	0.75	Power $1-\beta$ 0.80	0.85	0.90	0.95	0.99
						$\pi_1=0.15$					
0.20	0.01	0.005	767	1013	1110	1220	1348	1506	1717	2055	2770
0.20	0.02	0.010	626	849	938	1040	1158	1305	1502	1819	2495
0.20	0.05	0.025	444	635	713	801	906	1036	1212	1498	2118
0.20	0.10	0.050	313	476	543	621	714	829	988	1248	1817
0.20	0.20	0.100	190	321	377	442	520	620	758	987	1500
0.25	0.01	0.005	213	281	307	337	373	416	474	567	764
0.25	0.02	0.010	174	235	260	288	320	361	415	502	688
0.25	0.05	0.025	123	176	197	222	250	286	335	413	583
0.25	0.10	0.050	87	132	151	172	197	229	273	344	501
0.25	0.20	0.100	53	89	104	122	144	171	209	272	413
0.30	0.01	0.005	103	136	149	163	180	201	229	273	367
0.30	0.02	0.010	84	114	126	139	155	174	200	242	331
0.30	0.05	0.025	60	85	96	107	121	138	161	199	280
0.30	0.10	0.050	42	64	73	83	95	111	131	166	240
0.30	0.20	0.100	26	43	51	59	69	83	101	131	198
0.35	0.01	0.005	63	82	90	98	109	121	137	164	220
0.35	0.02	0.010	51	69	76	84	93	105	120	145	198
0.35	0.05	0.025	37	52	58	65	73	83	97	119	168
0.35	0.10	0.050	26	39	44	50	57	66	79	99	144
0.35	0.20	0.100	16	26	31	36	42	50	60	78	118
0.40	0.01	0.005	43	56	61	67	74	82	93	111	148
0.40	0.02	0.010	35	47	52	57	63	71	81	98	133
0.40	0.05	0.025	25	35	39	44	49	56	65	80	113
0.40	0.10	0.050	18	26	30	34	39	45	53	67	96
0.40	0.20	0.100	11	18	21	24	28	34	41	53	79
0.45	0.01	0.005	31	41	45	49	54	59	67	80	107
0.45	0.02	0.010	26	34	38	41	46	51	59	71	96
0.45	0.05	0.025	18	26	29	32	36	41	47	58	81
0.45	0.10	0.050	13	19	22	25	28	33	39	48	69
0.45	0.20	0.100	8	13	15	18	21	24	29	38	57
0.50	0.01	0.005	24	31	34	37	41	45	51	61	81
0.50	0.02	0.010	20	26	29	32	35	39	45	54	73
0.50	0.05	0.025	14	20	22	24	27	31	36	44	61
0.50	0.10	0.050	10	15	17	19	22	25	29	37	52
0.50	0.20	0.100	6	10	12	14	16	19	22	29	43
0.55	0.01	0.005	19	25	27	29	32	36	40	48	63
0.55	0.02	0.010	16	21	23	25	28	31	35	42	57
0.55	0.05	0.025	11	16	17	19	22	24	28	34	48
0.55	0.10	0.050	8	12	13	15	17	20	23	29	41
0.55	0.20	0.100	5	8	9	11	12	15	17	22	33
0.60	0.01	0.005	16	20	22	24	26	29	32	38	50
0.60	0.02	0.010	13	17	19	20	22	25	28	34	45
0.60	0.05	0.025	9	13	14	16	17	20	23	28	38
0.60	0.10	0.050	7	10	11	12	14	16	18	23	32
0.60	0.20	0.100	4	7	8	9	10	12	14	18	26
0.65	0.01	0.005	13	17	18	20	21	24	26	31	41
0.65	0.02	0.010	11	14	15	17	18	20	23	27	36
0.65	0.05	0.025	8	11	12	13	14	16	19	22	31
0.65	0.10	0.050	6	8	9	10	11	13	15	18	26
0.65	0.20	0.100	4	5	6	7	8	10	11	14	21
0.70	0.01	0.005	11	14	15	16	18	20	22	26	33
0.70	0.02	0.010	9	12	13	14	15	17	19	23	30
0.70	0.05	0.025	7	9	10	11	12	13	15	18	25
0.70	0.10	0.050	5	7	7	8	9	11	12	15	21
0.70	0.20	0.100	3	5	5	6	7	8	9	12	17

Table 3.1   22

Table 3.1 Continued.

π_2	α 2-sided	1-sided	Power $1-\beta$ 0.50	0.65	0.70	0.75	0.80	0.85	0.90	0.95	0.99
					$\pi_1 = 0.15$						
0.75	0.01	0.005	10	12	13	14	15	16	18	21	28
0.75	0.02	0.010	8	10	11	12	13	14	16	19	25
0.75	0.05	0.025	6	8	8	9	10	11	13	15	21
0.75	0.10	0.050	4	6	6	7	8	9	10	13	17
0.75	0.20	0.100	3	4	4	5	6	7	8	10	14
0.80	0.01	0.005	8	10	11	12	13	14	15	18	23
0.80	0.02	0.010	7	9	9	10	11	12	13	16	20
0.80	0.05	0.025	5	6	7	8	8	9	11	13	17
0.80	0.10	0.050	4	5	5	6	7	7	9	10	14
0.80	0.20	0.100	2	3	4	4	5	6	7	8	11
0.85	0.01	0.005	7	9	9	10	11	12	13	15	19
0.85	0.02	0.010	6	7	8	9	9	10	11	13	17
0.85	0.05	0.025	4	6	6	7	7	8	9	11	14
0.85	0.10	0.050	3	4	5	5	6	6	7	9	12
0.85	0.20	0.100	2	3	3	4	4	5	5	7	9
0.90	0.01	0.005	6	8	8	9	9	10	11	12	15
0.90	0.02	0.010	5	6	7	7	8	9	9	11	14
0.90	0.05	0.025	4	5	5	6	6	7	7	9	11
0.90	0.10	0.050	3	4	4	4	5	5	6	7	9
0.90	0.20	0.100	2	3	3	3	3	4	5	5	8
0.95	0.01	0.005	6	7	7	7	8	8	9	10	13
0.95	0.02	0.010	5	6	6	6	7	7	8	9	11
0.95	0.05	0.025	3	4	4	5	5	6	6	7	9
0.95	0.10	0.050	3	3	3	4	4	4	5	6	8
0.95	0.20	0.100	2	2	2	3	3	3	4	4	6

Table 3.1

Table 3.1 Continued.

π_2	α 2-sided	1-sided	0.50	0.65	0.70	Power $1-\beta$ 0.75	0.80	0.85	0.90	0.95	0.99
						$\pi_1 = 0.20$					
0.25	0.01	0.005	926	1223	1340	1473	1628	1819	2074	2482	3347
0.25	0.02	0.010	755	1026	1133	1256	1399	1576	1814	2197	3015
0.25	0.05	0.025	536	767	861	968	1094	1251	1464	1810	2558
0.25	0.10	0.050	378	575	656	750	862	1002	1193	1507	2196
0.25	0.20	0.100	230	388	455	534	628	749	915	1193	1812
0.30	0.01	0.005	249	329	360	396	437	488	556	665	896
0.30	0.02	0.010	203	276	305	337	376	423	486	589	807
0.30	0.05	0.025	145	206	231	260	294	336	392	485	684
0.30	0.10	0.050	102	155	176	201	231	269	320	404	587
0.30	0.20	0.100	62	104	122	143	169	201	245	319	484
0.35	0.01	0.005	118	155	170	187	206	230	262	313	421
0.35	0.02	0.010	96	130	144	159	177	199	229	277	379
0.35	0.05	.025	69	98	109	123	138	158	185	228	321
0.35	0.10	0.050	48	73	83	95	109	127	150	190	275
0.35	0.20	0.100	30	49	58	68	79	95	115	150	227
0.40	0.01	0.005	70	92	101	110	122	136	154	184	247
0.40	0.02	0.010	57	77	85	94	105	117	135	163	222
0.40	0.05	0.025	41	58	65	72	82	93	109	134	188
0.40	0.10	0.050	29	43	49	56	64	75	89	111	161
0.40	0.20	0.100	18	29	34	40	47	56	68	88	133
0.45	0.01	0.005	47	61	67	74	81	90	102	122	163
0.45	0.02	0.010	38	52	57	63	70	78	90	108	147
0.45	0.05	0.025	27	39	43	48	54	62	72	89	124
0.45	0.10	0.050	19	29	33	37	43	50	59	74	107
0.45	0.20	0.100	12	20	23	27	31	37	45	58	88
0.50	0.01	0.005	34	44	48	53	58	65	73	87	116
0.50	0.02	0.010	28	37	41	45	50	56	64	77	104
0.50	0.05	0.025	20	28	31	35	39	44	52	63	88
0.50	0.10	0.050	14	21	24	27	31	35	42	52	76
0.50	0.20	0.100	9	14	17	19	22	26	32	41	62
0.55	0.01	0.005	26	33	36	40	44	49	55	65	87
0.55	0.02	0.010	21	28	31	34	38	42	48	58	78
0.55	0.05	0.025	15	21	23	26	29	33	39	47	66
0.55	0.10	0.050	11	16	18	20	23	27	31	39	56
0.55	0.20	0.100	7	11	12	14	17	20	24	31	46
0.60	0.01	0.005	20	26	28	31	34	38	43	50	67
0.60	0.02	0.010	17	22	24	26	29	33	37	44	60
0.60	0.05	0.025	12	17	18	20	23	26	30	36	51
0.60	0.10	0.050	9	12	14	16	18	21	24	30	43
0.60	0.20	0.100	5	9	10	11	13	15	19	24	35
0.65	0.01	0.005	17	21	23	25	27	30	34	40	53
0.65	0.02	0.010	14	18	19	21	23	26	30	35	47
0.65	0.05	0.025	10	13	15	16	18	21	24	29	40
0.65	0.10	0.050	7	10	11	13	14	16	19	24	34
0.65	0.20	0.100	4	7	8	9	10	12	15	19	28
0.70	0.01	0.005	14	17	19	20	22	24	27	32	42
0.70	0.02	0.010	11	15	16	17	19	21	24	28	38
0.70	0.05	0.025	8	11	12	13	15	17	19	23	32
0.70	0.10	0.050	6	8	9	10	12	13	16	19	27
0.70	0.20	0.100	4	6	6	7	8	10	12	15	22
0.75	0.01	0.005	11	14	15	17	18	20	22	26	34
0.75	0.02	0.010	9	12	13	14	16	17	20	23	31
0.75	0.05	0.025	7	9	10	11	12	14	16	19	26
0.75	0.10	0.050	5	7	8	9	10	11	13	16	22
0.75	0.20	0.100	3	5	5	6	7	8	10	12	18

Table 3.1

Table 3.1 Continued.

π_2	α 2-sided	1-sided	Power $1-\beta$ 0.50	0.65	0.70	0.75	0.80	0.85	0.90	0.95	0.99
						$\pi_1 = 0.20$					
0.80	0.01	0.005	10	12	13	14	15	17	19	22	28
0.80	0.02	0.010	8	10	11	12	13	14	16	19	25
0.80	0.05	0.025	6	8	8	9	10	11	13	15	21
0.80	0.10	0.050	4	6	6	7	8	9	10	13	18
0.80	0.20	0.100	3	4	5	5	6	7	8	10	14
0.85	0.01	0.005	8	10	11	12	13	14	15	18	23
0.85	0.02	0.010	7	9	9	10	11	12	13	16	20
0.85	0.05	0.025	5	6	7	8	8	9	11	13	17
0.85	0.10	0.050	4	5	5	6	7	7	9	10	14
0.85	0.20	0.100	2	3	4	4	5	6	7	8	11
0.90	0.01	0.005	7	9	9	10	11	12	13	15	19
0.90	0.02	0.010	6	7	8	8	9	10	11	13	16
0.90	0.05	0.025	4	6	6	7	7	8	9	10	14
0.90	0.10	0.050	3	4	5	5	6	6	7	8	11
0.90	0.20	0.100	2	3	3	4	4	5	5	7	9
0.95	0.01	0.005	6	7	8	8	9	10	11	12	15
0.95	0.02	0.010	5	6	7	7	8	8	9	11	13
0.95	0.05	0.025	4	5	5	6	6	7	7	8	11
0.95	0.10	0.050	3	4	4	4	5	5	6	7	9
0.95	0.20	0.100	2	3	3	3	3	4	4	5	7

Table 3.1

Table 3.1 Continued.

π_2	2-sided	1-sided	0.50	0.65	0.70	0.75	0.80	0.85	0.90	0.95	0.99
	α						Power $1-\beta$				

$\pi_1 = 0.25$

π_2	2-sided	1-sided	0.50	0.65	0.70	0.75	0.80	0.85	0.90	0.95	0.99
0.30	0.01	0.005	1059	1398	1533	1684	1862	2080	2371	2838	3828
0.30	0.02	0.010	864	1173	1296	1436	1600	1802	2074	2513	3448
0.30	0.05	0.025	613	877	984	1107	1251	1431	1674	2070	2926
0.30	0.10	0.050	432	658	750	858	986	1146	1365	1724	2511
0.30	0.20	0.100	262	443	520	610	719	856	1047	1364	2073
0.35	0.01	0.005	279	368	403	443	490	547	623	745	1004
0.35	0.02	0.010	228	309	341	378	421	474	545	660	904
0.35	0.05	0.025	162	231	259	291	329	376	440	543	767
0.35	0.10	0.050	114	173	198	226	259	301	358	452	658
0.35	0.20	0.100	69	117	137	161	189	225	275	358	543
0.40	0.01	0.005	130	171	187	205	227	253	288	344	463
0.40	0.02	0.010	106	143	158	175	195	219	252	305	417
0.40	0.05	0.025	75	107	120	135	152	174	203	251	354
0.40	0.10	0.050	53	80	92	105	120	139	166	209	303
0.40	0.20	0.100	33	54	64	74	88	104	127	165	250
0.45	0.01	0.005	76	100	109	120	132	147	167	200	268
0.45	0.02	0.010	62	84	92	102	113	127	146	177	241
0.45	0.05	0.025	44	63	70	79	89	101	118	145	204
0.45	0.10	0.050	31	47	53	61	70	81	96	121	175
0.45	0.20	0.100	19	32	37	43	51	60	74	95	144
0.50	0.01	0.005	50	66	72	79	87	96	110	131	175
0.50	0.02	0.010	41	55	61	67	74	84	96	115	157
0.50	0.05	0.025	29	41	46	52	58	66	77	95	133
0.50	0.10	0.050	21	31	35	40	46	53	63	79	114
0.50	0.20	0.100	13	21	24	29	33	40	48	62	94
0.55	0.01	0.005	36	47	51	56	61	68	77	92	123
0.55	0.02	0.010	29	39	43	47	53	59	68	81	110
0.55	0.05	0.025	21	29	33	37	41	47	54	67	93
0.55	0.10	0.050	15	22	25	28	32	37	44	55	80
0.55	0.20	0.100	9	15	17	20	24	28	34	44	66
0.60	0.01	0.005	27	35	38	42	46	51	57	68	91
0.60	0.02	0.010	22	29	32	35	39	44	50	60	81
0.60	0.05	0.025	16	22	24	27	31	35	40	49	69
0.60	0.10	0.050	11	17	19	21	24	28	33	41	59
0.60	0.20	0.100	7	11	13	15	18	21	25	32	48
0.65	0.01	0.005	21	27	29	32	35	39	44	52	69
0.65	0.02	0.010	17	23	25	27	30	34	38	46	62
0.65	0.05	0.025	12	17	19	21	24	27	31	38	52
0.65	0.10	0.050	9	13	14	16	19	21	25	31	45
0.65	0.20	0.100	6	9	10	12	14	16	19	25	37
0.70	0.01	0.005	17	21	23	25	28	31	35	41	54
0.70	0.02	0.010	14	18	20	22	24	27	30	36	48
0.70	0.05	0.025	10	14	15	17	19	21	24	29	41
0.70	0.10	0.050	7	10	11	13	15	17	20	24	35
0.70	0.20	0.100	5	7	8	9	11	12	15	19	28
0.75	0.01	0.005	14	17	19	20	22	25	28	33	43
0.75	0.02	0.010	11	15	16	17	19	21	24	29	38
0.75	0.05	0.025	8	11	12	13	15	17	19	23	32
0.75	0.10	0.050	6	8	9	10	12	13	16	19	27
0.75	0.20	0.100	4	6	7	7	9	10	12	15	22
0.80	0.01	0.005	11	14	15	17	18	20	22	26	34
0.80	0.02	0.010	9	12	13	14	16	17	20	23	31
0.80	0.05	0.025	7	9	10	11	12	14	16	19	26
0.80	0.10	0.050	5	7	8	9	10	11	13	16	22
0.80	0.20	0.100	3	5	5	6	7	8	10	12	18

Table 3.1

Table 3.1 Continued.

π_2	α 2-sided	1-sided	Power $1-\beta$ 0.50	0.65	0.70	0.75	0.80	0.85	0.90	0.95	0.99
						$\pi_1 = 0.25$					
0.85	0.01	0.005	10	12	13	14	15	16	18	21	28
0.85	0.02	0.010	8	10	11	12	13	14	16	19	25
0.85	0.05	0.025	6	8	8	9	10	11	13	15	21
0.85	0.10	0.050	4	6	6	7	8	9	10	13	17
0.85	0.20	0.100	3	4	4	5	6	7	8	10	14
0.90	0.01	0.005	8	10	11	12	12	14	15	17	22
0.90	0.02	0.010	7	8	9	10	11	12	13	15	20
0.90	0.05	0.025	5	6	7	8	8	9	10	12	16
0.90	0.10	0.050	4	5	5	6	7	7	8	10	14
0.90	0.20	0.100	2	3	4	4	5	5	6	8	11
0.95	0.01	0.005	7	8	9	10	10	11	12	14	18
0.95	0.02	0.010	6	7	8	8	9	10	11	12	16
0.95	0.05	0.025	4	5	6	6	7	8	8	10	13
0.95	0.10	0.050	3	4	4	5	5	6	7	8	11
0.95	0.20	0.100	2	3	3	4	4	4	5	6	9
						$\pi_1 = 0.30$					
0.35	0.01	0.005	1165	1539	1686	1853	2049	2289	2609	3123	4212
0.35	0.02	0.010	950	1290	1426	1580	1760	1983	2283	2765	3794
0.35	0.05	0.025	675	965	1083	1218	1377	1575	1842	2278	3220
0.35	0.10	0.050	475	723	826	944	1084	1261	1502	1897	2764
0.35	0.20	0.100	289	488	572	671	791	942	1152	1501	2281
0.40	0.01	0.005	302	399	437	480	530	592	675	808	1088
0.40	0.02	0.010	247	335	370	409	456	513	590	715	980
0.40	0.05	0.025	175	250	281	315	356	407	477	589	831
0.40	0.10	0.050	124	188	214	244	281	326	388	490	713
0.40	0.20	0.100	75	127	148	174	205	244	298	388	589
0.45	0.01	0.005	139	183	200	219	242	270	308	368	495
0.45	0.02	0.010	113	153	169	187	208	234	269	326	446
0.45	0.05	0.025	81	115	128	144	163	186	217	268	378
0.45	0.10	0.050	57	86	98	112	128	149	177	223	324
0.45	0.20	0.100	35	58	68	80	94	111	136	177	267
0.50	0.01	0.005	80	105	115	126	139	155	177	211	283
0.50	0.02	0.010	65	88	97	108	120	134	154	186	255
0.50	0.05	0.025	47	66	74	83	93	107	124	153	216
0.50	0.10	0.050	33	50	56	64	74	85	101	128	185
0.50	0.20	0.100	20	34	39	46	54	64	78	101	152
0.55	0.01	0.005	52	68	75	82	90	101	114	136	183
0.55	0.02	0.010	43	57	63	70	78	87	100	121	164
0.55	0.05	0.025	31	43	48	54	61	69	81	99	139
0.55	0.10	0.050	22	32	37	42	48	55	66	82	119
0.55	0.20	0.100	13	22	26	30	35	41	50	65	98
0.60	0.01	0.005	37	48	53	57	63	70	80	95	127
0.60	0.02	0.010	30	40	44	49	54	61	70	84	114
0.60	0.05	0.025	22	30	34	38	42	48	56	69	97
0.60	0.10	0.050	15	23	26	29	33	39	46	57	83
0.60	0.20	0.100	10	15	18	21	24	29	35	45	68
0.65	0.01	0.005	28	36	39	42	47	52	59	69	93
0.65	0.02	0.010	23	30	33	36	40	45	51	61	83
0.65	0.05	0.025	16	22	25	28	31	35	41	50	70
0.65	0.10	0.050	12	17	19	22	25	28	33	42	60
0.65	0.20	0.100	7	11	13	15	18	21	26	33	49

27

Table 3.1

Table 3.1 Continued.

π_2	α 2-sided	1-sided	Power $1-\beta$ 0.50	0.65	0.70	0.75	0.80	0.85	0.90	0.95	0.99
						$\pi_1 = 0.30$					
0.70	0.01	0.005	21	27	30	32	36	39	44	53	70
0.70	0.02	0.010	17	23	25	28	30	34	39	46	63
0.70	0.05	0.025	13	17	19	21	24	27	31	38	53
0.70	0.10	0.050	9	13	15	17	19	22	25	32	45
0.70	0.20	0.100	6	9	10	12	14	16	19	25	37
0.75	0.01	0.005	17	21	23	25	28	31	35	41	54
0.75	0.02	0.010	14	18	20	22	24	27	30	36	48
0.75	0.05	0.025	10	14	15	17	19	21	24	29	41
0.75	0.10	0.050	7	10	11	13	15	17	20	24	35
0.75	0.20	0.100	5	7	8	9	11	12	15	19	28
0.80	0.01	0.005	14	17	19	20	22	24	27	32	42
0.80	0.02	0.010	11	15	16	17	19	21	24	28	38
0.80	0.05	0.025	8	11	12	13	15	17	19	23	32
0.80	0.10	0.050	6	8	9	10	12	13	16	19	27
0.80	0.20	0.100	4	6	6	7	8	10	12	15	22
0.85	0.01	0.005	11	14	15	16	18	20	22	26	33
0.85	0.02	0.010	9	12	13	14	15	17	19	23	30
0.85	0.05	0.025	7	9	10	11	12	13	15	18	25
0.85	0.10	0.050	5	7	7	8	9	11	12	15	21
0.85	0.20	0.100	3	5	5	6	7	8	9	12	17
0.90	0.01	0.005	9	12	12	13	15	16	18	21	26
0.90	0.02	0.010	8	10	11	11	12	14	15	18	24
0.90	0.05	0.025	6	7	8	9	10	11	12	15	20
0.90	0.10	0.050	4	6	6	7	8	9	10	12	17
0.90	0.20	0.100	3	4	4	5	6	6	8	9	13
0.95	0.01	0.005	8	10	10	11	12	13	14	16	21
0.95	0.02	0.010	7	8	9	9	10	11	12	14	19
0.95	0.05	0.025	5	6	7	7	8	9	10	12	16
0.95	0.10	0.050	4	5	5	6	6	7	8	10	13
0.95	0.20	0.100	2	3	4	4	5	5	6	7	11

Table 3.1

28

Table 3.1 Continued.

	α					Power 1−β					
π_2	2-sided	1-sided	0.50	0.65	0.70	0.75	0.80	0.85	0.90	0.95	0.99

<table>
<tr><td colspan="12" align="center">$\pi_1=0.35$</td></tr>
</table>

π_2	2-sided	1-sided	0.50	0.65	0.70	0.75	0.80	0.85	0.90	0.95	0.99
0.40	0.01	0.005	1245	1644	1802	1980	2189	2445	2788	3337	4501
0.40	0.02	0.010	1015	1379	1524	1688	1881	2119	2439	2954	4054
0.40	0.05	0.025	721	1031	1157	1301	1471	1682	1969	2434	3440
0.40	0.10	0.050	508	773	882	1008	1159	1347	1604	2027	2953
0.40	0.20	0.100	308	521	612	717	845	1007	1231	1604	2437
0.45	0.01	0.005	319	421	461	507	560	625	712	852	1148
0.45	0.02	0.010	260	353	390	432	481	542	623	754	1034
0.45	0.05	0.025	185	264	296	333	376	430	503	621	877
0.45	0.10	0.050	130	198	226	258	296	344	410	517	753
0.45	0.20	0.100	79	134	157	183	216	257	314	409	621
0.50	0.01	0.005	145	190	208	229	253	282	321	384	517
0.50	0.02	0.010	118	160	176	195	217	244	281	340	465
0.50	0.05	0.025	84	120	134	150	170	194	227	280	395
0.50	0.10	0.050	59	90	02	117	134	155	185	233	338
0.50	0.20	0.100	36	61	71	83	98	116	142	184	279
0.55	0.01	0.005	83	108	119	130	144	160	182	217	292
0.55	0.02	0.010	67	91	100	111	123	139	159	192	263
0.55	0.05	0.025	48	68	76	85	96	110	128	158	223
0.55	0.10	0.050	34	51	58	66	76	88	105	132	191
0.55	0.20	0.100	21	35	40	47	55	66	80	104	157
0.60	0.01	0.005	53	70	76	84	92	103	117	139	187
0.60	0.02	0.010	44	59	65	71	79	89	102	123	168
0.60	0.05	0.025	31	44	49	55	62	71	82	101	142
0.60	0.10	0.050	22	33	37	43	49	56	67	84	122
0.60	0.20	0.100	14	22	26	30	36	42	51	66	100
0.65	0.01	0.005	37	49	53	58	64	71	81	96	128
0.65	0.02	0.010	31	41	45	49	55	62	70	85	115
0.65	0.05	0.025	22	31	34	38	43	49	57	70	98
0.65	0.10	0.050	16	23	26	30	34	39	46	58	83
0.65	0.20	0.100	10	16	18	21	25	29	35	46	69
0.70	0.01	0.005	28	36	39	42	47	52	59	69	93
0.70	0.02	0.010	23	30	33	36	40	45	51	61	83
0.70	0.05	0.025	16	22	25	28	31	35	41	50	70
0.70	0.10	0.050	12	17	19	22	25	28	33	42	60
0.70	0.20	0.100	7	11	13	15	18	21	26	33	49
0.75	0.01	0.005	21	27	29	32	35	39	44	52	69
0.75	0.02	0.010	17	23	25	27	30	34	38	46	62
0.75	0.05	0.025	12	17	19	21	24	27	31	38	52
0.75	0.10	0.050	9	13	14	16	19	21	25	31	45
0.75	0.20	0.100	6	9	10	12	14	16	19	25	37
0.80	0.01	0.005	17	21	23	25	27	30	34	40	53
0.80	0.02	0.010	14	18	19	21	23	26	30	35	47
0.80	0.05	0.025	10	13	15	16	18	21	24	29	40
0.80	0.10	0.050	7	10	11	13	14	16	19	24	34
0.80	0.20	0.100	4	7	8	9	10	12	15	19	28
0.85	0.01	0.005	13	17	18	20	21	24	26	31	41
0.85	0.02	0.010	11	14	15	17	18	20	23	27	36
0.85	0.05	0.025	8	11	12	13	14	16	19	22	31
0.85	0.10	0.050	6	8	9	10	11	13	15	18	26
0.85	0.20	0.100	4	5	6	7	8	10	11	14	21
0.90	0.01	0.005	11	13	15	16	17	19	21	24	32
0.90	0.02	0.010	9	11	12	13	15	16	18	21	28
0.90	0.05	0.025	6	9	9	10	11	13	15	18	24
0.90	0.10	0.050	5	6	7	8	9	10	12	14	20
0.90	0.20	0.100	3	4	5	6	7	8	9	11	16

Table 3.1

Table 3.1 Continued.

π_2	α 2-sided	1-sided	0.50	0.65	0.70	Power $1-\beta$ 0.75	0.80	0.85	0.90	0.95	0.99
					$\pi_1 = 0.35$						
0.95	0.01	0.005	9	11	12	13	14	15	17	19	25
0.95	0.02	0.010	7	9	10	11	12	13	14	17	22
0.95	0.05	0.025	5	7	8	8	9	10	12	14	18
0.95	0.10	0.050	4	5	6	6	7	8	9	11	16
0.95	0.20	0.100	3	4	4	5	5	6	7	9	13

Table 3.1

30

Table 3.1 Continued.

π_2	α 2-sided	1-sided	Power $1-\beta$ 0.50	0.65	0.70	0.75	0.80	0.85	0.90	0.95	0.99
						$\pi_1 = 0.40$					
0.45	0.01	0.005	1298	1714	1879	2065	2282	2550	2907	3480	4693
0.45	0.02	0.010	1059	1438	1589	1760	1961	2210	2543	3080	4227
0.45	0.05	0.025	752	1075	1206	1356	1534	1754	2053	2538	3587
0.45	0.10	0.050	529	806	920	1051	1208	1405	1673	2114	3079
0.45	0.20	0.100	322	543	638	748	881	1050	1283	1672	2541
0.50	0.01	0.005	329	434	475	522	577	645	735	879	1184
0.50	0.02	0.010	268	364	402	445	496	559	643	778	1067
0.50	0.05	0.025	191	272	305	343	388	443	519	641	905
0.50	0.10	0.050	134	204	233	266	305	355	423	533	777
0.50	0.20	0.100	82	138	161	189	223	265	324	422	641
0.55	0.01	0.005	148	194	213	234	258	288	328	392	527
0.55	0.02	0.010	120	163	180	199	222	249	287	347	475
0.55	0.05	0.025	86	122	137	153	173	198	231	286	403
0.55	0.10	0.050	60	91	104	119	136	158	188	238	345
0.55	0.20	0.100	37	62	72	85	100	118	144	188	285
0.60	0.01	0.005	83	110	120	131	145	162	184	220	295
0.60	0.02	0.010	68	92	101	112	125	140	161	194	266
0.60	0.05	0.025	49	69	77	86	97	111	130	160	225
0.60	0.10	0.050	34	52	59	67	77	89	106	133	193
0.60	0.20	0.100	21	35	41	48	56	66	81	105	159
0.65	0.01	0.005	53	70	76	84	92	103	117	139	187
0.65	0.02	0.010	44	59	65	71	79	89	102	123	168
0.65	0.05	0.025	31	44	49	55	62	71	82	101	142
0.65	0.10	0.050	22	33	37	43	49	56	67	84	122
0.65	0.20	0.100	14	22	26	30	36	42	51	66	100
0.70	0.01	0.005	37	48	53	57	63	70	80	95	127
0.70	0.02	0.010	30	40	44	49	54	61	70	84	114
0.70	0.05	0.025	22	30	34	38	42	48	56	69	97
0.70	0.10	0.050	15	23	26	29	33	39	46	57	83
0.70	0.20	0.100	10	15	18	21	24	29	35	45	68
0.75	0.01	0.005	27	35	38	42	46	51	57	68	91
0.75	0.02	0.010	22	29	32	35	39	44	50	60	81
0.75	0.05	0.025	16	22	24	27	31	35	40	49	69
0.75	0.10	0.050	11	17	19	21	24	28	33	41	59
0.75	0.20	0.100	7	11	13	15	18	21	25	32	48
0.80	0.01	0.005	20	26	28	31	34	38	43	50	67
0.80	0.02	0.010	17	22	24	26	29	33	37	44	60
0.80	0.05	0.025	12	17	18	20	23	26	30	36	51
0.80	0.10	0.050	9	12	14	16	18	21	24	30	43
0.80	0.20	0.100	5	9	10	11	13	15	19	24	35
0.85	0.01	0.005	16	20	22	24	26	29	32	38	50
0.85	0.02	0.010	13	17	19	20	22	25	28	34	45
0.85	0.05	0.025	9	13	14	16	17	20	23	28	38
0.85	0.10	0.050	7	10	11	12	14	16	18	23	32
0.85	0.20	0.100	4	7	8	9	10	12	14	18	26
0.90	0.01	0.005	13	16	17	19	20	22	25	29	38
0.90	0.02	0.010	10	13	14	16	17	19	22	26	34
0.90	0.05	0.025	7	10	11	12	14	15	17	21	29
0.90	0.10	0.050	5	8	8	9	11	12	14	17	24
0.90	0.20	0.100	3	5	6	7	8	9	11	14	20
0.95	0.01	0.005	10	13	14	15	16	17	19	23	29
0.95	0.02	0.010	8	11	11	12	14	15	17	20	26
0.95	0.05	0.025	6	8	9	10	11	12	14	16	22
0.95	0.10	0.050	4	6	7	7	8	9	11	13	19
0.95	0.20	0.100	3	4	5	5	6	7	8	10	15

Table 3.1

Table 3.1 Continued.

π_2	α 2-sided	1-sided	0.50	0.65	0.70	Power $1-\beta$ 0.75	0.80	0.85	0.90	0.95	0.99
						$\pi_1 = 0.45$					
0.50	0.01	0.005	1324	1749	1917	2107	2329	2602	2966	3551	4789
0.50	0.02	0.010	1080	1467	1621	1796	2001	2255	2595	3143	4314
0.50	0.05	0.025	767	1097	1231	1384	1565	1790	2095	2590	3661
0.50	0.10	0.050	540	822	939	1073	1233	1433	1707	2157	3142
0.50	0.20	0.100	328	554	651	763	899	1071	1309	1707	2593
0.55	0.01	0.005	332	438	480	528	583	651	742	888	1196
0.55	0.02	0.010	271	368	406	450	501	564	649	786	1077
0.55	0.05	0.025	193	275	308	347	392	448	524	647	914
0.55	0.10	0.050	136	206	235	269	309	359	427	539	784
0.55	0.20	0.100	83	139	163	191	225	268	327	426	647
0.60	0.01	0.005	148	194	213	234	258	288	328	392	527
0.60	0.02	0.010	120	163	180	199	222	249	287	347	475
0.60	0.05	0.025	86	122	137	153	173	198	231	286	403
0.60	0.10	0.050	60	91	104	119	136	158	188	238	345
0.60	0.20	0.100	37	62	72	85	100	118	144	188	285
0.65	0.01	0.005	83	108	119	130	144	160	182	217	292
0.65	0.02	0.010	67	91	100	111	123	139	159	192	263
0.65	0.05	0.025	48	68	76	85	96	110	128	158	223
0.65	0.10	0.050	34	51	58	66	76	88	105	132	191
0.65	0.20	0.100	21	35	40	47	55	66	80	104	157
0.70	0.01	0.005	52	68	75	82	90	101	114	136	183
0.70	0.02	0.010	43	57	63	70	78	87	100	121	164
0.70	0.05	0.025	31	43	48	54	61	69	81	99	139
0.70	0.10	0.050	22	32	37	42	48	55	66	82	119
0.70	0.20	0.100	13	22	26	30	35	41	50	65	98
0.75	0.01	0.005	36	47	51	56	61	68	77	92	123
0.75	0.02	0.010	29	39	43	47	53	59	68	81	110
0.75	0.05	0.025	21	29	33	37	41	47	54	67	93
0.75	0.10	0.050	15	22	25	28	32	37	44	55	80
0.75	0.20	0.100	9	15	17	20	24	28	34	44	66
0.80	0.01	0.005	26	33	36	40	44	49	55	65	87
0.80	0.02	0.010	21	28	31	34	38	42	48	58	78
0.80	0.05	0.025	15	21	23	26	29	33	39	47	66
0.80	0.10	0.050	11	16	18	20	23	27	31	39	56
0.80	0.20	0.100	7	11	12	14	17	20	24	31	46
0.85	0.01	0.005	19	25	27	29	32	36	40	48	63
0.85	0.02	0.010	16	21	23	25	28	31	35	42	57
0.85	0.05	0.025	11	16	17	19	22	24	28	34	48
0.85	0.10	0.050	8	12	13	15	17	20	23	29	41
0.85	0.20	0.100	5	8	9	11	12	15	17	22	33
0.90	0.01	0.005	15	19	20	22	24	27	30	35	47
0.90	0.02	0.010	12	16	17	19	21	23	26	31	42
0.90	0.05	0.025	9	12	13	15	16	18	21	26	35
0.90	0.10	0.050	6	9	10	11	13	15	17	21	30
0.90	0.20	0.100	4	6	7	8	9	11	13	17	24
0.95	0.01	0.005	12	15	16	17	19	20	23	27	35
0.95	0.02	0.010	10	12	13	15	16	18	20	24	31
0.95	0.05	0.025	7	9	10	11	12	14	16	19	26
0.95	0.10	0.050	5	7	8	9	10	11	13	16	22
0.95	0.20	0.100	3	5	5	6	7	8	10	12	18

Table 3.1   32

Table 3.1 Continued.

π_2	α 2-sided	1-sided	Power $1-\beta$ 0.50	0.65	0.70	0.75	0.80	0.85	0.90	0.95	0.99
						$\pi_1=0.50$					
0.55	0.01	0.005	1324	1749	1917	2107	2329	2602	2966	3551	4789
0.55	0.02	0.010	1080	1467	1621	1796	2001	2255	2595	3143	4314
0.55	0.05	0.025	767	1097	1231	1384	1565	1790	2095	2590	3661
0.55	0.10	0.050	540	822	939	1073	1233	1433	1707	2157	3142
0.55	0.20	0.100	328	554	651	763	899	1071	1309	1707	2593
0.60	0.01	0.005	329	434	475	522	577	645	735	879	1184
0.60	0.02	0.010	268	364	402	445	496	559	643	778	1067
0.60	0.05	0.025	191	272	305	343	388	443	519	641	905
0.60	0.10	0.050	134	204	233	266	305	355	423	533	777
0.60	0.20	0.100	82	138	161	189	223	265	324	422	641
0.65	0.01	0.005	145	190	208	229	253	282	321	384	517
0.65	0.02	0.010	118	160	176	195	217	244	281	340	465
0.65	0.05	0.025	84	120	134	150	170	194	227	280	395
0.65	0.10	0.050	59	90	102	117	134	155	185	233	338
0.65	0.20	0.100	36	61	71	83	98	116	142	184	279
0.70	0.01	0.005	80	105	115	126	139	155	177	211	283
0.70	0.02	0.010	65	88	97	108	120	134	154	186	255
0.70	0.05	0.025	47	66	74	83	93	107	124	153	216
0.70	0.10	0.050	33	50	56	64	74	85	101	128	185
0.70	0.20	0.100	20	34	39	46	54	64	78	101	152
0.75	0.01	0.005	50	66	72	79	87	96	110	131	175
0.75	0.02	0.010	41	55	61	67	74	84	96	115	157
0.75	0.05	0.025	29	41	46	52	58	66	77	95	133
0.75	0.10	0.050	21	31	35	40	46	53	63	79	114
0.75	0.20	0.100	13	21	24	29	33	40	48	62	94
0.80	0.01	0.005	34	44	48	53	58	65	73	87	116
0.80	0.02	0.010	28	37	41	45	50	56	64	77	104
0.80	0.05	0.025	20	28	31	35	39	44	52	63	88
0.80	0.10	0.050	14	21	24	27	31	35	42	52	76
0.80	0.20	0.100	9	14	17	19	22	26	32	41	62
0.85	0.01	0.005	24	31	34	37	41	45	51	61	81
0.85	0.02	0.010	20	26	29	32	35	39	45	54	73
0.85	0.05	0.025	14	20	22	24	27	31	36	44	61
0.85	0.10	0.050	10	15	17	19	22	25	29	37	52
0.85	0.20	0.100	6	10	12	14	16	19	22	29	43
0.90	0.01	0.005	18	23	25	27	30	33	37	44	58
0.90	0.02	0.010	15	19	21	23	25	28	32	39	52
0.90	0.05	0.025	11	14	16	18	20	22	26	32	44
0.90	0.10	0.050	8	11	12	14	16	18	21	26	37
0.90	0.20	0.100	5	7	9	10	11	13	16	21	30
0.95	0.01	0.005	14	17	19	20	22	24	27	32	42
0.95	0.02	0.010	11	14	16	17	19	21	24	28	38
0.95	0.05	0.025	8	11	12	13	15	17	19	23	32
0.95	0.10	0.050	6	8	9	10	12	13	15	19	27
0.95	0.20	0.100	4	6	6	7	8	10	12	15	22

Table 3.1

Table 3.2 Multiplying factor for comparing proportions when using chi-squared test with continuity correction.

m	0.05	0.10	0.15	0.20	0.25	0.30	0.35	0.40	$\pi_i - \pi_1$ 0.45	0.50	0.55	0.60	0.65	0.70	0.75	0.80	0.85	0.90	0.95
10	4.000	2.618	2.124	1.866	1.706	1.597	1.518	1.457	1.409	1.371	1.339	1.312	1.289	1.270	1.252	1.237	1.224	1.212	1.201
20	2.618	1.866	1.597	1.457	1.371	1.312	1.270	1.237	1.212	1.192	1.175	1.161	1.149	1.138	1.129	1.122	1.115	1.108	1.103
30	2.124	1.597	1.409	1.312	1.252	1.212	1.183	1.161	1.143	1.129	1.118	1.108	1.100	1.093	1.087	1.082	1.077	1.073	1.069
40	1.866	1.457	1.312	1.237	1.192	1.161	1.138	1.122	1.108	1.098	1.089	1.082	1.076	1.070	1.066	1.062	1.058	1.055	1.052
50	1.706	1.371	1.252	1.192	1.154	1.129	1.111	1.098	1.087	1.079	1.071	1.066	1.061	1.056	1.053	1.049	1.047	1.044	1.042
60	1.597	1.312	1.212	1.161	1.129	1.108	1.093	1.082	1.073	1.066	1.060	1.055	1.051	1.047	1.044	1.041	1.039	1.037	1.035
70	1.518	1.270	1.183	1.138	1.111	1.093	1.080	1.070	1.063	1.056	1.051	1.047	1.043	1.040	1.038	1.035	1.033	1.032	1.030
80	1.457	1.237	1.161	1.122	1.098	1.082	1.070	1.062	1.055	1.049	1.045	1.041	1.038	1.035	1.033	1.031	1.029	1.028	1.026
90	1.409	1.212	1.143	1.108	1.087	1.073	1.063	1.055	1.049	1.044	1.040	1.037	1.034	1.032	1.029	1.028	1.026	1.025	1.023
100	1.371	1.192	1.129	1.098	1.079	1.066	1.056	1.049	1.044	1.040	1.036	1.033	1.031	1.028	1.026	1.025	1.023	1.022	1.021
150	1.252	1.129	1.087	1.066	1.053	1.044	1.038	1.033	1.029	1.026	1.024	1.022	1.020	1.019	1.018	1.017	1.016	1.015	1.014
200	1.192	1.098	1.066	1.049	1.040	1.033	1.028	1.025	1.022	1.020	1.018	1.017	1.015	1.014	1.013	1.012	1.012	1.011	1.010
250	1.154	1.079	1.053	1.040	1.032	1.026	1.023	1.020	1.018	1.016	1.014	1.013	1.012	1.011	1.011	1.010	1.009	1.009	1.008
300	1.129	1.066	1.044	1.033	1.026	1.022	1.019	1.016	1.017	1.015	1.013	1.012	1.011	1.010	1.010	1.009	1.008	1.008	1.007
350	1.111	1.056	1.038	1.028	1.023	1.019	1.014	1.013	1.011	1.010	1.010	1.009	1.008	1.008	1.007	1.007	1.006	1.006	
400	1.098	1.049	1.033	1.025	1.020	1.017	1.014	1.012	1.011	1.010	1.009	1.008	1.008	1.007	1.007	1.006	1.006	1.006	1.005
450	1.087	1.044	1.029	1.022	1.018	1.015	1.013	1.011	1.010	1.009	1.008	1.007	1.007	1.006	1.006	1.006	1.005	1.005	1.005
500	1.079	1.040	1.026	1.020	1.016	1.013	1.011	1.010	1.009	1.008	1.007	1.007	1.006	1.006	1.005	1.005	1.005	1.004	1.004
550	1.071	1.036	1.024	1.018	1.014	1.012	1.010	1.009	1.008	1.007	1.007	1.006	1.006	1.005	1.005	1.005	1.004	1.004	1.004
600	1.066	1.033	1.022	1.017	1.013	1.011	1.010	1.008	1.007	1.007	1.006	1.006	1.005	1.005	1.004	1.004	1.004	1.004	1.004
650	1.061	1.031	1.020	1.015	1.012	1.010	1.009	1.008	1.007	1.006	1.006	1.005	1.005	1.004	1.004	1.004	1.004	1.003	1.003
700	1.056	1.028	1.019	1.014	1.011	1.010	1.008	1.007	1.006	1.006	1.005	1.005	1.004	1.004	1.004	1.004	1.003	1.003	1.003
750	1.053	1.026	1.018	1.013	1.011	1.009	1.008	1.007	1.006	1.005	1.005	1.004	1.004	1.004	1.004	1.003	1.003	1.003	1.003
800	1.049	1.025	1.017	1.012	1.010	1.008	1.007	1.006	1.006	1.005	1.005	1.004	1.004	1.004	1.003	1.003	1.003	1.003	1.003
850	1.047	1.023	1.016	1.012	1.009	1.008	1.007	1.006	1.005	1.005	1.004	1.004	1.004	1.003	1.003	1.003	1.003	1.003	1.002
900	1.044	1.022	1.015	1.011	1.009	1.007	1.006	1.006	1.005	1.004	1.004	1.004	1.003	1.003	1.003	1.003	1.003	1.002	1.002
950	1.042	1.021	1.014	1.010	1.008	1.007	1.006	1.005	1.005	1.004	1.004	1.004	1.003	1.003	1.003	1.003	1.002	1.002	1.002
1000	1.040	1.020	1.013	1.010	1.008	1.007	1.006	1.005	1.004	1.004	1.004	1.003	1.003	1.003	1.003	1.002	1.002	1.002	1.002
1500	1.026	1.013	1.009	1.007	1.005	1.004	1.004	1.003	1.003	1.003	1.002	1.002	1.002	1.002	1.002	1.002	1.002	1.001	1.001
2000	1.020	1.010	1.007	1.005	1.004	1.003	1.003	1.002	1.002	1.002	1.002	1.002	1.002	1.001	1.001	1.001	1.001	1.001	1.001
2500	1.016	1.008	1.005	1.004	1.003	1.003	1.002	1.002	1.002	1.002	1.001	1.001	1.001	1.001	1.001	1.001	1.001	1.001	1.001
3000	1.013	1.007	1.004	1.003	1.003	1.002	1.002	1.002	1.001	1.001	1.001	1.001	1.001	1.001	1.001	1.001	1.001	1.001	1.001
3500	1.011	1.006	1.004	1.003	1.002	1.002	1.002	1.001	1.001	1.001	1.001	1.001	1.001	1.001	1.001	1.001	1.001	1.001	1.001
4000	1.010	1.005	1.003	1.002	1.002	1.002	1.001	1.001	1.001	1.001	1.001	1.001	1.001	1.001	1.001	1.001	1.001	1.001	1.001
4500	1.009	1.004	1.003	1.002	1.002	1.001	1.001	1.001	1.001	1.001	1.001	1.001	1.001	1.001	1.001	1.001	1.001	1.000	1.000
5000	1.008	1.004	1.003	1.002	1.002	1.001	1.001	1.001	1.001	1.001	1.001	1.001	1.001	1.001	1.001	1.001	1.001	1.000	1.000

Table 3.2
34

Chapter 4
Testing Equivalence of Two Binomial Proportions

4.1 INTRODUCTION

If a reasonably effective therapy exists for a disease, then when a new treatment is introduced it may not be appropriate simply to compare proportions, as described in Chapter 3. Instead of attempting to find a significant difference between two treatments we may wish to show that a new treatment is equivalent in efficacy to the standard treatment, but the new treatment may have less side effects, may be socially more acceptable or be cheaper and less time-consuming. For example we may wish to compare radical mastectomy versus tumour resection in the treatment of breast cancer.

Failure to find a significant difference does not mean that the two treatments are equivalent. With a finite number of subjects we can never prove that two treatments are exactly equivalent. Having conducted an experiment to compare two treatments, we can calculate the proportion of successes under each one, and a confidence interval for the difference between them. With a given probability, $1-\alpha$, this confidence interval will cover the true difference in proportions. We now need to specify a limit within which we would accept that the two proportions are equivalent, and to specify a probability $1-\beta$ that the confidence limit will not exceed this specified value.

In general we are trying to show that the new treatment is not significantly *worse* than the standard treatment. It is not usually a serious error to claim that the two treatments are significantly different, when in fact they are not, since the consequence would be to keep the patients on the standard treatment. Thus α the type 1 error can be quite large, i.e. 0.1 or 0.2. We will usually be considering one-sided tests, since we are not trying to prove that the new treatment is better than the standard.

4.2 THEORY AND FORMULAE

Sample Size

Assume that outcome of the trial can be measured by one of two possibilities, e.g. 'cured' or 'not-cured', as in Chapter 3 and the true probability of success under treatment 1 is π_1 and under treatment 2 is π_2. After testing for equivalence of treatment we would assume that $\pi_1 = \pi_2$ for all practical purposes although we might have evidence that they in fact differ by a small amount. In conducting such a test for equivalence we wish to show, with a given probability $1-\beta$, that the trial will rule out any difference in proportions greater than some prespecified, usually small, difference ε. In addition, we specify the probability that the test might indicate that the two proportions are significantly different when there is, in fact, no difference. The number of patients required to be randomized to each treatment is given by

$$m = \frac{(z_{1-\alpha} + z_{1-\beta})^2 \{\pi_1(1-\pi_1) + \pi_2(1-\pi_2)\}}{\{\varepsilon - (\pi_1 - \pi_2)\}^2}. \tag{4.1}$$

This can be contrasted with equation (3.3) of Chapter 3. For the special case $\pi_1 = \pi_2 = \pi$ equation (4.1) simplifies to

$$m = \frac{2(z_{1-\alpha} + z_{1-\beta})^2 \pi(1-\pi)}{\varepsilon^2}. \tag{4.2}$$

Equations (4.1) and (4.2) can be easily evaluated using Table 2.3 which gives $(z_{1-\alpha} + z_{1-\beta})^2$ for different values of α and β. In practice values of ε between 0.05 and 0.2 are likely to be usual.

Power

In contrast, if one were asked for the power of a trial for given π_1, π_2, α, m and ε, then equation (4.1) can be inverted to give

$$z_{1-\beta} = \{\varepsilon - (\pi_1 - \pi_2)\} \sqrt{\left[\frac{m}{\pi_1(1-\pi_1) + \pi_2(1-\pi_2)} \right]} - z_{1-\alpha}. \tag{4.3}$$

4.3 BIBLIOGRAPHY

The formulae (4.1) and (4.2) are given by Makuch & Simon (1978) and are based on the normal approximation to the binomial distribution. Dunnett & Gent (1977) discuss alternative procedures which, in general, would yield similar sample sizes. A review of this approach has been given by Blackwelder (1982) and sample size graphs by Blackwelder & Chang (1984).

4.4 DESCRIPTION OF THE TABLE

Table 4.1

The table gives sample sizes required for the testing of equivalence of two binomial proportions for a range of two-sided test sizes of α from 0.02 to 0.40, one sided values from 0.01 to 0.20 and power $1-\beta$ ranging from 0.50 to 0.99. In contrast to Table 3.1, the larger binomial proportion is taken as that corresponding to treatment 1 and is denoted π_1. The smaller proportion is taken as π_2. In the table π_1 varies from 0.1 (0.05) 0.95, the value of π_2 increases in steps of 0.05 from $\pi_1 - \varepsilon$ to π_1. The values for ε are 0.05 (0.05) 0.20. The entries in the table are calculated using equations (4.1) and are rounded upwards to the nearest whole number.

4.5 THE USE OF THE TABLE

Table 4.1

Example 4.1

Consider a clinical trial to compare two treatments for stage I breast cancer. The treatments are mastectomy or simple removal of the lump, but leaving the remainder of the breast. We would like to show that, at worst, lump removal is only 10% inferior to mastectomy. Assuming the 5 year survival rate of stage I breast cancer after mastectomy is 60%, how large a trial would be needed to show that the 5 year survival rate for lump removal was at least 50%?

Equivalence is here defined as not more than 10% inferior, thus $\varepsilon = 0.1$. If in addition we set $\alpha = 0.1$ (one-sided test) then, on average, in one trial in ten we would say that mastectomy had a significantly lower 5 year survival rate than lump removal when in fact their rates were the same. If we choose a power $1-\beta = 0.8$, then we require a probability of 0.8 that our $100(1-\alpha)\%$ or 90% confidence limit does not exceed $\varepsilon = 0.1$. Then, from Table 4.1 with $\pi_1 = 0.60$, maximum allowable difference in proportions $\varepsilon = 0.10$, assuming π_2 is also equal to 0.60 and power $1-\beta = 0.80$, we would require at least 217 patients on each treatment.

If we were prepared to concede that lump removal did not in fact have the same survival rate as mastectomy, i.e. $\pi_2 < \pi_1$, but say the survival rate $\pi_2 = 0.55$, then from Table 4.1 the number of patients required for each treatment is 880. The number has increased considerably because now we wish to rule out a difference in proportions of 10%, when the true difference is in fact 5%.

Example 4.2

If we knew, before the trial planned in Example 4.1, that 150 patients for each treatment was the maximum that we would be able to recruit, then given that the expected survival rate is 60%, what sort of difference in proportions can we rule out?

Let us assume that we are considering a one-sided test, at $\alpha = 0.05$. Then, if the maximum allowable difference is set as $\varepsilon = 0.2$, with $\pi_1 = \pi_2 = 0.6$, then the corresponding entries in Table 4.1 are 130 patients with $1 - \beta = 0.95$ and 190 patients with $1 - \beta = 0.99$. Thus recruiting 150 patients per treatment would allow us to rule out a maximum allowable difference $\varepsilon = 0.2$ with a probability somewhere between 0.95 and 0.99.

Exact calculations using equation (4.3) give power $1 - \beta = 0.988$. If the maximum allowance difference were set at $\varepsilon = 0.1$, then we can rule out this difference with a probability between 0.65 and 0.70; more exactly power $1 - \beta = 0.685$.

Clearly such examples can only give an approximate guide to the sort of differences one is likely to be able to rule out. After any trial one should estimate the confidence interval for the difference in proportions to give a better idea as to how close the two treatments really are.

Example 4.3

Bennett, Dismukes, Duma *et al.* (1979) designed a clinical trial to test whether a combination chemotherapy for a shorter period would be at least as good as conventional therapy for patients with cryptococcal meningitis. They recruited 39 patients to each treatment arm and wished to conclude that a difference of less than 20% in response rate between the treatments would indicate equivalence. Assuming a one sided test size of 10% and an overall response rate of 50%, what is the power of the trial?

Here the maximum allowable difference $\varepsilon = 0.20$, $\pi_1 = \pi_2 = 0.50$ and one sided $\alpha = 0.1$. From Table 4.1 a trial of 35 patients has a power $1 - \beta = 0.65$ and a trial of 41 patients a power $1 - \beta = 0.70$, thus their trial with 39 patients per arm has a power between 65 and 70%. More exactly, using (4.3) $z_{1-\beta} = 0.4848$ which gives from Table 2.1 power $1 - \beta = 0.686$ or 68.6%.

4.5 REFERENCES

Bennett J. E., Dismukes W. E., Duma R. J., Medoff G., Sande M. A., Gallis H., Leonard J., Fields B., Bradshaw M., Haywood H., McGee Z., Cate T. R., Cobbs C. G., Warner J. F. & Alling D. W. (1979) A comparison of amphotericin B alone and combined with flucytosine in the treatment of cryptococcal meningitis. *New. Engl. J. Med.*, **301**, 126–131.

Blackwelder W. C. (1982) "Proving the null hypothesis" in clinical trials. *Controlled Clinical Trials,* **3**, 345–353.

Blackwelder W.C. & Chang M.A. (1984) Sample size graphs for 'Proving the null hypothesis'. *Controlled Clinical Trials,* **5**, 97–105.

Dunnett C. W. & Gent M. (1977) Significance testing to establish equivalence between treatments, with special reference to data in the form of 2 × 2 tables. *Biometrics,* **33**, 593–602.

Makuch R. & Simon R. (1978) Sample size requirements for evaluating a conservative therapy. *Cancer Treatment Reports,* **62**, 1037–1040.

Table 4.1 Sample sizes for testing equivalence of proportions.

ε	π_2	α 2-sided	α 1-sided	0.50	0.65	0.70	0.75	0.80	0.85	0.90	0.95	0.99
							Power $1-\beta$					
								$\pi_1 = 0.10$				
0.05	0.10	0.02	0.010	390	530	586	649	723	815	938	1136	1559
0.05	0.10	0.05	0.025	277	397	445	500	566	647	757	936	1323
0.05	0.10	0.10	0.050	195	297	339	388	446	518	617	780	1136
0.05	0.10	0.20	0.100	119	201	235	276	325	387	474	617	938
0.05	0.10	0.40	0.200	51	109	135	166	204	254	325	446	723
								$\pi_1 = 0.15$				
0.05	0.15	0.02	0.010	553	751	829	919	1024	1154	1328	1609	2209
0.05	0.15	0.05	0.025	392	562	630	708	801	916	1072	1326	1874
0.05	0.15	0.10	0.050	276	421	480	549	631	734	874	1104	1609
0.05	0.15	0.20	0.100	168	284	333	391	460	549	671	874	1328
0.05	0.15	0.40	0.200	73	154	191	235	289	360	460	631	1024
0.10	0.10	0.02	0.010	471	640	708	784	874	984	1133	1373	1884
0.10	0.10	0.05	0.025	335	479	537	604	683	782	915	1131	1599
0.10	0.10	0.10	0.050	236	359	410	469	538	626	746	942	1373
0.10	0.10	0.20	0.100	143	242	284	333	393	468	572	746	1133
0.10	0.10	0.40	0.200	62	131	163	200	247	307	393	538	874
0.10	0.15	0.02	0.010	139	188	208	230	256	289	332	403	553
0.10	0.15	0.05	0.025	98	141	158	177	201	229	268	332	469
0.10	0.15	0.10	0.050	69	106	120	138	158	184	219	276	403
0.10	0.15	0.20	0.100	42	71	84	98	115	138	168	219	332
0.10	0.15	0.40	0.200	19	39	48	59	73	90	115	158	256
								$\pi_1 = 0.20$				
0.05	0.20	0.02	0.010	693	942	1041	1153	1285	1448	1667	2019	2771
0.05	0.20	0.05	0.025	492	705	791	889	1005	1150	1345	1664	2352
0.05	0.20	0.10	0.050	347	528	603	689	792	921	1097	1386	2019
0.05	0.20	0.20	0.100	211	356	418	490	578	688	841	1097	1667
0.05	0.20	0.40	0.200	91	193	239	295	363	452	578	792	1285
0.10	0.15	0.02	0.010	623	846	935	1036	1155	1301	1497	1814	2490
0.10	0.15	0.05	0.025	442	633	710	799	903	1033	1209	1495	2113
0.10	0.15	0.10	0.050	312	474	542	619	711	827	985	1245	1814
0.10	0.15	0.20	0.100	189	320	376	441	519	618	756	985	1497
0.10	0.15	0.40	0.200	82	174	215	265	326	406	519	711	1155
0.10	0.20	0.02	0.010	174	236	261	289	322	362	417	505	693
0.10	0.20	0.05	0.025	123	177	198	223	252	288	337	416	588
0.10	0.20	0.10	0.050	87	132	151	173	198	231	275	347	505
0.10	0.20	0.20	0.100	53	89	105	123	145	172	211	275	417
0.10	0.20	0.40	0.200	23	49	60	74	91	113	145	198	322
0.15	0.10	0.02	0.010	542	736	813	901	1004	1131	1302	1578	2165
0.15	0.10	0.05	0.025	385	551	618	695	785	898	1051	1300	1838
0.15	0.10	0.10	0.050	271	413	471	538	619	719	857	1083	1578
0.15	0.10	0.20	0.100	165	278	327	383	451	538	657	857	1302
0.15	0.10	0.40	0.200	71	151	187	230	284	353	451	619	1004
0.15	0.15	0.02	0.010	156	212	234	259	289	326	375	454	623
0.15	0.15	0.05	0.025	111	159	178	200	226	259	303	374	529
0.15	0.15	0.10	0.050	78	119	136	155	178	207	247	312	454
0.15	0.15	0.20	0.100	48	80	94	111	130	155	189	247	375
0.15	0.15	0.40	0.200	21	44	54	67	82	102	130	178	289
0.15	0.20	0.02	0.010	77	105	116	129	143	161	186	225	308
0.15	0.20	0.05	0.025	55	79	88	99	112	128	150	185	262
0.15	0.20	0.10	0.050	39	59	67	77	88	103	122	154	225
0.15	0.20	0.20	0.100	24	40	47	55	65	77	94	122	186
0.15	0.20	0.40	0.200	11	22	27	33	41	51	65	88	143

Table 4.1

Table 4.1 Continued.

ε	π_2	α 2-sided	1-sided	Power $1-\beta$ 0.50	0.65	0.70	0.75	0.80	0.85	0.90	0.95	0.99
							$\pi_1 = 0.25$					
0.05	0.25	0.02	0.010	812	1103	1220	1351	1506	1697	1953	2366	3248
0.05	0.25	0.05	0.025	577	826	926	1042	1178	1347	1577	1950	2756
0.05	0.25	0.10	0.050	406	619	706	807	928	1079	1285	1624	2366
0.05	0.25	0.20	0.100	247	417	490	574	677	806	986	1285	1953
0.05	0.25	0.40	0.200	107	226	280	345	425	530	677	928	1506
0.10	0.20	0.02	0.010	753	1023	1130	1252	1396	1572	1810	2193	3010
0.10	0.20	0.05	0.025	534	765	858	965	1091	1248	1461	1807	2554
0.10	0.20	0.10	0.050	377	573	655	748	860	1000	1191	1505	2193
0.10	0.20	0.20	0.100	229	387	454	532	627	747	914	1191	1810
0.10	0.20	0.40	0.200	99	210	260	320	394	491	627	860	1396
0.10	0.25	0.02	0.010	203	276	305	338	377	425	489	592	812
0.10	0.25	0.05	0.025	145	207	232	261	295	337	395	488	689
0.10	0.25	0.10	0.050	102	155	177	202	232	270	322	406	592
0.10	0.25	0.20	0.100	62	105	123	144	170	202	247	322	489
0.10	0.25	0.40	0.200	27	57	70	87	107	133	170	232	377
0.15	0.15	0.02	0.010	682	927	1024	1135	1265	1425	1641	1988	2728
0.15	0.15	0.05	0.025	485	694	778	875	989	1132	1324	1638	2315
0.15	0.15	0.10	0.050	341	520	593	678	780	906	1080	1364	1988
0.15	0.15	0.20	0.100	207	351	411	483	568	678	828	1080	1641
0.15	0.15	0.40	0.200	90	190	236	290	357	445	568	780	1265
0.15	0.20	0.02	0.010	189	256	283	313	349	393	453	549	753
0.15	0.20	0.05	0.025	134	192	215	242	273	312	366	452	639
0.15	0.20	0.10	0.050	95	144	164	187	215	250	298	377	549
0.15	0.20	0.20	0.100	58	97	114	133	157	187	229	298	453
0.15	0.20	0.40	0.200	25	53	65	80	99	123	157	215	349
0.15	0.25	0.02	0.010	91	123	136	151	168	189	217	263	361
0.15	0.25	0.05	0.025	65	92	103	116	131	150	176	217	307
0.15	0.25	0.10	0.050	46	69	79	90	104	120	143	181	263
0.15	0.25	0.20	0.100	28	47	55	64	76	90	110	143	217
0.15	0.25	0.40	0.200	12	26	32	39	48	59	76	104	168
0.20	0.10	0.02	0.010	601	817	903	1000	1114	1256	1445	1751	2403
0.20	0.10	0.05	0.025	427	611	686	771	872	997	1167	1443	2040
0.20	0.10	0.10	0.050	301	458	523	598	687	799	951	1202	1751
0.20	0.10	0.20	0.100	183	309	363	425	501	597	730	951	1445
0.20	0.10	0.40	0.200	79	168	208	256	315	392	501	687	1114
0.20	0.15	0.02	0.010	171	232	256	284	317	357	411	497	682
0.20	0.15	0.05	0.025	122	174	195	219	248	283	331	410	579
0.20	0.15	0.10	0.050	86	130	149	170	195	227	270	341	497
0.20	0.15	0.20	0.100	52	88	103	121	142	170	207	270	411
0.20	0.15	0.40	0.200	23	48	59	73	90	112	142	195	317
0.20	0.20	0.02	0.010	84	114	126	140	156	175	202	244	335
0.20	0.20	0.05	0.025	60	85	96	108	122	139	163	201	284
0.20	0.20	0.10	0.050	42	64	73	84	96	112	133	168	244
0.20	0.20	0.20	0.100	26	43	51	60	70	83	102	133	202
0.20	0.20	0.40	0.200	11	24	29	36	44	55	70	96	156
0.20	0.25	0.02	0.010	51	69	77	85	95	107	123	148	203
0.20	0.25	0.05	0.025	37	52	58	66	74	85	99	122	173
0.20	0.25	0.10	0.050	26	39	45	51	58	68	81	102	148
0.20	0.25	0.20	0.100	16	27	31	36	43	51	62	81	123
0.20	0.25	0.40	0.200	7	15	18	22	27	34	43	58	95

Table 4.1

Table 4.1 Continued.

ε	π_2	α 2-sided	1-sided	0.50	0.65	0.70	0.75	Power $1-\beta$ 0.80	0.85	0.90	0.95	0.99
							$\pi_1 = 0.35$					
0.05	0.35	0.02	0.010	985	1339	1480	1639	1827	2059	2370	2871	3940
0.05	0.35	0.05	0.025	700	1002	1124	1264	1429	1635	1913	2366	3344
0.05	0.35	0.10	0.050	493	751	857	980	1126	1309	1559	1970	2871
0.05	0.35	0.20	0.100	299	506	594	697	821	978	1196	1559	2370
0.05	0.35	0.40	0.200	129	274	340	419	516	642	821	1126	1827
0.10	0.30	0.02	0.010	948	1287	1423	1576	1757	1979	2278	2760	3789
0.10	0.30	0.05	0.025	673	963	1081	1215	1374	1572	1839	2275	3216
0.10	0.30	0.10	0.050	474	722	824	942	1082	1259	1499	1894	2760
0.10	0.30	0.20	0.100	288	487	571	670	789	941	1150	1499	2278
0.10	0.30	0.40	0.200	124	264	327	403	496	618	789	1082	1757
0.10	0.35	0.02	0.010	247	335	370	410	457	515	593	718	985
0.10	0.35	0.05	0.025	175	251	281	316	358	409	479	592	836
0.10	0.35	0.10	0.050	124	188	215	245	282	328	390	493	718
0.10	0.35	0.20	0.100	75	127	149	175	206	245	299	390	593
0.10	0.35	0.40	0.200	33	69	85	105	129	161	206	282	457
0.15	0.25	0.02	0.010	899	1221	1350	1495	1666	1878	2161	2618	3594
0.15	0.25	0.05	0.025	638	914	1025	1153	1303	1491	1745	2158	3050
0.15	0.25	0.10	0.050	450	685	782	893	1027	1194	1422	1797	2618
0.15	0.25	0.20	0.100	273	462	542	636	749	892	1091	1422	2161
0.15	0.25	0.40	0.200	118	250	310	382	471	586	749	1027	1666
0.15	0.30	0.02	0.010	237	322	356	394	440	495	570	690	948
0.15	0.30	0.05	0.025	169	241	271	304	344	393	460	569	804
0.15	0.30	0.10	0.050	119	181	206	236	271	315	375	474	690
0.15	0.30	0.20	0.100	72	122	143	168	198	236	288	375	570
0.15	0.30	0.40	0.200	31	66	82	101	124	155	198	271	440
0.15	0.35	0.02	0.010	110	149	165	183	203	229	264	319	438
0.15	0.35	0.05	0.025	78	112	125	141	159	182	213	263	372
0.15	0.35	0.10	0.050	55	84	96	109	126	146	174	219	319
0.15	0.35	0.20	0.100	34	57	66	78	92	109	133	174	264
0.15	0.35	0.40	0.200	15	31	38	47	58	72	92	126	203
0.20	0.20	0.02	0.010	839	1140	1260	1396	1556	1753	2018	2445	3356
0.20	0.20	0.05	0.025	596	853	957	1076	1217	1392	1629	2015	2848
0.20	0.20	0.10	0.050	420	639	730	834	959	1115	1328	1678	2445
0.20	0.20	0.20	0.100	255	431	506	594	699	833	1019	1328	2018
0.20	0.20	0.40	0.200	110	234	290	357	440	547	699	959	1556
0.20	0.25	0.02	0.010	225	306	338	374	417	470	541	655	899
0.20	0.25	0.05	0.025	160	229	257	289	326	373	437	540	763
0.20	0.25	0.10	0.050	113	172	196	224	257	299	356	450	655
0.20	0.25	0.20	0.100	69	116	136	159	188	223	273	356	541
0.20	0.25	0.40	0.200	30	63	78	96	118	147	188	257	417
0.20	0.30	0.02	0.010	106	143	159	176	196	220	254	307	421
0.20	0.30	0.05	0.025	75	107	121	135	153	175	205	253	358
0.20	0.30	0.10	0.050	53	81	92	105	121	140	167	211	307
0.20	0.30	0.20	0.100	32	55	64	75	88	105	128	167	254
0.20	0.30	0.40	0.200	14	30	37	45	56	69	88	121	196
0.20	0.35	0.02	0.010	62	84	93	103	115	129	149	180	247
0.20	0.35	0.05	0.025	44	63	71	79	90	103	120	148	209
0.20	0.35	0.10	0.050	31	47	54	62	71	82	98	124	180
0.20	0.35	0.20	0.100	19	32	38	44	52	62	75	98	149
0.20	0.35	0.40	0.200	9	18	22	27	33	41	52	71	115

Table 4.1

Table 4.1 Continued.

ε	π_2	α 2-sided	1-sided	Power $1-\beta$ 0.50	0.65	0.70	0.75	0.80	0.85	0.90	0.95	0.99
							$\pi_1 = 0.40$					
0.05	0.40	0.02	0.010	1040	1412	1561	1729	1927	2172	2500	3028	4157
0.05	0.40	0.05	0.025	738	1057	1186	1333	1507	1724	2018	2495	3528
0.05	0.40	0.10	0.050	520	792	904	1033	1188	1381	1645	2078	3028
0.05	0.40	0.20	0.100	316	534	627	735	866	1032	1262	1645	2500
0.05	0.40	0.40	0.200	136	290	359	442	544	678	866	1188	1927
0.10	0.35	0.02	0.010	1013	1376	1520	1684	1877	2115	2435	2950	4049
0.10	0.35	0.05	0.025	719	1029	1155	1298	1468	1679	1965	2431	3436
0.10	0.35	0.10	0.050	506	771	880	1006	1157	1345	1602	2024	2950
0.10	0.35	0.20	0.100	308	520	610	716	843	1005	1229	1602	2435
0.10	0.35	0.40	0.200	133	282	349	430	530	660	843	1157	1877
0.10	0.40	0.02	0.010	260	353	391	433	482	543	625	757	1040
0.10	0.40	0.05	0.025	185	265	297	334	377	431	505	624	882
0.10	0.40	0.10	0.050	130	198	226	259	297	346	412	520	757
0.10	0.40	0.20	0.100	79	134	157	184	217	258	316	412	625
0.10	0.40	0.40	0.200	34	73	90	111	136	170	217	297	482
0.15	0.30	0.02	0.010	975	1324	1463	1621	1807	2036	2344	2839	3897
0.15	0.30	0.05	0.025	692	991	1111	1250	1413	1617	1892	2340	3308
0.15	0.30	0.10	0.050	487	742	848	969	1113	1295	1542	1948	2839
0.15	0.30	0.20	0.100	296	501	588	689	812	968	1183	1542	2344
0.15	0.30	0.40	0.200	128	271	336	414	510	635	812	1113	1807
0.15	0.35	0.02	0.010	254	344	380	421	470	529	609	738	1013
0.15	0.35	0.05	0.025	180	258	289	325	367	420	492	608	859
0.15	0.35	0.10	0.050	127	193	220	252	290	337	401	506	738
0.15	0.35	0.20	0.100	77	130	153	179	211	252	308	401	609
0.15	0.35	0.40	0.200	34	71	88	108	133	165	211	290	470
0.15	0.40	0.02	0.010	116	157	174	193	215	242	278	337	462
0.15	0.40	0.05	0.025	82	118	132	149	168	192	225	278	392
0.15	0.40	0.10	0.050	58	88	101	115	132	154	183	231	337
0.15	0.40	0.20	0.100	36	60	70	82	97	115	141	183	278
0.15	0.40	0.40	0.200	16	33	40	50	61	76	97	132	215
0.20	0.25	0.02	0.010	926	1258	1390	1540	1717	1934	2226	2697	3702
0.20	0.25	0.05	0.025	657	941	1056	1187	1343	1536	1797	2223	3142
0.20	0.25	0.10	0.050	463	705	805	920	1058	1230	1465	1851	2697
0.20	0.25	0.20	0.100	281	476	558	655	771	919	1124	1465	2226
0.20	0.25	0.40	0.200	122	258	320	394	485	604	771	1058	1717
0.20	0.30	0.02	0.010	244	331	366	406	452	509	586	710	975
0.20	0.30	0.05	0.025	173	248	278	313	354	405	473	585	827
0.20	0.30	0.10	0.050	122	186	212	243	279	324	386	487	710
0.20	0.30	0.20	0.100	74	126	147	173	203	242	296	386	586
0.20	0.30	0.40	0.200	32	68	84	104	128	159	203	279	452
0.20	0.35	0.02	0.010	113	153	169	188	209	235	271	328	450
0.20	0.35	0.05	0.025	80	115	129	145	164	187	219	271	382
0.20	0.35	0.10	0.050	57	86	98	112	129	150	178	225	328
0.20	0.35	0.20	0.100	35	58	68	80	94	112	137	178	271
0.20	0.35	0.40	0.200	15	32	39	48	59	74	94	129	209
0.20	0.40	0.02	0.010	65	89	98	109	121	136	157	190	260
0.20	0.40	0.05	0.025	47	67	75	84	95	108	127	156	221
0.20	0.40	0.10	0.050	33	50	57	65	75	87	103	130	190
0.20	0.40	0.20	0.100	20	34	40	46	55	65	79	103	157
0.20	0.40	0.40	0.200	9	19	23	28	34	43	55	75	121

Table 4.1

Table 4.1 Continued.

ε	π₂	α 2-sided	α 1-sided	Power 1−β 0.50	0.65	0.70	0.75	0.80	0.85	0.90	0.95	0.99
							π₁ = 0.30					
0.05	0.30	0.02	0.010	910	1236	1366	1513	1687	1900	2187	2650	3637
0.05	0.30	0.05	0.025	646	925	1037	1166	1319	1509	1766	2184	3087
0.05	0.30	0.10	0.050	455	693	791	904	1039	1208	1439	1819	2650
0.05	0.30	0.20	0.100	276	467	548	643	758	903	1104	1439	2187
0.05	0.30	0.40	0.200	119	253	314	387	476	593	758	1039	1687
0.10	0.25	0.02	0.010	861	1170	1293	1432	1596	1799	2070	2508	3442
0.10	0.25	0.05	0.025	611	875	982	1104	1248	1428	1671	2067	2922
0.10	0.25	0.10	0.050	431	656	749	856	984	1144	1362	1721	2508
0.10	0.25	0.20	0.100	262	442	519	609	717	855	1045	1362	2070
0.10	0.25	0.40	0.200	113	240	297	366	451	561	717	984	1596
0.10	0.30	0.02	0.010	228	309	342	379	422	475	547	663	910
0.10	0.30	0.05	0.025	162	232	260	292	330	378	442	546	772
0.10	0.30	0.10	0.050	114	174	198	226	260	302	360	455	663
0.10	0.30	0.20	0.100	69	117	137	161	190	226	276	360	547
0.10	0.30	0.40	0.200	30	64	79	97	119	149	190	260	422
0.15	0.20	0.02	0.010	801	1089	1203	1333	1486	1674	1927	2335	3204
0.15	0.20	0.05	0.025	569	815	914	1028	1162	1329	1556	1924	2720
0.15	0.20	0.10	0.050	401	610	697	797	916	1065	1268	1602	2335
0.15	0.20	0.20	0.100	244	412	483	567	668	796	973	1268	1927
0.15	0.20	0.40	0.200	105	223	277	341	420	523	668	916	1486
0.15	0.25	0.02	0.010	216	293	324	358	399	450	518	627	861
0.15	0.25	0.05	0.025	153	219	246	276	312	357	418	517	731
0.15	0.25	0.10	0.050	108	164	188	214	246	286	341	431	627
0.15	0.25	0.20	0.100	66	111	130	153	180	214	262	341	518
0.15	0.25	0.40	0.200	29	60	75	92	113	141	180	246	399
0.15	0.30	0.02	0.010	102	138	152	169	188	212	243	295	405
0.15	0.30	0.05	0.025	72	103	116	130	147	168	197	243	343
0.15	0.30	0.10	0.050	51	77	88	101	116	135	160	203	295
0.15	0.30	0.20	0.100	31	52	61	72	85	101	123	160	243
0.15	0.30	0.40	0.200	14	29	35	43	53	66	85	116	188
0.20	0.15	0.02	0.010	731	993	1098	1216	1355	1527	1758	2130	2923
0.20	0.15	0.05	0.025	519	743	834	937	1060	1213	1419	1755	2481
0.20	0.15	0.10	0.050	366	557	636	727	835	971	1157	1461	2130
0.20	0.15	0.20	0.100	222	376	441	517	609	726	887	1157	1758
0.20	0.15	0.40	0.200	96	204	252	311	383	477	609	835	1355
0.20	0.20	0.02	0.010	201	273	301	334	372	419	482	584	801
0.20	0.20	0.05	0.025	143	204	229	257	291	333	389	481	680
0.20	0.20	0.10	0.050	101	153	175	200	229	267	317	401	584
0.20	0.20	0.20	0.100	61	103	121	142	167	199	244	317	482
0.20	0.20	0.40	0.200	27	56	70	86	105	131	167	229	372
0.20	0.25	0.02	0.010	96	130	144	160	178	200	230	279	383
0.20	0.25	0.05	0.025	68	98	110	123	139	159	186	230	325
0.20	0.25	0.10	0.050	48	73	84	96	110	128	152	192	279
0.20	0.25	0.20	0.100	30	50	58	68	80	95	117	152	230
0.20	0.25	0.40	0.200	13	27	33	41	51	63	80	110	178
0.20	0.30	0.02	0.010	57	78	86	95	106	119	137	166	228
0.20	0.30	0.05	0.025	41	58	65	73	83	95	111	137	193
0.20	0.30	0.10	0.050	29	44	50	57	65	76	90	114	166
0.20	0.30	0.20	0.100	18	30	35	41	48	57	69	90	137
0.20	0.30	0.40	0.200	8	16	20	25	30	38	48	65	106

Table 4.1   42

Table 4.1 Continued.

ε	π_2	α 2-sided	α 1-sided	Power $1-\beta$ 0.50	0.65	0.70	0.75	0.80	0.85	0.90	0.95	0.99
							$\pi_1 = 0.45$					
0.05	0.45	0.02	0.010	1072	1456	1610	1783	1988	2240	2578	3123	4287
0.05	0.45	0.05	0.025	761	1090	1223	1375	1555	1778	2081	2573	3638
0.05	0.45	0.10	0.050	536	817	932	1066	1225	1424	1696	2143	3123
0.05	0.45	0.20	0.100	326	551	646	758	893	1064	1301	1696	2578
0.05	0.45	0.40	0.200	141	299	370	456	561	699	893	1225	1988
0.10	0.40	0.02	0.010	1056	1434	1585	1756	1958	2206	2539	3076	4222
0.10	0.40	0.05	0.025	750	1073	1204	1354	1531	1751	2049	2534	3583
0.10	0.40	0.10	0.050	528	804	918	1049	1206	1402	1670	2111	3076
0.10	0.40	0.20	0.100	321	542	636	747	880	1048	1282	1670	2539
0.10	0.40	0.40	0.200	139	294	364	449	553	688	880	1206	1958
0.10	0.45	0.02	0.010	268	364	403	446	497	560	645	781	1072
0.10	0.45	0.05	0.025	191	273	306	344	389	445	521	644	910
0.10	0.45	0.10	0.050	134	205	233	267	307	356	424	536	781
0.10	0.45	0.20	0.100	82	138	162	190	224	266	326	424	645
0.10	0.45	0.40	0.200	36	75	93	114	141	175	224	307	497
0.15	0.35	0.02	0.010	1029	1398	1545	1711	1907	2149	2474	2997	4114
0.15	0.35	0.05	0.025	730	1046	1173	1319	1492	1706	1997	2469	3491
0.15	0.35	0.10	0.050	515	784	895	1023	1175	1366	1628	205	2997
0.15	0.35	0.20	0.100	313	528	620	727	857	1021	1249	1628	2474
0.15	0.35	0.40	0.200	135	287	355	437	539	671	857	1175	1907
0.15	0.40	0.02	0.010	264	359	397	439	490	552	635	769	1056
0.15	0.40	0.05	0.025	188	269	301	339	383	438	513	634	896
0.15	0.40	0.10	0.050	132	201	230	263	302	351	418	528	769
0.15	0.40	0.20	0.100	81	136	159	187	220	262	321	418	635
0.15	0.40	0.40	0.200	35	74	91	113	139	172	220	302	490
0.15	0.45	0.02	0.010	120	162	179	199	221	249	287	347	477
0.15	0.45	0.05	0.025	85	122	136	153	173	198	232	286	405
0.15	0.45	0.10	0.050	60	91	104	119	137	159	189	239	347
0.15	0.45	0.20	0.100	37	62	72	85	100	119	145	189	287
0.15	0.45	0.40	0.200	16	34	42	51	63	78	100	137	221
0.20	0.30	0.02	0.010	991	1346	1488	1648	1837	2070	2383	2886	3962
0.20	0.30	0.05	0.025	703	1007	1130	1271	1437	1644	1923	2379	3363
0.20	0.30	0.10	0.050	496	755	862	985	1132	1316	1568	1981	2886
0.20	0.30	0.20	0.100	301	509	597	701	825	984	1203	1568	2383
0.20	0.30	0.40	0.200	130	276	342	421	519	646	825	1132	1837
0.20	0.35	0.02	0.010	258	350	387	428	477	538	619	750	1029
0.20	0.35	0.05	0.025	183	262	294	330	373	427	500	618	873
0.20	0.35	0.10	0.050	129	196	224	256	294	342	407	515	350
0.20	0.35	0.20	0.100	79	132	155	182	215	256	313	407	619
0.20	0.35	0.40	0.200	34	72	89	110	135	168	215	294	477
0.20	0.40	0.02	0.010	118	160	177	196	218	246	283	342	470
0.20	0.40	0.05	0.025	84	120	134	151	171	195	228	282	399
0.20	0.40	0.10	0.050	59	90	102	117	134	156	186	235	342
0.20	0.40	0.20	0.100	36	61	71	83	98	117	143	186	283
0.20	0.40	0.40	0.200	16	33	41	50	62	77	98	134	218
0.20	0.45	0.02	0.010	67	91	101	112	125	140	162	196	268
0.20	0.45	0.05	0.025	48	69	77	86	98	112	131	161	228
0.20	0.45	0.10	0.050	34	52	59	67	77	89	106	134	196
0.20	0.45	0.20	0.100	21	35	41	48	56	67	82	106	162
0.20	0.45	0.40	0.200	9	19	24	29	36	44	56	77	125

Table 4.1

Table 4.1 Continued.

ε	π_2	α 2-sided	1-sided	Power $1-\beta$ 0.50	0.65	0.70	0.75	0.80	0.85	0.90	0.95	0.99
							$\pi_1 = 0.50$					
0.05	0.50	0.02	0.010	1083	1471	1626	1802	2008	2262	2604	3155	4330
0.05	0.50	0.05	0.025	769	1101	1235	1389	1570	1796	2102	2599	3675
0.05	0.50	0.10	0.050	542	825	942	1076	1237	1438	1713	2165	3155
0.05	0.50	0.20	0.100	329	556	653	766	902	1075	1314	1713	2604
0.05	0.50	0.40	0.200	142	302	374	460	567	706	902	1237	2008
0.10	0.45	0.02	0.010	1077	1464	1618	1793	1998	2251	2591	3139	4308
0.10	0.45	0.05	0.025	765	1095	1229	1382	1562	1787	2091	2586	3657
0.10	0.45	0.10	0.050	539	821	937	1071	1231	1431	1705	2154	3139
0.10	0.45	0.20	0.100	327	553	650	762	898	1070	1308	1705	2591
0.10	0.45	0.40	0.200	141	300	372	458	564	702	898	1231	1998
0.10	0.50	0.02	0.010	271	368	407	451	502	566	651	789	1083
0.10	0.50	0.05	0.025	193	276	309	348	393	449	526	650	919
0.10	0.50	0.10	0.050	136	207	236	269	310	360	429	542	789
0.10	0.50	0.20	0.100	83	139	164	192	226	269	329	429	651
0.10	0.50	0.40	0.200	36	76	94	115	142	177	226	310	502
0.15	0.40	0.02	0.010	1061	1442	1593	1765	1968	2217	2552	3092	4243
0.15	0.40	0.05	0.025	753	1079	1210	1361	1539	1760	2060	2547	3602
0.15	0.40	0.10	0.050	531	808	923	1055	1212	1410	1679	2122	3092
0.15	0.40	0.20	0.100	322	545	640	750	884	1054	1288	1679	2552
0.15	0.40	0.40	0.200	139	296	366	451	556	692	884	1212	1968
0.15	0.45	0.02	0.010	270	366	405	449	500	563	648	785	1077
0.15	0.45	0.05	0.025	192	274	308	346	391	447	523	647	915
0.15	0.45	0.10	0.050	135	206	235	268	308	358	427	539	785
0.15	0.45	0.20	0.100	82	139	163	191	225	268	327	427	648
0.15	0.45	0.40	0.200	36	75	93	115	141	176	225	308	500
0.15	0.50	0.02	0.010	121	164	181	201	224	252	290	351	482
0.15	0.50	0.05	0.025	86	123	138	155	175	200	234	289	409
0.15	0.50	0.10	0.050	61	92	105	120	138	160	191	241	351
0.15	0.50	0.20	0.100	37	62	73	86	101	120	146	191	290
0.15	0.50	0.40	0.200	16	34	42	52	63	79	101	138	224
0.20	0.35	0.02	0.010	1034	1405	1553	1720	1917	2160	2487	3013	4135
0.20	0.35	0.05	0.025	734	1051	1179	1326	1500	1715	2007	2482	3510
0.20	0.35	0.10	0.050	517	788	899	1028	1181	1374	1636	2068	3013
0.20	0.35	0.20	0.100	314	531	623	731	862	1027	1255	1636	2487
0.20	0.35	0.40	0.200	136	288	357	440	542	674	862	1181	1917
0.20	0.40	0.02	0.010	266	361	399	442	492	555	638	773	1061
0.20	0.40	0.05	0.025	189	270	303	341	385	440	515	637	901
0.20	0.40	0.10	0.050	133	202	231	264	303	353	420	531	773
0.20	0.40	0.20	0.100	81	137	160	188	221	264	322	420	638
0.20	0.40	0.40	0.200	35	74	92	113	139	173	221	303	492
0.20	0.45	0.02	0.010	120	163	180	200	222	251	288	349	479
0.20	0.45	0.05	0.025	85	122	137	154	174	199	233	288	407
0.20	0.45	0.10	0.050	60	92	105	119	137	159	190	240	349
0.20	0.45	0.20	0.100	37	62	73	85	100	119	146	190	288
0.20	0.45	0.40	0.200	16	34	42	51	63	78	100	137	222
0.20	0.50	0.02	0.010	68	92	102	113	126	142	163	198	271
0.20	0.50	0.05	0.025	49	69	78	87	99	113	132	163	230
0.20	0.50	0.10	0.050	34	52	59	68	78	90	108	136	198
0.20	0.50	0.20	0.100	21	35	41	48	57	68	83	108	163
0.20	0.50	0.40	0.200	9	19	24	29	36	45	57	78	126

Table 4.1   44

Table 4.1 Continued.

ε	π_2	2-sided	1-sided	0.50	0.65	0.70	0.75	0.80	0.85	0.90	0.95	0.99
						$\pi_1 = 0.55$						
0.05	0.55	0.02	0.010	1072	1456	1610	1783	1988	2240	2578	3123	4287
0.05	0.55	0.05	0.025	761	1090	1223	1375	1555	1778	2081	2573	3638
0.05	0.55	0.10	0.050	536	817	932	1066	1225	1424	1696	2143	3123
0.05	0.55	0.20	0.100	326	551	646	758	893	1064	1301	1696	2578
0.05	0.55	0.40	0.200	141	299	370	456	561	699	893	1225	1988
0.10	0.50	0.02	0.010	1077	1464	1618	1793	1998	2251	2591	3139	4308
0.10	0.50	0.05	0.025	765	1095	1229	1382	1562	1787	2091	2586	3657
0.10	0.50	0.10	0.050	539	821	937	1071	1231	1431	1705	2154	3139
0.10	0.50	0.20	0.100	327	553	650	762	898	1070	1308	1705	2591
0.10	0.50	0.40	0.200	141	300	372	458	564	702	898	1231	1998
0.10	0.55	0.02	0.010	268	364	403	446	497	560	645	781	1072
0.10	0.55	0.05	0.025	191	273	306	344	389	445	521	644	910
0.10	0.55	0.10	0.050	134	205	233	267	307	356	424	536	781
0.10	0.55	0.20	0.100	82	138	162	190	224	266	326	424	645
0.10	0.55	0.40	0.200	36	75	93	114	141	175	224	307	497
0.15	0.45	0.02	0.010	1072	1456	1610	1783	1988	2240	2578	3123	4287
0.15	0.45	0.05	0.025	761	1090	1223	1375	1555	1778	2081	2573	3638
0.15	0.45	0.10	0.050	536	817	932	1066	1225	1424	1696	2143	3123
0.15	0.45	0.20	0.100	326	551	646	758	893	1064	1301	1696	2578
0.15	0.45	0.40	0.200	141	299	370	456	561	699	893	1225	1988
0.15	0.50	0.02	0.010	270	366	405	449	500	563	648	785	1077
0.15	0.50	0.05	0.025	192	274	308	346	391	447	523	647	915
0.15	0.50	0.10	0.050	135	206	235	268	308	358	427	539	785
0.15	0.50	0.20	0.100	82	139	163	191	225	268	327	427	648
0.15	0.50	0.40	0.200	36	75	93	115	141	176	225	308	500
0.15	0.55	0.02	0.010	120	162	179	199	221	249	287	347	477
0.15	0.55	0.05	0.025	85	122	136	153	173	198	232	286	405
0.15	0.55	0.10	0.050	60	91	104	119	137	159	189	239	347
0.15	0.55	0.20	0.100	37	62	72	85	100	119	145	189	287
0.15	0.55	0.40	0.200	16	34	42	51	63	78	100	137	221
0.20	0.40	0.02	0.010	1056	1434	1585	1756	1958	2206	2539	3076	4222
0.20	0.40	0.05	0.025	750	1073	1204	1354	1531	1751	2049	2534	3583
0.20	0.40	0.10	0.050	528	804	918	1049	1206	1402	1670	2111	3076
0.20	0.40	0.20	0.100	321	542	636	747	880	1048	1282	1670	2539
0.20	0.40	0.40	0.200	139	294	364	449	553	688	880	1206	1958
0.20	0.45	0.02	0.010	268	364	403	446	497	560	645	781	1072
0.20	0.45	0.05	0.025	191	273	306	344	389	445	521	644	910
0.20	0.45	0.10	0.050	134	205	233	267	307	356	424	536	781
0.20	0.45	0.20	0.100	82	138	162	190	224	266	326	424	645
0.20	0.45	0.40	0.200	36	75	93	114	141	175	224	307	497
0.20	0.50	0.02	0.010	120	163	180	200	222	251	288	349	479
0.20	0.50	0.05	0.025	85	122	137	154	174	199	233	288	407
0.20	0.50	0.10	0.050	60	92	105	119	137	159	190	240	349
0.20	0.50	0.20	0.100	37	62	73	85	100	119	146	190	288
0.20	0.50	0.40	0.200	16	34	42	51	63	78	100	137	222
0.20	0.55	0.02	0.010	67	91	101	112	125	140	162	196	268
0.20	0.55	0.05	0.025	48	69	77	86	98	112	131	161	228
0.20	0.55	0.10	0.050	34	52	59	67	77	89	106	134	196
0.20	0.55	0.20	0.100	21	35	41	48	56	67	82	106	162
0.20	0.55	0.40	0.200	9	19	24	29	36	44	56	77	125

Table 4.1

Table 4.1 Continued.

ε	π_2	α 2-sided	α 1-sided	0.50	0.65	0.70	Power $1-\beta$ 0.75	0.80	0.85	0.90	0.95	0.99
						$\pi_1 = 0.60$						
0.05	0.60	0.02	0.010	1040	1412	1561	1729	1927	2172	2500	3028	4157
0.05	0.60	0.05	0.025	738	1057	1186	1333	1507	1724	2018	2495	3528
0.05	0.60	0.10	0.050	520	792	904	1033	1188	1381	1645	2078	3028
0.05	0.60	0.20	0.100	316	534	627	735	866	1032	1262	1645	2500
0.05	0.60	0.40	0.200	136	290	359	442	544	678	866	1188	1927
0.10	0.55	0.02	0.010	1056	1434	1585	1756	1958	2206	2539	3076	4222
0.10	0.55	0.05	0.025	750	1073	1204	1354	1531	1751	2049	2534	3583
0.10	0.55	0.10	0.050	528	804	918	1049	1206	1402	1670	2111	3076
0.10	0.55	0.20	0.100	321	542	636	747	880	1048	1282	1670	2539
0.10	0.55	0.40	0.200	139	294	364	449	553	688	880	1206	1958
0.10	0.60	0.02	0.010	260	353	391	433	482	543	625	757	1040
0.10	0.60	0.05	0.025	185	265	297	334	377	431	505	624	882
0.10	0.60	0.10	0.050	130	198	226	259	297	346	412	520	757
0.10	0.60	0.20	0.100	79	134	157	184	217	258	316	412	625
0.10	0.60	0.40	0.200	34	73	90	111	136	170	217	297	482
0.15	0.50	0.02	0.010	1061	1442	1593	1765	1968	2217	2552	3092	4243
0.15	0.50	0.05	0.025	753	1079	1210	1361	1539	1760	2060	2547	3602
0.15	0.50	0.10	0.050	531	808	923	1055	1212	1410	1679	2122	3092
0.15	0.50	0.20	0.100	322	545	640	750	884	1054	1288	1679	2552
0.15	0.50	0.40	0.200	139	296	366	451	556	692	884	1212	1968
0.15	0.55	0.02	0.010	264	359	397	439	490	552	635	769	1056
0.15	0.55	0.05	0.025	188	269	301	339	383	438	513	634	896
0.15	0.55	0.10	0.050	132	201	230	263	302	351	418	528	769
0.15	0.55	0.20	0.100	81	136	159	187	220	262	321	418	635
0.15	0.55	0.40	0.200	35	74	91	113	139	172	220	302	490
0.15	0.60	0.02	0.010	116	157	174	193	215	242	278	337	462
0.15	0.60	0.05	0.025	82	118	132	149	168	192	225	278	392
0.15	0.60	0.10	0.050	58	88	101	115	132	154	183	231	337
0.15	0.60	0.20	0.100	36	60	70	82	97	115	141	183	278
0.15	0.60	0.40	0.200	16	33	40	50	61	76	97	132	215
0.20	0.45	0.02	0.010	1056	1434	1585	1756	1958	2206	2539	3076	4222
0.20	0.45	0.05	0.025	750	1073	1204	1354	1531	1751	2049	2534	3583
0.20	0.45	0.10	0.050	528	804	918	1049	1206	1402	1670	2111	3076
0.20	0.45	0.20	0.100	321	542	636	747	880	1048	1282	1670	2539
0.20	0.45	0.40	0.200	139	294	364	449	553	688	880	1206	1958
0.20	0.50	0.02	0.010	266	361	399	442	492	555	638	773	1061
0.20	0.50	0.05	0.025	189	270	303	341	385	440	515	637	901
0.20	0.50	0.10	0.050	133	202	231	264	303	353	420	531	773
0.20	0.50	0.20	0.100	81	137	160	188	221	264	322	420	638
0.20	0.50	0.40	0.200	35	74	92	113	139	173	221	303	492
0.20	0.55	0.02	0.010	118	160	177	196	218	246	283	342	470
0.20	0.55	0.05	0.025	84	120	134	151	171	195	228	282	399
0.20	0.55	0.10	0.050	59	90	102	117	134	156	186	235	342
0.20	0.55	0.20	0.100	36	61	71	83	98	117	143	186	283
0.20	0.55	0.40	0.200	16	33	41	50	62	77	98	134	218
0.20	0.60	0.02	0.010	65	89	98	109	121	136	157	190	260
0.20	0.60	0.05	0.025	47	67	75	84	95	108	127	156	221
0.20	0.60	0.10	0.050	33	50	57	65	75	87	103	130	190
0.20	0.60	0.20	0.100	20	34	40	46	55	65	79	103	157
0.20	0.60	0.40	0.200	9	19	23	28	34	43	55	75	121

Table 4.1   46

Table 4.1 Continued.

ε	π_2	α 2-sided	1-sided	Power $1-\beta$ 0.50	0.65	0.70	0.75	0.80	0.85	0.90	0.95	0.99
						$\pi_1 = 0.65$						
0.05	0.65	0.02	0.010	985	1339	1480	1639	1827	2059	2370	2871	3940
0.05	0.65	0.05	0.025	700	1002	1124	1264	1429	1635	1913	2366	3344
0.05	0.65	0.10	0.050	493	751	857	980	1126	1309	1559	1970	2871
0.05	0.65	0.20	0.100	299	506	594	697	821	978	1196	1559	2370
0.05	0.65	0.40	0.200	129	274	340	419	516	642	821	1126	1827
0.10	0.60	0.02	0.010	1013	1376	1520	1684	1877	2115	2435	2950	4049
0.10	0.60	0.05	0.025	719	1029	1155	1298	1468	1679	1965	2431	3436
0.10	0.60	0.10	0.050	506	771	880	1006	1157	1345	1602	2024	2950
0.10	0.60	0.20	0.100	308	520	610	716	843	1005	1229	1602	2435
0.10	0.60	0.40	0.200	133	282	349	430	530	660	843	1157	1877
0.10	0.65	0.02	0.010	247	335	370	410	457	515	593	718	985
0.10	0.65	0.05	0.025	175	251	281	316	358	409	479	592	836
0.10	0.65	0.10	0.050	124	188	215	245	282	328	390	493	718
0.10	0.65	0.20	0.100	75	127	149	175	206	245	299	390	593
0.10	0.65	0.40	0.200	33	69	85	105	129	161	206	282	457
0.15	0.55	0.02	0.010	1029	1398	1545	1711	1907	2149	2474	2997	4114
0.15	0.55	0.05	0.025	730	1046	1173	1319	1492	1706	1997	2469	3491
0.15	0.55	0.10	0.050	515	784	895	1023	1175	1366	1628	2057	2997
0.15	0.55	0.20	0.100	313	528	620	727	857	1021	1249	1628	2474
0.15	0.55	0.40	0.200	135	287	355	437	539	671	857	1175	1907
0.15	0.60	0.02	0.010	254	344	380	421	470	529	609	738	1013
0.15	0.60	0.05	0.025	180	258	289	325	367	420	492	608	859
0.15	0.60	0.10	0.050	127	193	220	252	290	337	401	506	738
0.15	0.60	0.20	0.100	77	130	153	179	211	252	308	401	609
0.15	0.60	0.40	0.200	34	71	88	108	133	165	211	290	470
0.15	0.65	0.02	0.010	110	149	165	183	203	229	264	319	438
0.15	0.65	0.05	0.025	78	112	125	141	159	182	213	263	372
0.15	0.65	0.10	0.050	55	84	96	109	126	146	174	219	319
0.15	0.65	0.20	0.100	34	57	66	78	92	109	133	174	264
0.15	0.65	0.40	0.200	15	31	38	47	58	72	92	126	203
0.20	0.50	0.02	0.010	1034	1405	1553	1720	1917	2160	2487	3013	4135
0.20	0.50	0.05	0.025	734	1051	1179	1326	1500	1715	2007	2482	3510
0.20	0.50	0.10	0.050	517	788	899	1028	1181	1374	1636	2068	3013
0.20	0.50	0.20	0.100	314	531	623	731	862	1027	1255	1636	2487
0.20	0.50	0.40	0.200	136	288	357	440	542	674	862	1181	1917
0.20	0.55	0.02	0.010	258	350	387	428	477	538	619	750	1029
0.20	0.55	0.05	0.025	183	262	294	330	373	427	500	618	873
0.20	0.55	0.10	0.050	129	196	224	256	294	342	407	515	750
0.20	0.55	0.20	0.100	79	132	155	182	215	256	313	407	619
0.20	0.55	0.40	0.200	34	72	89	110	135	168	215	294	477
0.20	0.60	0.02	0.010	113	153	169	188	209	235	271	328	450
0.20	0.60	0.05	0.025	80	115	129	145	164	187	219	271	382
0.20	0.60	0.10	0.050	57	86	98	112	129	150	178	225	328
0.20	0.60	0.20	0.100	35	58	68	80	94	112	137	178	271
0.20	0.60	0.40	0.200	15	32	39	48	59	74	94	129	209
0.20	0.65	0.02	0.010	62	84	93	103	115	129	149	180	247
0.20	0.65	0.05	0.025	44	63	71	79	90	103	120	148	209
0.20	0.65	0.10	0.050	31	47	54	62	71	82	98	124	180
0.20	0.65	0.20	0.100	19	32	38	44	52	62	75	98	149
0.20	0.65	0.40	0.200	9	18	22	27	33	41	52	71	115

Table 4.1

Table 4.1 Continued.

ε	π_2	α 2-sided	α 1-sided	Power $1-\beta$ 0.50	0.65	0.70	0.75	0.80	0.85	0.90	0.95	0.99
						$\pi_1 = 0.70$						
0.05	0.70	0.02	0.010	910	1236	1366	1513	1687	1900	2187	2650	3637
0.05	0.70	0.05	0.025	646	925	1037	1166	1319	1509	1766	2184	3087
0.05	0.70	0.10	0.050	455	693	791	904	1039	1208	1439	1819	2650
0.05	0.70	0.20	0.100	276	467	548	643	758	903	1104	1439	2187
0.05	0.70	0.40	0.200	119	253	314	387	476	593	758	1039	1687
0.10	0.65	0.02	0.010	948	1287	1423	1576	1757	1979	2278	2760	3789
0.10	0.65	0.05	0.025	673	963	1081	1215	1374	1572	1839	2275	3216
0.10	0.65	0.10	0.050	474	722	824	942	1082	1259	1499	1894	2760
0.10	0.65	0.20	0.100	288	487	571	670	789	941	1150	1499	2278
0.10	0.65	0.40	0.200	124	264	327	403	496	618	789	1082	1757
0.10	0.70	0.02	0.010	228	309	342	379	422	475	547	663	910
0.10	0.70	0.05	0.025	162	232	260	292	330	378	442	546	772
0.10	0.70	0.10	0.050	114	174	198	226	260	302	360	455	663
0.10	0.70	0.20	0.100	69	117	137	161	190	226	276	360	547
0.10	0.70	0.40	0.200	30	64	79	97	119	149	190	260	422
0.15	0.60	0.02	0.010	975	1324	1463	1621	1807	2036	2344	2839	3897
0.15	0.60	0.05	0.025	692	991	1111	1250	1413	1617	1892	2340	3308
0.15	0.60	0.10	0.050	487	742	848	969	1113	1295	1542	1948	2839
0.15	0.60	0.20	0.100	296	501	588	689	812	968	1183	1542	2344
0.15	0.60	0.40	0.200	128	271	336	414	510	635	812	1113	1807
0.15	0.65	0.02	0.010	237	322	356	394	440	495	570	690	948
0.15	0.65	0.05	0.025	169	241	271	304	344	393	460	569	804
0.15	0.65	0.10	0.050	119	181	206	236	271	315	375	474	690
0.15	0.65	0.20	0.100	72	122	143	168	198	236	288	375	570
0.15	0.65	0.40	0.200	31	66	82	101	124	155	198	271	440
0.15	0.70	0.02	0.010	102	138	152	169	188	212	243	295	405
0.15	0.70	0.05	0.025	72	103	116	130	147	168	197	243	343
0.15	0.70	0.10	0.050	51	77	88	101	116	135	160	203	295
0.15	0.70	0.20	0.100	31	52	61	72	85	101	123	160	243
0.15	0.70	0.40	0.200	14	29	35	43	53	66	85	116	188
0.20	0.55	0.02	0.010	991	1346	1488	1648	1837	2070	2383	2886	3962
0.20	0.55	0.05	0.025	703	1007	1130	1271	1437	1644	1923	2379	3363
0.20	0.55	0.10	0.050	496	755	862	985	1132	1316	1568	1981	2886
0.20	0.55	0.20	0.100	301	509	597	701	825	984	1203	1568	2383
0.20	0.55	0.40	0.200	130	276	342	421	519	646	825	1132	1837
0.20	0.60	0.02	0.010	244	331	366	406	452	509	586	710	975
0.20	0.60	0.05	0.025	173	248	278	313	354	405	473	585	827
0.20	0.60	0.10	0.050	122	186	212	243	279	324	386	487	710
0.20	0.60	0.20	0.100	74	126	147	173	203	242	296	386	586
0.20	0.60	0.40	0.200	32	68	84	104	128	159	203	279	452
0.20	0.65	0.02	0.010	106	143	159	176	196	220	254	307	421
0.20	0.65	0.05	0.025	75	107	121	135	153	175	205	253	358
0.20	0.65	0.10	0.050	53	81	92	105	121	140	167	211	307
0.20	0.65	0.20	0.100	32	55	64	75	88	105	128	167	254
0.20	0.65	0.40	0.200	14	30	37	45	56	69	88	121	196
0.20	0.70	0.02	0.010	57	78	86	95	106	119	137	166	228
0.20	0.70	0.05	0.025	41	58	65	73	83	95	111	137	193
0.20	0.70	0.10	0.050	29	44	50	57	65	76	90	114	166
0.20	0.70	0.20	0.100	18	30	35	41	48	57	69	90	137
0.20	0.70	0.40	0.200	8	16	20	25	30	38	48	65	106

Table 4.1

Table 4.1 Continued.

		α		Power $1-\beta$								
ε	π_2	2-sided	1-sided	0.50	0.65	0.70	0.75	0.80	0.85	0.90	0.95	0.99
						$\pi_1 = 0.75$						
0.05	0.75	0.02	0.010	812	1103	1220	1351	1506	1697	1953	2366	3248
0.05	0.75	0.05	0.025	577	826	926	1042	1178	1347	1577	1950	2756
0.05	0.75	0.10	0.050	406	619	706	807	928	1079	1285	1624	2366
0.05	0.75	0.20	0.100	247	417	490	574	677	806	986	1285	1953
0.05	0.75	0.40	0.200	107	226	280	345	425	530	677	928	1506
0.10	0.70	0.02	0.010	861	1170	1293	1432	1596	1799	2070	2508	3442
0.10	0.70	0.05	0.025	611	875	982	1104	1248	1428	1671	2067	2922
0.10	0.70	0.10	0.050	431	656	749	856	984	1144	1362	1721	2508
0.10	0.70	0.20	0.100	262	442	519	609	717	855	1045	1362	2070
0.10	0.70	0.40	0.200	113	240	297	366	451	561	717	984	1596
0.10	0.75	0.02	0.010	203	276	305	338	377	425	489	592	812
0.10	0.75	0.05	0.025	145	207	232	261	295	337	395	488	689
0.10	0.75	0.10	0.050	102	155	177	202	232	270	322	406	592
0.10	0.75	0.20	0.100	62	105	123	144	170	202	247	322	489
0.10	0.75	0.40	0.200	27	57	70	87	107	133	170	232	377
0.15	0.65	0.02	0.010	899	1221	1350	1495	1666	1878	2161	2618	3594
0.15	0.65	0.05	0.025	638	914	1025	1153	1303	1491	1745	2158	3050
0.15	0.65	0.10	0.050	450	685	782	893	1027	1194	1422	1797	2618
0.15	0.65	0.20	0.100	273	462	542	636	749	892	1091	1422	2161
0.15	0.65	0.40	0.200	118	250	310	382	471	586	749	1027	1666
0.15	0.70	0.02	0.010	216	293	324	358	399	450	518	627	861
0.15	0.70	0.05	0.025	153	219	246	276	312	357	418	517	731
0.15	0.70	0.10	0.050	108	164	188	214	246	286	341	431	627
0.15	0.70	0.20	0.100	66	111	130	153	180	214	262	341	518
0.15	0.70	0.40	0.200	29	60	75	92	113	141	180	246	399
0.15	0.75	0.02	0.010	91	123	136	151	168	189	217	263	361
0.15	0.75	0.05	0.025	65	92	103	116	131	150	176	217	307
0.15	0.75	0.10	0.050	46	69	79	90	104	120	143	181	263
0.15	0.75	0.20	0.100	28	47	55	64	76	90	110	143	217
0.15	0.75	0.40	0.200	12	26	32	39	48	59	76	104	168
0.20	0.60	0.02	0.010	926	1258	1390	1540	1717	1934	2226	2697	3702
0.20	0.60	0.05	0.025	657	941	1056	1187	1343	1536	1797	2223	3142
0.20	0.60	0.10	0.050	463	705	805	920	1058	1230	1465	1851	2697
0.20	0.60	0.20	0.100	281	476	558	655	771	919	1124	1465	2226
0.20	0.60	0.40	0.200	122	258	320	394	485	604	771	1058	1717
0.20	0.65	0.02	0.010	225	306	338	374	417	470	541	655	899
0.20	0.65	0.05	0.025	160	229	257	289	326	373	437	540	763
0.20	0.65	0.10	0.050	113	172	196	224	257	299	356	450	655
0.20	0.65	0.20	0.100	69	116	136	159	188	223	273	356	541
0.20	0.65	0.40	0.200	30	63	78	96	118	147	188	257	417
0.20	0.70	0.02	0.010	96	130	144	160	178	200	230	279	383
0.20	0.70	0.05	0.025	68	98	110	123	139	159	186	230	325
0.20	0.70	0.10	0.050	48	73	84	96	110	128	152	192	279
0.20	0.70	0.20	0.100	30	50	58	68	80	95	117	152	230
0.20	0.70	0.40	0.200	13	27	33	41	51	63	80	110	178
0.20	0.75	0.02	0.010	51	69	77	85	95	107	123	148	203
0.20	0.75	0.05	0.025	37	52	58	66	74	85	99	122	173
0.20	0.75	0.10	0.050	26	39	45	51	58	68	81	102	148
0.20	0.75	0.20	0.100	16	27	31	36	43	51	62	81	123
0.20	0.75	0.40	0.200	7	15	18	22	27	34	43	58	95

Table 4.1

Table 4.1 Continued.

ε	π_2	2-sided	1-sided	0.50	0.65	0.70	0.75	0.80	0.85	0.90	0.95	0.99
						$\pi_1 = 0.80$						
0.05	0.80	0.02	0.010	693	942	1041	1153	1285	1448	1667	2019	2771
0.05	0.80	0.05	0.025	492	705	791	889	1005	1150	1345	1664	2352
0.05	0.80	0.10	0.050	347	528	603	689	792	921	1097	1386	2019
0.05	0.80	0.20	0.100	211	356	418	490	578	688	841	1097	1667
0.05	0.80	0.40	0.200	91	193	239	295	363	452	578	792	1285
0.10	0.75	0.02	0.010	753	1023	1130	1252	1396	1572	1810	2193	3010
0.10	0.75	0.05	0.025	534	765	858	965	1091	1248	1461	1807	2554
0.10	0.75	0.10	0.050	377	573	655	748	860	1000	1191	1505	2193
0.10	0.75	0.20	0.100	229	387	454	532	627	747	914	1191	1810
0.10	0.75	0.40	0.200	99	210	260	320	394	491	627	860	1396
0.10	0.80	0.02	0.010	174	236	261	289	322	362	417	505	693
0.10	0.80	0.05	0.025	123	177	198	223	252	288	337	416	588
0.10	0.80	0.10	0.050	87	132	151	173	198	231	275	347	505
0.10	0.80	0.20	0.100	53	89	105	123	145	172	211	275	417
0.10	0.80	0.40	0.200	23	49	60	74	91	113	145	198	322
0.15	0.70	0.02	0.010	801	1089	1203	1333	1486	1674	1927	2335	3204
0.15	0.70	0.05	0.025	569	815	914	1028	1162	1329	1556	1924	2720
0.15	0.70	0.10	0.050	401	610	697	797	916	1065	1268	1602	2335
0.15	0.70	0.20	0.100	244	412	483	567	668	796	973	1268	1927
0.15	0.70	0.40	0.200	105	223	277	341	420	523	668	916	1486
0.15	0.75	0.02	0.010	189	256	283	313	349	393	453	549	753
0.15	0.75	0.05	0.025	134	192	215	242	273	312	366	452	639
0.15	0.75	0.10	0.050	95	144	164	187	215	250	298	377	549
0.15	0.75	0.20	0.100	58	97	114	133	157	187	229	298	453
0.15	0.75	0.40	0.200	25	53	65	80	99	123	157	215	349
0.15	0.80	0.02	0.010	77	105	116	129	143	161	186	225	308
0.15	0.80	0.05	0.025	55	79	88	99	112	128	150	185	262
0.15	0.80	0.10	0.050	39	59	67	77	88	103	122	154	225
0.15	0.80	0.20	0.100	24	40	47	55	65	77	94	122	186
0.15	0.80	0.40	0.200	11	22	27	33	41	51	65	88	143
0.20	0.65	0.02	0.010	839	1140	1260	1396	1556	1753	2018	2445	3356
0.20	0.65	0.05	0.025	596	853	957	1076	1217	1392	1629	2015	2848
0.20	0.65	0.10	0.050	420	639	730	834	959	1115	1328	1678	2445
0.20	0.65	0.20	0.100	255	431	506	594	699	833	1019	1328	2018
0.20	0.65	0.40	0.200	110	234	290	357	440	547	699	959	1556
0.20	0.70	0.02	0.010	201	273	301	334	372	419	482	584	801
0.20	0.70	0.05	0.025	143	204	229	257	291	333	389	481	680
0.20	0.70	0.10	0.050	101	153	175	200	229	267	317	401	584
0.20	0.70	0.20	0.100	61	103	121	142	167	199	244	317	482
0.20	0.70	0.40	0.200	27	56	70	86	105	131	167	229	372
0.20	0.75	0.02	0.010	84	114	126	140	156	175	202	244	335
0.20	0.75	0.05	0.025	60	85	96	108	122	139	163	201	284
0.20	0.75	0.10	0.050	42	64	73	84	96	112	133	168	244
0.20	0.75	0.20	0.100	26	43	51	60	70	83	102	133	202
0.20	0.75	0.40	0.200	11	24	29	36	44	55	70	96	156
0.20	0.80	0.02	0.010	44	59	66	73	81	91	105	127	174
0.20	0.80	0.05	0.025	31	45	50	56	63	72	85	104	147
0.20	0.80	0.10	0.050	22	33	38	44	50	58	69	87	127
0.20	0.80	0.20	0.100	14	23	27	31	37	43	53	69	105
0.20	0.80	0.40	0.200	6	13	15	19	23	29	37	50	81

Table 4.1

Table 4.1 Continued.

ε	π_2	α 2-sided	1-sided	0.50	0.65	0.70	Power $1-\beta$ 0.75	0.80	0.85	0.90	0.95	0.99
							$\pi_1 = 0.85$					
0.05	0.85	0.02	0.010	553	751	829	919	1024	1154	1328	1609	2209
0.05	0.85	0.05	0.025	392	562	630	708	801	916	1072	1326	1874
0.05	0.85	0.10	0.050	276	421	480	549	631	734	874	1104	1609
0.05	0.85	0.20	0.100	168	284	333	391	460	549	671	874	1328
0.05	0.85	0.40	0.200	73	154	191	235	289	360	460	631	1024
0.10	0.80	0.02	0.010	623	846	935	1036	1155	1301	1497	1814	2490
0.10	0.80	0.05	0.025	442	633	710	799	903	1033	1209	1495	2113
0.10	0.80	0.10	0.050	312	474	542	619	711	827	985	1245	1814
0.10	0.80	0.20	0.100	189	320	376	441	519	618	756	985	1497
0.10	0.80	0.40	0.200	82	174	215	265	326	406	519	711	1155
0.10	0.85	0.02	0.010	139	188	208	230	256	289	332	403	553
0.10	0.85	0.05	0.025	98	141	158	177	201	229	268	332	469
0.10	0.85	0.10	0.050	69	106	120	138	158	184	219	276	403
0.10	0.85	0.20	0.100	42	71	84	98	115	138	168	219	332
0.10	0.85	0.40	0.200	19	39	48	59	73	90	115	158	256
0.15	0.75	0.02	0.010	682	927	1024	1135	1265	1425	1641	1988	2728
0.15	0.75	0.05	0.025	485	694	778	875	989	1132	1324	1638	2315
0.15	0.75	0.10	0.050	341	520	593	678	780	906	1080	1364	1988
0.15	0.75	0.20	0.100	207	351	411	483	568	678	828	1080	1641
0.15	0.75	0.40	0.200	90	190	236	290	357	445	568	780	1265
0.15	0.80	0.02	0.010	156	212	234	259	289	326	375	454	623
0.15	0.80	0.05	0.025	111	159	178	200	226	259	303	374	529
0.15	0.80	0.10	0.050	78	119	136	155	178	207	247	312	454
0.15	0.80	0.20	0.100	48	80	94	111	130	155	189	247	375
0.15	0.80	0.40	0.200	21	44	54	67	82	102	130	178	289
0.15	0.85	0.02	0.010	62	84	93	103	114	129	148	179	246
0.15	0.85	0.05	0.025	44	63	70	79	89	102	120	148	209
0.15	0.85	0.10	0.050	31	47	54	61	71	82	98	123	179
0.15	0.85	0.20	0.100	19	32	37	44	52	61	75	98	148
0.15	0.85	0.40	0.200	9	18	22	27	33	40	52	71	114
0.20	0.70	0.02	0.010	731	993	1098	1216	1355	1527	1758	2130	2923
0.20	0.70	0.05	0.025	519	743	834	937	1060	1213	1419	1755	2481
0.20	0.70	0.10	0.050	366	557	636	727	835	971	1157	1461	2130
0.20	0.70	0.20	0.100	222	376	441	517	609	726	887	1157	1758
0.20	0.70	0.40	0.200	96	204	252	311	383	477	609	835	1355
0.20	0.75	0.02	0.010	171	232	256	284	317	357	411	497	682
0.20	0.75	0.05	0.025	122	174	195	219	248	283	331	410	579
0.20	0.75	0.10	0.050	86	130	149	170	195	227	270	341	497
0.20	0.75	0.20	0.100	52	88	103	121	142	170	207	270	411
0.20	0.75	0.40	0.200	23	48	59	73	90	112	142	195	317
0.20	0.80	0.02	0.010	70	94	104	116	129	145	167	202	277
0.20	0.80	0.05	0.025	50	71	79	89	101	115	135	167	235
0.20	0.80	0.10	0.050	35	53	61	69	79	92	110	139	202
0.20	0.80	0.20	0.100	21	36	42	49	58	69	84	110	167
0.20	0.80	0.40	0.200	10	20	24	30	37	46	58	79	129
0.20	0.85	0.02	0.010	35	47	52	58	64	73	83	101	139
0.20	0.85	0.05	0.025	25	36	40	45	51	58	67	83	118
0.20	0.85	0.10	0.050	18	27	30	35	40	46	55	69	101
0.20	0.85	0.20	0.100	11	18	21	25	29	35	42	55	83
0.20	0.85	0.40	0.200	5	10	12	15	19	23	29	40	64

Table 4.1

Table 4.1 Continued.

ε	π_2	α 2-sided	1-sided	0.50	0.65	0.70	Power $1-\beta$ 0.75	0.80	0.85	0.90	0.95	0.99
							$\pi_1 = 0.90$					
0.05	0.90	0.02	0.010	390	530	586	649	723	815	938	1136	1559
0.05	0.90	0.05	0.025	277	397	445	500	566	647	757	936	1323
0.05	0.90	0.10	0.050	195	297	339	388	446	518	617	780	1136
0.05	0.90	0.20	0.100	119	201	235	276	325	387	474	617	938
0.05	0.90	0.40	0.200	51	109	135	166	204	254	325	446	723
0.10	0.85	0.02	0.010	471	640	708	784	874	984	1133	1373	1884
0.10	0.85	0.05	0.025	335	479	537	604	683	782	915	1131	1599
0.10	0.85	0.10	0.050	236	359	410	469	538	626	746	942	1373
0.10	0.85	0.20	0.100	143	242	284	333	393	468	572	746	1133
0.10	0.85	0.40	0.200	62	131	163	200	247	307	393	538	874
0.10	0.90	0.02	0.010	98	133	147	163	181	204	235	284	390
0.10	0.90	0.05	0.025	70	100	112	125	142	162	190	234	331
0.10	0.90	0.10	0.050	49	75	85	97	112	130	155	195	284
0.10	0.90	0.20	0.100	30	51	59	69	82	97	119	155	235
0.10	0.90	0.40	0.200	13	28	34	42	51	64	82	112	181
0.15	0.80	0.02	0.010	542	736	813	901	1004	1131	1302	1578	2165
0.15	0.80	0.05	0.025	385	551	618	695	785	898	1051	1300	1838
0.15	0.80	0.10	0.050	271	413	471	538	619	719	857	1083	1578
0.15	0.80	0.20	0.100	165	278	327	383	451	538	657	857	1302
0.15	0.80	0.40	0.200	71	151	187	230	284	353	451	619	1004
0.15	0.85	0.02	0.010	118	160	177	196	219	246	284	344	471
0.15	0.85	0.05	0.025	84	120	135	151	171	196	229	283	400
0.15	0.85	0.10	0.050	59	90	103	118	135	157	187	236	344
0.15	0.85	0.20	0.100	36	61	71	84	99	117	143	187	284
0.15	0.85	0.40	0.200	16	33	41	50	62	77	99	135	219
0.15	0.90	0.02	0.010	44	59	66	73	81	91	105	127	174
0.15	0.90	0.05	0.025	31	45	50	56	63	72	85	104	147
0.15	0.90	0.10	0.050	22	33	38	44	50	58	69	87	127
0.15	0.90	0.20	0.100	14	23	27	31	37	43	53	69	105
0.15	0.90	0.40	0.200	6	13	15	19	23	29	37	50	81
0.20	0.75	0.02	0.010	601	817	903	1000	1114	1256	1445	1751	2403
0.20	0.75	0.05	0.025	427	611	686	771	872	997	1167	1443	2040
0.20	0.75	0.10	0.050	301	458	523	598	687	799	951	1202	1751
0.20	0.75	0.20	0.100	183	309	363	425	501	597	730	951	1445
0.20	0.75	0.40	0.200	79	168	208	256	315	392	501	687	1114
0.20	0.80	0.02	0.010	136	184	204	226	251	283	326	395	542
0.20	0.80	0.05	0.025	97	138	155	174	197	225	263	325	460
0.20	0.80	0.10	0.050	68	104	118	135	155	180	215	271	395
0.20	0.80	0.20	0.100	42	70	82	96	113	135	165	215	326
0.20	0.80	0.40	0.200	18	38	47	58	71	89	113	155	251
0.20	0.85	0.02	0.010	53	72	79	88	98	110	126	153	210
0.20	0.85	0.05	0.025	38	54	60	68	76	87	102	126	178
0.20	0.85	0.10	0.050	27	40	46	53	60	70	83	105	153
0.20	0.85	0.20	0.100	16	27	32	37	44	52	64	83	126
0.20	0.85	0.40	0.200	7	15	19	23	28	35	44	60	98
0.20	0.90	0.02	0.010	25	34	37	41	46	51	59	71	98
0.20	0.90	0.05	0.025	18	25	28	32	36	41	48	59	83
0.20	0.90	0.10	0.050	13	19	22	25	28	33	39	49	71
0.20	0.90	0.20	0.100	8	13	15	18	21	25	30	39	59
0.20	0.90	0.40	0.200	4	7	9	11	13	16	21	28	46

Table 4.1

Table 4.1 Continued.

ε	π_2	α 2-sided	1-sided	0.50	0.65	0.70	Power $1-\beta$ 0.75	0.80	0.85	0.90	0.95	0.99
						$\pi_1 = 0.95$						
0.05	0.95	0.02	0.010	206	280	309	343	382	430	495	600	823
0.05	0.95	0.05	0.025	146	210	235	264	299	342	400	494	699
0.05	0.95	0.10	0.050	103	157	179	205	235	274	326	412	600
0.05	0.95	0.20	0.100	63	106	124	146	172	205	250	326	495
0.05	0.95	0.40	0.200	27	58	71	88	108	135	172	235	382
0.10	0.90	0.02	0.010	298	405	447	496	552	622	716	868	1191
0.10	0.90	0.05	0.025	212	303	340	382	432	494	578	715	1011
0.10	0.90	0.10	0.050	149	227	259	296	341	396	472	596	868
0.10	0.90	0.20	0.100	91	153	180	211	248	296	362	472	716
0.10	0.90	0.40	0.200	39	83	103	127	156	194	248	341	552
0.10	0.95	0.02	0.010	52	70	78	86	96	108	124	150	206
0.10	0.95	0.05	0.025	37	53	59	66	75	86	100	124	175
0.10	0.95	0.10	0.050	26	40	45	52	59	69	82	103	150
0.10	0.95	0.20	0.100	16	27	31	37	43	52	63	82	124
0.10	0.95	0.40	0.200	7	15	18	22	27	34	43	59	96
0.15	0.85	0.02	0.010	379	515	569	631	703	792	912	1104	1516
0.15	0.85	0.05	0.025	269	386	433	486	550	629	736	910	1287
0.15	0.85	0.10	0.050	190	289	330	377	433	504	600	758	1104
0.15	0.85	0.20	0.100	115	195	229	268	316	377	460	600	912
0.15	0.85	0.40	0.200	50	106	131	161	199	247	316	433	703
0.15	0.90	0.02	0.010	75	102	112	124	138	156	179	217	298
0.15	0.90	0.05	0.025	53	76	85	96	108	124	145	179	253
0.15	0.90	0.10	0.050	38	57	65	74	86	99	118	149	217
0.15	0.90	0.20	0.100	23	39	45	53	62	74	91	118	179
0.15	0.90	0.40	0.200	10	21	26	32	39	49	62	86	138
0.15	0.95	0.02	0.010	23	32	35	39	43	48	55	67	92
0.15	0.95	0.05	0.025	17	24	27	30	34	38	45	55	78
0.15	0.95	0.10	0.050	12	18	20	23	27	31	37	46	67
0.15	0.95	0.20	0.100	7	12	14	17	20	23	28	37	55
0.15	0.95	0.40	0.200	3	7	8	10	12	15	20	27	43
0.20	0.80	0.02	0.010	450	611	675	748	833	939	1081	1309	1797
0.20	0.80	0.05	0.025	319	457	513	577	652	746	873	1079	1525
0.20	0.80	0.10	0.050	225	343	391	447	514	597	711	899	1309
0.20	0.80	0.20	0.100	137	231	271	318	375	446	546	711	1081
0.20	0.80	0.40	0.200	59	125	155	191	236	293	375	514	833
0.20	0.85	0.02	0.010	95	129	143	158	176	198	228	276	379
0.20	0.85	0.05	0.025	68	97	109	122	138	158	184	228	322
0.20	0.85	0.10	0.050	48	73	83	95	109	126	150	190	276
0.20	0.85	0.20	0.100	29	49	58	67	79	95	115	150	228
0.20	0.85	0.40	0.200	13	27	33	41	50	62	79	109	176
0.20	0.90	0.02	0.010	34	45	50	56	62	70	80	97	133
0.20	0.90	0.05	0.025	24	34	38	43	48	55	65	80	113
0.20	0.90	0.10	0.050	17	26	29	33	38	44	53	67	97
0.20	0.90	0.20	0.100	11	17	20	24	28	33	41	53	80
0.20	0.90	0.40	0.200	5	10	12	15	18	22	28	38	62
0.20	0.95	0.02	0.010	13	18	20	22	24	27	31	38	52
0.20	0.95	0.05	0.025	10	14	15	17	19	22	25	31	44
0.20	0.95	0.10	0.050	7	10	12	13	15	18	21	26	38
0.20	0.95	0.20	0.100	4	7	8	10	11	13	16	21	31
0.20	0.95	0.40	0.200	2	4	5	6	7	9	11	15	24

Table 4.1

Chapter 5
Confidence Limits for a Binomial Proportion

5.1 INTRODUCTION

As part of an epidemiological survey, or as a prelude to conducting a clinical trial one might require the prevalence of a particular disease in a population. For example, in a clinical trial for the treatment of asthma in young children, a knowledge of the prevalence of the disease in the area surrounding his hospital would enable a clinician to gauge the need for treatment and the maximum numbers available for trial entry. Given that the disease is fairly well defined, the clinician is likely to ask how many subjects do I need to examine in order to assess prevalence with a reasonable degree of accuracy? Providing that the sample is selected by simple random sampling (described in Chapter 13), then the required sample size can be calculated.

The sample size is determined by a number of factors. Firstly, how precise should the estimate be? If the investigator can only allow a small margin of error, then he will need a large sample. Often the investigator will express his error as a percentage of the estimate to within, say, 5%.

The second factor is the probability that our estimate is close to the population parameter we are trying to estimate. The actual prevalence can be determined only by examining the entire population. It is possible that the limits obtained from a sample do not contain the actual prevalence but one can estimate the probability that the interval contains it. We often strive for a 95% confidence that the limits we have obtained contain the actual prevalence. This means that if we conducted a large number of fixed size surveys, then we could claim that 95% of them contained the true prevalence. Sometimes it is desirable to set the limits so that 99% of all samples will contain the true prevalence. For a fixed sample size this clearly means wider limits, alternatively for a prespecified set of confidence limits one may have to increase the sample size in order to achieve them.

Finally we need to have some idea of the prevalence in the population under study. Clearly we need to examine larger numbers of people to estimate the prevalence of a rare event than the prevalence of one that is relatively common. Sometimes we may have an idea of the prevalence from previous studies, in other cases it may merely be an intelligent guess.

5.2 THEORY AND FORMULAE

Infinite Population

Let π be the true prevalence and n be the number of subjects required to estimate π to within $\varepsilon\%$. Provided that neither $n\pi$ nor $n(1-\pi)$ is not too small (a good guide is that they are both greater than 10) then we can use the normal approximation to the binomial distribution and obtain approximate $100(1-\alpha)\%$ confidence limits for π as

$$p + z_{1-\alpha}\sqrt{\frac{\pi(1-\pi)}{n}}, \tag{5.1}$$

where p is the estimate of π based on the n subjects. It is the right handside of expression (5.1) that we require to be $\varepsilon\%$ of π. At the planning stage, however, we have not observed p and only have an informed guess for π. This leads to an estimate of the required sample size as

$$n = \frac{10000(1-\pi)z_{1-\alpha}^2}{\pi\varepsilon^2}. \tag{5.2}$$

On completion of the study an estimate p of π is obtained and the confidence interval for π corresponding to equation (5.1) can be expressed as

$$p - \frac{\varepsilon p}{100} \text{ to } p + \frac{\varepsilon p}{100} \tag{5.3}$$

If $n\pi$ or $n(1 - \pi)$ is small, then the normal approximation used in (5.1) breaks down, and the confidence interval is no longer symmetric about p. The upper and lower limits π_U and π_L for a 95% confidence interval are the values for which the observed proportion p is significant on a one-sided test at the 2½% level.

Let r be the number of cases found in a sample of size n. Then, for a $100(1 - \alpha)\%$ confidence interval π_L is chosen as the value of π that satisfies

$$\sum_{s=r}^{n} \binom{n}{s} \pi^S (1 - \pi)^{n-s} = \frac{\alpha}{2} \tag{5.4}$$

while π_U is chosen as the value of π that satisfies

$$\sum_{s=0}^{r} \binom{n}{s} {}^s (1 - \pi)^{n;-1} = \frac{\alpha}{2}. \tag{5.5}$$

These values can be obtained from tables of the binomial distribution. They are not symmetric about $p = r/n$; the asymmetry is most pronounced when either ε is large or p is close to one and in these situations π_L is further from p than is π_U. Tables giving solutions to equations (5.3) and (5.4) for values of n from 10 to 1000 are available, and from these values of n and r were chosen so that r/n is approximately the required proportion p, and 100 $\{(r/n) - \pi_L\}/(r/n)$ is approximately equal to ε.

Finite Population

In many cases the population from which the researcher is sampling is known and of limited size. For example he may wish to assess the prevalence of impotence amongst diabetics on a diabetic register. In this case it is clear that if one sampled 60 out of a total population of 100 one would have a more accurate assessment of the prevalence than if one took a sample of 60 out a population of size 1000. Given that we have estimated p by the ratio of r cases from a sample of size n diabetics from a population of size N, then, if the true prevalence is π an approximate $100(1-\alpha)\%$ confidence interval for π is

$$p \pm z_{1-\alpha}\sqrt{\left\{\frac{\pi(1-\pi)}{n}\left(\frac{N-n}{N}\right)\right\}}. \tag{5.6}$$

Given that n is the sample size given by equation (5.2), which in fact assumes that the population is infinitely large, it can be shown that for a finite population of size N the actual sample size required, n', is given by

$$n' = n/(1 + n/N). \tag{5.7}$$

Thus we can obtain the sample size required when sampling from a finite population simply by multiplying n, calculated as though the population were infinite, by the correction factor $1/(1 + n/N)$.

The purpose of sampling is to reduce the number of observations required by the investigator. However, if the required sample size is a major proportion of the population, say 80%, then it may be sensible to examine the whole population rather than a sample of it.

5.3 BIBLIOGRAPHY

Confidence intervals for a binomial proportion were discussed by Armitage (1971, p.115–118) and by Altman & Gore (1982, p.73-75). *Documenta Geigy* (1962) gives confidence limits for an observed proportion as do Mainland, Herrern & Sutcliffe (1956).

5.4 DESCRIPTION OF THE TABLES

Table 5.1

Table 5.1 differs in format from the tables described in Chapters 3 and 4 of this book in that only two significance levels 0.01 and 0.05 are given, and only two-sided confidence limits are considered. One rarely requires confidence intervals with different significance levels. The expected proportion π ranges from 0.05 to 0.95 in steps of 0.05. The columns are labelled 1, 5, 10, 15 or 20 being the value of ε, the required percentage of the expected value. For large values of p and ε the confidence intervals are no longer symmetric and so the figures below the dotted line are the number of subjects required to estimate the lower limit to the required precision as discussed in Section 5.2. This is the longer of the two limits. The numbers are given to two significant digits.

Table 5.2

Table 5.2 gives the correction factor required to obtain the size of a sample from a finite population when sampling without replacement. If n is the sample size indicated by Table 5.1, assuming an infinite population, and N is the actual population size then the first column gives the ratio n/N in steps of 0.05 from 0.05 to 2.00 and the second column gives $1/(1 + n/N)$. Note that the ratio n/N can exceed 1, i.e. the assumption of an infinite population, rather than a finite population, might suggest sampling more than the entire available subjects! However the correction factor will always reduce the required sample to be less than the available population.

5.4 USE OF THE TABLES

Table 5.1

Example 5.1

The prevalence rate of a disease among children in a particular area is believed to be about 30%. How many subjects are required if we wish to determine this prevalence to within about 3 percentage points with 95% confidence that this is true?

This is equivalent to determining the prevalence to within 10% of its true value. From Table 5.1, with $p = 0.30$, $\alpha = 0.05$ and $\varepsilon = 0.10$ it can be seen that the number of subjects required is approximately 900. If the investigator then decides that he would prefer to be 99% rather than 95% confident of being within 10% of the true value he will need 1600 children. Note that if he wishes to be within 5% of the true value, i.e. halving the confidence intervals, he needs to quadruple the number of subjects to approximately 3600 and 6200 for the 95% and 99% confidence intervals respectively.

Example 5.2

Suppose all that is known about the prevalence of the disease is that it might be somewhere between 10% and

40%. How does this effect the sample size calculation described in Example 5.1?

If it were 10% then from Table 5.1 the investigator would require about 3500 subjects, whereas if it were 40%, he needs only 580. If he decides to stick with the original estimate of 900 children from Example 5.1, then he will have a more precise estimate if the prevalence turns out to be near 40% than if it turns out to be near 10%.

Example 5.3

Campbell, Elwood, Abbas & Waters (1984) estimated the prevalence of angina in a group of 1400 women to be about 15%. What is the 95% confidence interval of this result?

From Table 5.1 with $p = 0.15$ and $n = 1400$ we can see that for a 95% confidence interval the value of ε lies between 10 and 15. Solving equation (5.1) for ε we obtain $\varepsilon = 0.12$, i.e. the 95% confidence interval is approximately $(15 - 0.12 \times 15, 15 + 0.12 \times 15)$ or $(13.2, 16.8)$.

Table 5.2

Example 5.4

There are 1000 diabetics on a register. If the estimated prevalence of impotence amongst diabetics is assumed to be 20%, how many subjects do we require if we are willing to allow a 95% confidence interval of 4 percentage points either side of the true prevalence?

Here $\pi = 0.2$ and four percentage points of this prevalence of 20% means that $\varepsilon = 4/20 = 0.20$. From Table 5.1 we would require about 390 subjects. This is approximately 40% of the diabetics on the register. Using Table 5.2 the correction factor is 0.71 and so the actual number of subjects required is $390 \times 0.71 = 277$.

REFERENCES

Altman D. G. & Gore S. M. (1982) *Statistics in Practice*. British Medical Association, London.
Armitage P. (1971) *Statistical Methods in Medical Research*. Blackwell Scientific Publication, Oxford.
Campbell M. J., Elwood P. C., Abass S., Waters W. E. (1984) Chest pain in women − a study of prevalence and mortality follow-up in South Wales. *J. Epid. and Community Health*, **38**, 17−20.
Documenta Geigy (1962) *Scientific Tables*, 6th edn. Manchester, Geigy.
Mainland D., Herrern L. & Sutcliffe M. I. (1956) *Statistical Tables for use with Binomial Samples − Contingency Tests, Confidence Limits and Sample Size Estimates*. University College of Medicine, New York.

Table 5.1 Number of subjects required to give approximate $100(1-\alpha)\%$ confidence intervals of $(p - \varepsilon\, p/100, p + \varepsilon\, p/100)$.

p	α 2-sided	0.01	0.05	0.10	0.15	0.20
0.05	0.01	1260900	50500	12700	5600	3200
0.05	0.05	730000	30000	7300	3300	1900
0.10	0.01	597300	23900	6000	2700	1500
0.10	0.05	345800	13900	3500	1600	870
0.15	0.01	376100	15100	3800	1700	940
0.15	0.05	217700	8800	2200	970	550
0.20	0.01	265500	10700	2700	1200	670
0.20	0.05	153700	6200	1600	690	390
0.25	0.01	199100	8000	2000	880	500
0.25	0.05	115300	4700	1200	520	290
0.30	0.01	154900	6200	1600	690	390
0.30	0.05	89700	3600	900	400	230
0.35	0.01	123300	5000	1300	550	310
0.35	0.05	71400	2900	720	320	180
0.40	0.01	99600	4000	1000	440	250
0.40	0.05	57700	2400	580	260	150
0.45	0.01	81200	3300	820	360	210
0.45	0.05	47000	1900	470	210	120
0.50	0.01	66400	2700	670	290	170
0.50	0.05	38500	1600	390	170	100
0.55	0.01	54300	2200	550	240	145
0.55	0.05	31500	1300	320	140	88
0.60	0.01	44300	1800	450	200	125
0.60	0.05	25700	1100	260	120	75
0.65	0.01	35800	1500	360	180	105
0.65	0.05	20700	830	210	110	63
0.70	0.01	28500	1200	290	145	87
0.70	0.05	16500	660	170	91	53
0.75	0.01	22200	890	250	120	73
0.75	0.05	12900	520	150	77	46
0.80	0.01	16600	670	200	100	63
0.80	0.05	9700	380	130	60	40
0.85	0.01	11800	470	160	81	52
0.85	0.05	6800	270	100	49	30
0.90	0.01	7400	300	120	62	41
0.90	0.05	4300	170	74	40	25
0.95	0.01	3500	220	83	51	30
0.95	0.05	2100	140	56	33	21

Below dotted line the confidence limits are no longer approximately symmetric and the table gives the number of subjects required to give the *lower* limit to within $\varepsilon\%$ of p.

Table 5.1

58

Table 5.2 Correction factor for sampling from a finite population.

n/N†	Correction factor $1/(1 + n/N)$	n/N	Correction factor $1/(1 + n/N)$
0.05	0.95	1.05	0.49
0.10	0.91	1.10	0.48
0.15	0.87	1.15	0.47
0.20	0.83	1.20	0.45
0.25	0.80	1.25	0.44
0.30	0.77	1.30	0.43
0.35	0.74	1.35	0.43
0.40	0.71	1.40	0.42
0.45	0.69	1.45	0.41
0.50	0.67	1.50	0.40
0.55	0.65	1.55	0.39
0.60	0.63	1.60	0.38
0.65	0.61	1.65	0.38
0.70	0.59	1.70	0.37
0.75	0.57	1.75	0.36
0.80	0.56	1.80	0.36
0.85	0.54	1.85	0.35
0.90	0.53	1.90	0.34
0.95	0.51	1.95	0.34
1.00	0.50	2.00	0.33

†Ratio of sample size assuming infinite population to actual population size.

Chapter 6
Post-Marketing Surveillance

6.1 INTRODUCTION

After a drug has been accepted for general use it is common practice to survey a large population of treated patients to identify adverse effects. In some circumstances it is possible that only one adverse reaction, such as a drug related death, would be necessary for the drug to be considered unacceptable and withdrawn from a prescription list. In other situations a single adverse occurrence of a particular event would be put down to chance and two or three occurrences required to confirm suspicion about the drug. In most situations the observed adverse reactions may occur without the drug, for example, in an elderly population deaths are likely to occur anyway. Many common adverse reactions such as nausea, drowsiness and headache are prevalent in the population, and we need to know whether the drug has provoked an increase in prevalence of such adverse reactions over this background rate. If the background incidence is known, then the sample size calculations are relatively straightforward. If, as is more usual, the incidence is not known, then a control population might also be monitored for comparison purposes. Often the control group would be the same size as the patient group. However, if one takes a control group approximately five times the size of the patient group, it can be shown that the number of patients actually on the drug who need to be monitored can be reduced to approximately the number required if the incidence were in fact known.

6.2 THEORY AND FORMULAE

Sample Size

No Background Incidence of Adverse Reactions

Suppose the expected incidence of adverse reactions is λ, the number of occurrences of a particular adverse reaction is a and the number of patients required to be monitored is m. If the incidence of adverse reactions is reasonably low then one might assume that they follow a Poisson distribution. With these assumptions and defining β to be the probability that, for given incidence λ, we will *not* find a reactions in a sample of m patients on the particular drug under study, it can be shown that m satisfies

$$\sum_{x=0}^{a-1} \frac{m \, \lambda^x \, e^{-m\lambda}}{x!} = \beta. \tag{6.1}$$

For the special case in which the particular adverse reaction need occur in only 1 patient equation (6.1) simplifies and

$$m = -\log\beta/\lambda. \tag{6.2}$$

For $a > 1$ there is no simple expression for the solution to equation (6.1) but the equation can be solved using numerical methods.

With a known Background Incidence of Adverse Reactions

If λ_0 is the background incidence of the adverse reaction, and λ_1 the additional incidence caused by use of the particular drug under study, then, when the background incidence is known, for given significance level α and power $1-\beta$ we would require

$$m = \frac{\{z_{1-\alpha}\sqrt{\lambda_o} + z_{1-\beta}\sqrt{(\lambda_o + \lambda_1)}\}^2}{\lambda_1^2} .$$ (6.3)

With an unknown Background Incidence of Adverse Reactions

If the background incidence is unknown, then a control group is needed. However, in order to estimate the number of subjects required in the study we still need to hazard a guess to the background incidence λ_0. If the control group is r times as big as the treated group then m satisfies

$$m = \frac{[z_{1-\alpha}\sqrt{\{(r+1)\bar{\lambda}(1-\bar{\lambda})\}} + z_{1-\beta}\sqrt{\{r\lambda_0(1-\lambda_0) + (\lambda_0 + \lambda_1)(1 - \lambda_0 - \lambda_1)\}}]^2}{r\lambda_1^2}$$ (6.4)

where $\bar{\lambda} = \{\lambda_0 + r(\lambda_0 + \lambda_1)\}/(r+1)$.

When λ_0 and λ_1 are both small and λ_1 much less than λ_0 equation (6.4) can be approximated by

$$m = \frac{[z_{1-\alpha}\sqrt{\{\lambda_o(r+1) + \lambda_1\}} + z_{1-\beta}\sqrt{\{\lambda_o(r+1) + r\lambda_1\}}]^2}{r\lambda_1^2} .$$ (6.5)

If the control group is the same size as the treated group, then $r = 1$ and (6.4) reduces to

$$m = \frac{(z_{1-\alpha} + z_{1-\beta})^2 \{\lambda_o(1-\lambda_o) + (\lambda_o + \lambda_1)(1-\lambda_o-\lambda_1)\}}{\lambda_1^2}$$ (6.6)

which in turn can be approximated by

$$m = \frac{(z_{1-\alpha} + z_{1-\beta})^2 (2\lambda_o + \lambda_1)}{\lambda_1^2} .$$ (6.7)

Equations (6.5) and (6.7) can more easily be evaluated by using Table 2.3 which gives $(z_{1-\alpha} + z_{1-\beta})^2$ for different values of α and β.

Several Independent Reactions

In practice several adverse reactions to a particular drug are often monitored simultaneously. For planning purposes these are often assumed all to have approximately the same incidence and to act independently of each other. If s reactions are being monitored simultaneously, then to avoid getting many false positive results the significance level of $\alpha = \alpha s$. Thus the only change required is to replace $z_{1-\alpha}$ by $z_{1-\alpha'}$ in these equations.

Power

In contrast, if one were asked for the power of a study for a given λ_0, λ_1, m and α then equations (6.3) and (6.4) could be inverted to give

$$z_{1-\beta} = \frac{\lambda_1\sqrt{m} - z_{1-\alpha}\sqrt{\lambda_o}}{\sqrt{(\lambda_o + \lambda_1)}}$$ (6.8)

and

$$z_{1-\beta} = \frac{\lambda_1\sqrt{(rm)} - z_{1-\alpha}\sqrt{\{r\lambda_0(1-\lambda_0) + (\lambda_0 + \lambda_1)(1-\lambda_0-\lambda_1)\}}}{\sqrt{\{(r+1)\bar{\lambda}(1-\bar{\lambda}\}}}$$ (6.9)

respectively.

6.2

6.3 BIBLIOGRAPHY

Post-marketing surveillance has been discussed by Lewis (1981, 1983) from the statistical point of view and also by Skegg & Doll (1977) and Wilson (1977).

6.4 DESCRIPTION OF THE TABLES

Table 6.1

The table gives the number of subjects to be monitored in order that either 1, 2, 3, 4 or 5 occurrences of the adverse side effect will be observed with probability $1-\beta$ ranging from 0.5 to 0.99 and for given incidence, λ, ranging from 0.0001 to 0.01.

Table 6.2

The table gives sample sizes required for detection of adverse reactions amongst the treated group of patients when compared with a known background incidence. The number of subjects to be monitored is given for background incidence λ_0 ranging from 0.001 to 0.1, increase in incidence λ_1 ranging from 0.0005 to 0.05, two-sided test sizes of α from 0.01 to 0.20, one sided values from 0.005 to 0.1 and power $1-\beta$ ranging from 0.5 to 0.99. The entries in the table are calculated using equation (6.3) and are rounded upwards.

Tables 6.3, 6.4 and 6.5

These tables are tabulated as Table 6.2 except that comparisons of the monitored group are with a simultaneously studied control group. Sample sizes are given for values of the ratio of controls to monitored subjects, $r = 1, 2$ and 5.

Table 6.6

This table is similar to Table 6.3 except that, instead of just one type of adverse reaction, we are now looking for 50 independent adverse reactions simultaneously.

Table 6.7

This table is similar to Table 6.3 except that, instead of just one type of adverse reaction, we are now looking for 100 independent adverse reactions simultaneously.

6.5 USE OF THE TABLES

Table 6.1

Example 6.1

In a previous survey, a hypertensive drug produced cardiac arrhythmias in about 1 in 10,000 people, i.e. an incidence of 0.0001. A researcher decides that if a new hypertensive drug produces three such arrhythmias then the drug will have to be withdrawn pending further research. He wishes to detect three events with a 99% probability of success.

Table 6.1, with $1-\beta = 0.99$, incidence $\lambda = 0.0001$ and $a = 3$, gives $m = 84,070$ subjects. If, on the other hand, the marketing division told him the maximum number of subjects was 30,000 then one can see by scanning Table 6.1 that he could detect 1 adverse reaction with a probability of success of 0.95 ($\lambda = 0.0001$, $a = 1$, $1-\beta = 0.95$) or 2 reactions with a probability of 0.80 ($\lambda = 0.0001$, $a = 2$, $1-\beta = 0.80$).

Table 6.2

Example 6.2

Suppose that a possible side-effect of a drug is an increased incidence of gastric cancer. In an elderly population, say that the annual incidence of gastric cancer is 1%, and the drug will be deemed unacceptable if this increases to 1.5%. What size of study is required?

If an experimenter is prepared to discount any result that states that the drug actually prevents gastric cancer, then this is a one-sided test, at say $\alpha = 0.05$, and if he requires a power $1-\beta = 0.9$ to detect this increase, then one can see from Table 6.2 with $\lambda_0 = 0.01$ and $\lambda_1 = 0.005$ that he would be required to study 4200 subjects receiving the drug.

Tables 6.3, 6.4 and 6.5

Example 6.3

If the experimenter of Example 6.2 did no know the annual incidence of gastric cancer, but was prepared to monitor a comparable population of equal size, how many subjects should be monitored in each group?

For one sided $\alpha = 0.05$, $1-\beta = 0.90$, $\lambda_0 = 0.01$, $\lambda_1 = 0.005$ and $r = 1$, Table 6.3 gives 8500 in each group. It should be emphasized that, although he does not know the actual incidence for the control group, he has to make some guess at its value in order to estimate the required number of patients. The patient number to be recruited is quite sensitive to the anticipated incidence λ_0. Thus in this example, if in fact $\lambda_0 = 0.005$ rather than 0.01 then the number of subjects in each group, for the same test size and power, would be 5100.

Example 6.4

Suppose that an investigator planning the study described in Example 6.3 has access to a cancer registry as a source for controls, and this enables him to monitor many more controls than patients on the drug. What affect does this have on the number of patients to be monitored?

As in Example 6.3, assuming one sided, $\alpha = 0.05$, $1-\beta = 0.90$, $\lambda_0 = 0.01$, $\lambda_1 = 0.005$, if we further assume $r = 5$ then Table 6.5 suggests recruiting 5100 patients receiving the drug and $5 \times 5100 = 25,500$ controls.

Example 6.5

A post-marketing surveillance study of 3000 patients and the same number of controls is planned. The incidence of an adverse reaction is estimated at about 0.5% in the control group and 1.0% in the treated group. What is the power of the proposed study for a one-sided significance test at the 5% level?

Table 6.3 with one-sided $\alpha = 0.05$, $\lambda_0 = 0.005$, $\lambda_1 = 0.005$, $r = 1$ gives $m = 2900$ patients for power

$1 - \beta = 0.70$, and $m = 3300$ patients for power $1 - \beta = 0.75$. Thus with 3000 patients there is a 70–75% chance of detecting a doubling of the incidence, if the true incidence is in fact 0.5%. Equation (6.9) with $m = 3000$ gives $z_{1-\beta} = 0.6003$, and from Table 2.1 the power, $1 - \beta = 0.73$.

Tables 6.6 and 6.7

Example 6.6

A post-marketing surveillance study is planned to determine the adverse reactions of a new analgesic. It is recognized that a large number of types of adverse reactions are possible and one might wish to look out for nausea, stomach-ache, sleepiness, skin rashes etc. It is anticipated that the average incidence of each of these reactions in an untreated population is about 1 in 100. How many patients receiving the drug ought the investigators to recruit to the surveillance in order to detect a doubling of the incidence?

We suppose the investigators are looking out for 100 different types of adverse reaction. Clearly when studying so many reactions a number are likely to be significantly raised by chance. If each reaction is tested at a nominal two-sided $\alpha = 0.0005$ then the overall significance level for testing 100 reactions is $\alpha = 100 \, \alpha' = 0.05$. To be 90% certain of detecting a doubling of the incidence rate, with $\lambda_0 = 0.01$, $\lambda_1 = 0.01$, overall significance level $\alpha = 0.05$ and $1 - \beta = 0.90$, from Table 6.7 we would require 2900 patients.

For 50 adverse reactions with the same significance level and power Table 6.6 gives 2700 patients. This contrasts with 1500 patients for one reaction given by Table 6.2. For numbers of adverse reactions between 1 and 50 the required number of patients will be between 1500 and 2700.

6.5 REFERENCES

Lewis J. A. (1981) Post-marketing surveillance: how many patients? *Tips* (April), 93–94.
Lewis J. A. (1983) Clinical Trials: Statistical developments of practical benefit to the Pharmaceutical Industry. *J. Roy. Statist. Soc. (A)*, **146**, 362–393.
Skegg D. C. G. & Doll R. (1977) The case for recording events in clinical trials. *Br. Med. J*, **2**, 1523–1524.
Wilson A. B. (1977) Post-marketing surveillance of adverse reactions to new medicines. *Br. Med. J*, **2**, 1001–1003.

Table 6.1 Numbers of subjects required to produce a adverse reactions with probability $1 - \beta$, given expected incidence λ and background incidence

λ	α	0.50	0.65	0.70	0.75	0.80	0.85	0.90	0.95	0.99
					$1-\beta$ Probability					
0.0100	1	80	110	130	140	170	200	240	310	470
0.0100	2	200	240	310	470	310	340	390	480	670
0.0100	3	340	390	480	670	430	480	540	640	850
0.0100	4	480	540	640	850	560	610	670	780	1010
0.0100	5	610	670	780	1010	680	730	800	92-	1170
0.0050	1	140	220	250	280	330	390	470	610	930
0.0050	2	390	470	610	930	600	680	780	950	1330
0.0050	3	680	780	950	1330	860	950	1070	1270	1690
0.0050	4	950	1070	1270	1690	1110	1210	1340	1560	2020
0.0050	5	1210	1340	1560	2020	1350	1460	1600	1840	2330
0.0010	1	700	1060	1210	1390	1620	1900	2310	3000	4610
0.0010	2	1900	2310	3000	4610	3000	3380	3900	4750	6640
0.0010	3	3380	3900	4750	6640	4230	4730	5330	6300	8410
0.0010	4	4730	5330	6300	8410	5520	6020	6690	7760	10050
0.0010	5	6020	6690	7760	10050	6730	7270	8000	9160	11610
0.0005	1	1390	2110	2410	2780	3220	3800	4610	6000	9220
0.0005	2	3800	4610	6000	9220	5990	6750	7790	9490	13280
0.0005	3	6750	7790	9490	13280	8560	9450	10650	12600	16820
0.0005	4	9450	10650	12600	16820	11040	12030	13370	15510	20100
0.0005	5	12030	13370	15510	20100	13450	14540	15990	18310	23210
0.0001	1	6940	10500	12050	13870	16100	18980	23030	29960	46060
0.0001	2	18980	23030	29960	46060	29950	33730	38900	47440	66390
0.0001	3	33730	38900	47440	66390	42800	47240	53230	62960	84070
0.0001	4	47240	53230	62960	84070	55150	60140	66810	77540	100460
0.0001	5	60140	66810	77540	100460	67210	72670	79940	91540	116050

Table 6.1

Table 6.2 Sample sizes for detection of adverse reactions: background incidence known.

λ_1	α 2-Sided	1-Sided	Power $1-\beta$ 0.50	0.65	0.70	0.75	0.80	0.85	0.90	0.95	0.99
						$\lambda_0=0.100$					
0.0500	0.01	0.005	270	380	420	470	530	600	690	850	1200
0.0500	0.02	0.010	220	320	360	400	460	520	610	760	1100
0.0500	0.05	0.025	160	240	280	320	360	420	500	640	930
0.0500	0.10	0.050	110	190	220	250	290	350	420	540	810
0.0500	0.20	0.100	70	130	150	180	220	270	330	440	690
0.0100	0.01	0.005	6700	8900	9800	11000	12000	14000	16000	19000	26000
0.0100	0.02	0.010	5500	7500	8300	9300	11000	12000	14000	17000	23000
0.0100	0.05	0.025	3900	5600	6400	7200	8100	9300	11000	14000	20000
0.0100	0.10	0.050	2800	4200	4900	5600	6400	7500	9000	12000	17000
0.0100	0.20	0.100	1700	2900	3400	4000	4700	5700	6900	9100	14000
0.0050	0.01	0.005	27000	36000	39000	43000	48000	53000	61000	73000	99000
0.0050	0.02	0.010	22000	30000	33000	37000	41000	46000	53000	65000	89000
0.0050	0.05	0.025	16000	23000	25000	29000	32000	37000	43000	54000	76000
0.0050	0.10	0.050	11000	17000	20000	22000	26000	30000	36000	45000	65000
0.0050	0.20	0.100	6600	12000	14000	16000	19000	22000	27000	36000	54000
0.0010	0.01	0.005	670000	880000	970000	1060000	1180000	1310000	1500000	1790000	2420000
0.0010	0.02	0.010	550000	740000	820000	910000	1010000	1140000	1310000	1590000	2180000
0.0010	0.05	0.025	390000	560000	620000	700000	790000	910000	1060000	1310000	1850000
0.0010	0.10	0.050	280000	420000	480000	540000	630000	730000	870000	1090000	1590000
0.0010	0.20	0.100	170000	280000	330000	390000	460000	540000	670000	870000	1320000
0.0005	0.01	0.005	2660000	3510000	3850000	4240000	4680000	5230000	5970000	7140000	9640000
0.0005	0.02	0.010	2170000	2950000	3260000	3610000	4020000	4540000	5220000	6330000	8690000
0.0005	0.05	0.025	1540000	2210000	2480000	2780000	3150000	3600000	4220000	5210000	7370000
0.0005	0.10	0.050	1090000	1660000	1890000	2160000	2480000	2890000	3440000	4340000	6330000
0.0005	0.20	0.100	660000	1120000	1310000	1540000	1810000	2160000	2640000	3440000	5230000
						$\lambda_0=0.050$					
0.0500	0.01	0.005	140	200	230	260	290	330	390	490	690
0.0500	0.02	0.010	110	170	190	220	250	290	350	440	640
0.0500	0.05	0.025	80	130	150	180	200	240	290	370	560
0.0500	0.10	0.050	60	100	120	140	170	200	250	320	490
0.0500	0.20	0.100	40	70	90	110	130	160	200	270	420
0.0100	0.01	0.005	3400	4500	5000	5500	6200	6900	8000	9600	14000
0.0100	0.02	0.010	2800	3800	4300	4700	5300	6000	7000	8600	12000
0.0100	0.05	0.025	2000	2900	3300	3700	4200	4800	5700	7100	11000
0.0100	0.10	0.050	1400	2200	2500	2900	3300	3900	4700	6000	8800
0.0100	0.20	0.100	830	1500	1800	2100	2500	3000	3700	4800	7400
0.0050	0.01	0.005	14000	18000	20000	22000	24000	27000	31000	37000	51000
0.0050	0.02	0.010	11000	15000	17000	19000	21000	24000	27000	33000	46000
0.0050	0.05	0.025	7700	12000	13000	15000	17000	19000	22000	28000	39000
0.0050	0.10	0.050	5500	8400	9700	12000	13000	15000	18000	23000	34000
0.0050	0.20	0.100	3300	5700	6800	8000	9400	12000	14000	19000	28000
0.0010	0.01	0.005	340000	440000	490000	540000	590000	660000	750000	900000	1220000
0.0010	0.02	0.010	280000	370000	410000	460000	510000	570000	660000	800000	1100000
0.0010	0.05	0.025	200000	280000	310000	350000	400000	460000	530000	660000	930000
0.0010	0.10	0.050	140000	210000	240000	280000	320000	370000	440000	550000	800000
0.0010	0.20	0.100	83000	140000	170000	200000	230000	280000	340000	440000	660000
0.0005	0.01	0.005	1330000	1760000	1930000	2120000	2350000	2620000	2990000	3580000	4830000
0.0005	0.02	0.010	1090000	1480000	1630000	1810000	2020000	2270000	2620000	3170000	4360000
0.0005	0.05	0.025	770000	1110000	1240000	1400000	1580000	1810000	2110000	2620000	3700000
0.0005	0.10	0.050	550000	830000	950000	1080000	1250000	1450000	1730000	2180000	3180000
0.0005	0.20	0.100	330000	560000	660000	770000	910000	1080000	1330000	1730000	2630000

Table 6.2

66

Table 6.2 Continued.

λ_1	α 2-Sided	1-Sided	0.50	0.65	0.70	0.75	0.80	0.85	0.90	0.95	0.99
						Power $1-\beta$					
					$\lambda_0 = 0.010$						
0.0100	0.01	0.005	670	980	1200	1300	1500	1700	2000	2500	3500
0.0100	0.02	0.010	550	830	950	1100	1300	1500	1800	2200	3200
0.0100	0.05	0.025	390	630	740	860	1000	1200	1500	1900	2800
0.0100	0.10	0.050	280	490	580	680	810	970	1200	1600	2500
0.0100	0.20	0.100	170	340	420	510	620	760	960	1400	2100
0.0050	0.01	0.005	2700	3800	4200	4700	5300	6000	6900	8500	12000
0.0050	0.02	0.010	2200	3200	3600	4000	4600	5200	6100	7600	11000
0.0050	0.05	0.025	1600	2400	2800	3200	3600	4200	500	6400	9300
0.0050	0.10	0.050	1100	1800	2100	2500	2900	3400	4200	5400	8100
0.0050	0.20	0.100	660	1300	1500	1800	2200	2700	3300	4400	6900
0.0010	0.01	0.005	67000	89000	98000	110000	120000	140000	160000	190000	260000
0.0010	0.02	0.010	55000	75000	83000	93000	110000	120000	140000	170000	230000
0.0010	0.05	0.025	39000	56000	63000	72000	81000	93000	110000	140000	200000
0.0010	0.10	0.050	28000	42000	49000	56000	64000	75000	90000	120000	170000
0.0010	0.20	0.100	17000	29000	34000	40000	47000	57000	69000	91000	140000
0.0005	0.01	0.005	270000	360000	390000	430000	480000	530000	610000	730000	990000
0.0005	0.02	0.010	220000	300000	330000	370000	410000	460000	530000	650000	890000
0.0005	0.05	0.025	160000	230000	250000	290000	320000	370000	430000	540000	760000
0.0005	0.10	110000	170000	200000	220000	260000	300000	360000	450000	650000	
0.0005	0.20	0.100	66000	120000	140000	160000	190000	220000	270000	360000	540000
					$\lambda_0 = 0.005$						
0.0050	0.01	0.005	1400	2000	2300	2500	2900	3300	3900	4900	6900
0.0050	0.02	0.010	1100	1700	1900	2200	2500	2900	3500	4400	6400
0.0050	0.05	0.025	770	1300	1500	1700	2000	2400	2900	3700	5600
0.0050	0.10	0.050	550	970	1200	1400	1700	2000	2400	3200	4900
0.0050	0.20	0.100	330	670	820	1010	1300	1600	2000	2700	4200
0.0010	0.01	0.005	34000	45000	50000	55000	62000	69000	80000	96000	140000
0.0010	0.02	0.010	28000	38000	43000	53000	60000	70000	86000	120000	
0.0010	0.05	0.025	20000	29000	37000	42000	48000	57000	71000	110000	
0.0010	0.10	0.050	14000	22000	25000	29000	33000	39000	47000	60000	88000
0.0010	0.20	0.100	8300	15000	18000	21000	25000	30000	37000	48000	74000
0.0005	0.01	0.005	140000	180000	200000	220000	240000	270000	310000	370000	510000
0.0005	0.02	0.010	110000	150000	170000	190000	210000	240000	270000	330000	460000
0.0005	0.05	0.025	77000	120000	130000	150000	170000	190000	220000	280000	390000
0.0005	0.10	0.050	55000	84000	97000	120000	130000	150000	180000	230000	340000
0.0005	0.20	0.100	33000	57000	68000	80000	94000	120000	140000	190000	280000
					$\lambda_0 = 0.001$						
0.0010	0.01	0.005	6700	12000	13000	15000	17000	20000	25000	35000	
0.0010	0.02	0.010	5500	8300	9500	11000	13000	15000	18000	22000	32000
0.0010	0.05	0.025	3900	6300	7300	8500	10000	12000	15000	19000	28000
0.0010	0.10	0.050	2800	4800	5700	6800	8100	9700	12000	16000	25000
0.0010	0.20	0.100	1700	3400	4100	5000	6200	7600	9600	14000	21000
0.0005	0.01	0.005	27000	38000	42000	47000	53000	60000	69000	85000	120000
0.0005	0.02	0.010	22000	32000	36000	40000	46000	52000	61000	76000	110000
0.0005	0.05	0.025	16000	24000	28000	32000	36000	42000	50000	64000	93000
0.0005	0.10	0.050	11000	18000	21000	25000	29000	34000	42000	54000	81000
0.0005	0.20	0.100	6600	13000	15000	18000	22000	27000	33000	44000	69000

Table 6.2

Table 6.3 Sample sizes for detection of adverse reactions: treated and control groups equal size.

λ_1	α 2-sided	1-sided	0.50	0.65	0.70	Power $1-\beta$ 0.75	0.80	0.85	0.90	0.95	0.99
					$\lambda_0 = 0.100$						
0.100	0.01	0.005	180	230	250	270	300	340	380	460	610
0.100	0.02	0.010	140	190	210	230	260	290	340	400	550
0.100	0.05	0.025	100	150	160	180	200	230	270	330	470
0.100	0.10	0.050	70	110	130	140	160	190	220	280	400
0.100	0.20	0.100	50	80	90	100	120	140	170	220	330
0.050	0.01	0.005	590	770	850	930	1100	1200	1400	1600	2100
0.050	0.02	0.010	480	650	720	790	880	990	1200	1400	1900
0.050	0.05	0.025	340	490	550	610	690	790	920	1200	1700
0.050	0.10	0.050	240	370	420	480	550	630	750	950	1400
0.050	0.20	0.100	150	250	290	340	400	470	580	750	1200
0.010	0.01	0.005	13000	17000	19000	20000	22000	25000	28000	34000	46000
0.010	0.02	0.010	11000	14000	16000	17000	19000	22000	25000	30000	41000
0.010	0.05	0.025	1300	11000	12000	14000	15000	17000	20000	25000	35000
0.010	0.10	0.050	5100	7800	8900	11000	12000	14000	17000	21000	30000
0.010	0.20	0.100	3100	5300	6200	7200	8500	11000	13000	17000	25000
0.005	0.01	0.005	49000	65000	71000	78000	86000	97000	110000	140000	180000
0.005	0.02	0.010	40000	55000	60000	67000	74000	84000	96000	120000	160000
0.005	0.05	0.025	29000	41000	46000	52000	58000	67000	78000	96000	140000
0.005	0.10	0.050	20000	31000	35000	40000	46000	53000	64000	80000	120000
0.005	0.20	0.100	13000	21000	25000	29000	34000	40000	49000	64000	96000
0.001	0.01	0.005	1200000	1590000	1740000	1920000	2120000	2360000	2700000	3230000	4350000
0.001	0.02	0.010	980000	1330000	1470000	1630000	1820000	2050000	2360000	2860000	3920000
0.001	0.05	0.025	700000	1000000	1120000	1260000	1420000	1630000	1900000	2350000	3330000
0.001	0.10	0.050	490000	750000	860000	980000	1120000	1300000	1550000	1960000	2860000
0.001	0.20	0.100	300000	510000	590000	700000	820000	980000	1190000	1550000	2360000
					$\lambda_0 = 0.050$						
0.100	0.01	0.005	130	160	180	200	210	240	270	320	430
0.100	0.02	0.010	100	140	150	170	190	210	240	290	390
0.100	0.05	0.025	80	100	120	130	150	170	190	240	330
0.100	0.10	0.050	50	80	90	100	120	130	160	200	290
0.100	0.20	0.100	40	60	60	70	90	100	120	160	240
0.050	0.01	0.005	370	490	540	590	650	730	830	990	1400
0.050	0.02	0.010	310	410	460	500	560	630	730	880	1200
0.050	0.05	0.025	220	310	350	390	440	500	590	720	1100
0.050	0.10	0.050	160	230	270	300	350	400	480	600	880
0.050	0.20	0.100	100	160	190	220	260	300	370	480	720
0.010	0.01	0.005	6900	9200	10000	11000	13000	14000	16000	19000	25000
0.010	0.02	0.010	5700	7700	8500	9400	11000	12000	14000	17000	23000
0.010	0.05	0.025	4000	5800	6500	7300	8200	9400	11000	14000	20000
0.010	0.10	0.050	2900	4300	4900	5600	6500	7500	9000	12000	17000
0.010	0.20	0.100	1800	2900	3400	4000	4700	5600	6900	9000	14000
0.005	0.01	0.005	27000	35000	39000	43000	47000	52000	60000	71000	96000
0.005	0.02	0.010	22000	30000	33000	36000	40000	46000	52000	63000	87000
0.005	0.05	0.025	16000	22000	25000	28000	32000	36000	42000	52000	74000
0.005	0.10	0.050	11000	17000	19000	22000	25000	29000	35000	44000	63000
0.005	0.20	0.100	6600	12000	13000	16000	18000	22000	27000	35000	52000
0.001	0.01	0.005	640000	850000	930000	1020000	1130000	1260000	1430000	1710000	2310000
0.001	0.02	0.010	520000	710000	780000	870000	970000	1090000	1250000	1520000	2080000
0.001	0.05	0.025	370000	530000	600000	670000	760000	870000	1010000	1250000	1770000
0.001	0.10	0.050	260000	400000	460000	520000	600000	690000	830000	1040000	1520000
0.001	0.20	0.100	160000	270000	320000	370000	440000	520000	640000	830000	1250000

Table 6.3

Table 6.3 Continued.

λ_1	α 2-sided	1-sided	0.50	0.65	0.70	0.75	Power $1-\beta$ 0.80	0.85	0.90	0.95	0.99
					$\lambda_0 = 0.010$						
0.100	0.01	0.005	80	100	110	120	140	150	170	200	270
0.100	0.02	0.010	70	90	100	110	120	130	150	180	240
0.100	0.05	0.025	50	70	70	80	90	110	120	150	210
0.100	0.10	0.050	40	50	60	70	70	90	100	130	180
0.100	0.20	0.100	20	40	40	50	60	70	80	100	150
0.050	0.01	0.005	190	240	260	290	320	360	410	480	650
0.050	0.02	0.010	150	200	220	250	280	310	360	430	590
0.050	0.05	0.025	110	150	170	190	220	250	290	350	500
0.050	0.10	0.050	80	120	130	150	170	200	240	300	430
0.050	0.20	0.100	50	80	90	110	130	150	180	230	350
0.010	0.01	0.005	2000	2600	2900	3200	3500	3900	4400	5300	7100
0.010	0.02	0.010	1700	2200	2500	2700	3000	3400	3900	4700	6400
0.010	0.05	0.025	1200	1700	1900	2100	2400	2700	3200	3900	5500
0.010	0.10	0.050	810	1300	1400	1600	1900	2200	2600	3200	4700
0.010	0.20	0.100	490	830	970	1200	1400	1600	2000	2600	3900
0.005	0.01	0.005	6600	8700	9500	11000	12000	13000	15000	18000	24000
0.005	0.02	0.010	5400	7300	8100	8900	10000	12000	13000	16000	22000
0.005	0.05	0.025	3800	5500	6100	6900	7800	8900	11000	13000	19000
0.005	0.10	0.050	2700	4100	4700	5400	6200	7100	8500	11000	16000
0.005	0.20	0.100	1700	2800	3300	3800	4500	5400	6500	8500	13000
0.001	0.01	0.005	140000	190000	200000	220000	250000	280000	310000	380000	500000
0.001	0.02	0.010	120000	160000	170000	190000	210000	240000	280000	330000	450000
0.001	0.05	0.025	80000	120000	130000	150000	170000	190000	220000	280000	390000
0.001	0.10	0.050	57000	86000	98000	120000	130000	150000	180000	230000	330000
0.001	0.20	0.100	35000	58000	68000	80000	94000	120000	140000	180000	280000
					$\lambda_0 = 0.005$						
0.100	0.01	0.005	70	100	110	110	130	140	160	190	250
0.100	0.02	0.010	60	80	90	100	110	120	140	170	230
0.100	0.05	0.025	50	60	70	80	90	100	110	140	190
0.100	0.10	0.050	30	50	50	60	70	80	90	120	170
0.100	0.20	0.100	20	30	40	50	50	60	70	90	140
0.050	0.01	0.005	160	210	230	250	280	310	350	420	560
0.050	0.02	0.010	130	180	190	210	240	270	310	370	500
0.050	0.05	0.025	100	130	150	170	190	210	250	310	430
0.050	0.10	0.050	70	100	110	130	150	170	200	260	370
0.050	0.20	0.100	40	70	80	90	110	130	160	200	300
0.010	0.01	0.005	1400	1800	2000	2100	2400	2600	3000	3600	4800
0.010	0.02	0.010	1100	1500	1700	1800	2000	2300	2600	3200	4300
0.010	0.05	0.025	770	1100	1300	1400	1600	1800	2100	2600	3700
0.010	0.10	0.050	540	820	940	1100	1300	1500	1700	2200	3200
0.010	0.20	0.100	330	560	650	760	900	1100	1400	1700	2600
0.005	0.01	0.005	4000	5300	5800	6300	7000	7800	8900	11000	15000
0.005	0.02	0.010	3300	4400	4900	5400	6000	6800	7800	9400	13000
0.005	0.05	0.025	2300	3300	3700	4200	4700	5400	6300	7800	11000
0.005	0.10	0.050	1700	2500	2900	3300	3700	4300	5100	6500	9400
0.005	0.20	0.100	980	1700	2000	2300	2700	3200	4000	5100	7800
0.001	0.01	0.005	73000	96000	110000	120000	130000	150000	170000	200000	270000
0.001	0.02	0.010	60000	81000	89000	99000	110000	130000	150000	180000	240000
0.001	0.05	0.025	43000	61000	68000	76000	86000	99000	120000	150000	210000
0.001	0.10	0.050	30000	46000	52000	59000	68000	79000	94000	120000	180000
0.001	0.20	0.100	18000	31000	36000	42000	50000	59000	72000	94000	150000

Table 6.3 Continued.

λ_1	α 2-sided	1-sided	Power $1-\beta$ 0.50	0.65	0.70	0.75	0.80	0.85	0.90	0.95	0.99
					$\lambda_0 = 0.001$						
0.100	0.01	0.005	70	90	100	110	120	130	150	170	230
0.100	0.02	0.010	60	80	80	90	100	110	130	160	210
0.100	0.05	0.025	40	60	70	70	80	90	110	130	180
0.100	0.10	0.050	30	50	50	60	60	70	90	110	150
0.100	0.20	0.100	20	30	40	40	50	60	70	90	130
0.050	0.01	0.005	140	180	200	220	240	270	300	360	490
0.050	0.02	0.010	120	150	170	190	210	230	270	320	440
0.050	0.05	0.025	80	120	130	150	160	190	220	270	370
0.050	0.10	0.050	60	90	100	110	130	150	180	220	320
0.050	0.20	0.100	40	60	70	80	100	110	140	180	270
0.010	0.01	0.005	800	1100	1200	1300	1400	1600	1800	2200	2900
0.010	0.02	0.010	650	880	970	1100	1200	1400	1600	1900	2600
0.010	0.05	0.025	460	660	740	830	940	1100	1300	1600	2200
0.010	0.10	0.050	330	500	570	650	740	860	1100	1300	1900
0.010	0.20	0.100	200	340	390	460	540	650	790	1100	1600
0.005	0.01	0.005	1900	2500	2700	3000	3300	3700	4200	5000	6800
0.005	0.02	0.010	1600	2100	2300	2600	2800	3200	3700	4400	6100
0.005	0.05	0.025	1100	1600	1800	2000	2200	2600	3000	3700	5200
0.005	0.10	0.050	760	1200	1400	1600	1800	2100	2400	3100	4400
0.005	0.20	0.100	460	780	920	1100	1300	1500	1900	2400	3700
0.001	0.01	0.005	20000	27000	29000	32000	35000	40000	45000	54000	72000
0.001	0.02	0.010	17000	23000	25000	27000	31000	34000	39000	48000	65000
0.001	0.05	0.025	12000	17000	19000	21000	24000	27000	32000	39000	56000
0.001	0.10	0.050	8200	13000	15000	17000	19000	22000	26000	33000	48000
0.001	0.20	0.100	5000	8400	9800	12000	14000	17000	20000	26000	39000
					$\lambda_0 = 0.100$						
0.0500	0.01	0.005	470	600	660	720	790	880	1000	1200	1600
0.0500	0.02	0.010	380	510	560	610	680	760	870	1100	1500
0.0500	0.05	0.025	270	380	420	470	530	600	700	860	1200
0.0500	0.10	0.050	190	290	320	370	420	480	570	710	1100
0.0500	0.20	0.100	120	190	220	260	300	360	440	560	840
0.0100	0.01	0.005	9500	13000	14000	16000	17000	19000	22000	26000	34000
0.0100	0.02	0.010	7800	11000	12000	13000	15000	17000	19000	23000	31000
0.0100	0.05	0.025	5500	7900	8800	9900	12000	13000	15000	19000	26000
0.0100	0.10	0.050	3900	5900	6700	7700	8800	11000	13000	16000	23000
0.0100	0.20	0.100	2400	4000	4700	5500	6400	7600	9300	13000	19000
0.0050	0.01	0.005	37000	49000	54000	59000	65000	73000	83000	99000	140000
0.0050	0.02	0.010	31000	41000	46000	50000	56000	63000	72000	88000	120000
0.0050	0.05	0.025	22000	31000	35000	39000	44000	50000	59000	72000	110000
0.0050	0.10	0.050	16000	23000	27000	30000	35000	40000	48000	60000	87000
0.0050	0.20	0.100	9200	16000	19000	22000	25000	30000	37000	48000	72000
0.0010	0.01	0.005	910000	1200000	1310000	1440000	1590000	1780000	2020000	2420000	3260000
0.0010	0.02	0.010	740000	1000000	1110000	1230000	1370000	1540000	1770000	2140000	2940000
0.0010	0.05	0.025	530000	750000	840000	950000	1070000	1220000	1430000	1770000	2500000
0.0010	0.10	0.050	370000	560000	640000	730000	840000	980000	1170000	1470000	2140000
0.0010	0.20	0.100	230000	380000	450000	520000	620000	730000	900000	1170000	1770000
0.0005	0.01	0.005	3600000	4750000	5210000	5730000	6330000	7070000	8060000	9650000	13010000
0.0005	0.02	0.010	2940000	3990000	4410000	4880000	5440000	6130000	7050000	8540000	11720000
0.0005	0.05	0.025	2090000	2980000	3350000	3760000	4250000	4870000	5690000	7040000	9950000
0.0005	0.10	0.050	1470000	2240000	2550000	2920000	3350000	3900000	4640000	5860000	8540000
0.0005	0.20	0.100	890000	1510000	1770000	2080000	2450000	2910000	3560000	4640000	7050000

Table 6.3

Table 6.4 Sample sizes for detection of adverse reactions: control group 2 times treated group.

λ_1	α 2-Sided	1-Sided	0.50	0.65	0.70	0.75	Power $1-\beta$ 0.80	0.85	0.90	0.95	0.99
						$\lambda_0 = 0.050$					
0.0500	0.01	0.005	310	400	430	470	570	640	760	1000	
0.0500	0.02	0.010	250	330	360	400	440	490	560	670	900
0.0500	0.05	0.025	180	250	280	310	340	390	450	550	760
0.0500	0.10	0.050	130	190	210	240	270	310	360	450	650
0.0500	0.20	0.100	80	130	150	170	200	230	280	350	530
0.0100	0.01	0.005	5400	7000	8400	9300	11000	12000	14000	19000	
0.0100	0.02	0.010	4400	5900	6500	7200	8000	9000	11000	13000	17000
0.0100	0.05	0.025	3100	4400	4900	5500	6200	7100	8300	11000	15000
0.0100	0.10	0.050	2200	3300	3800	4300	4900	5700	6700	8500	13000
0.0100	0.20	0.100	1400	2200	2600	3100	3600	4300	5200	6700	11000
0.0050	0.01	0.005	21000	27000	29000	32000	36000	40000	45000	54000	72000
0.0050	0.02	0.010	17000	23000	25000	28000	31000	34000	40000	48000	65000
0.0050	0.05	0.025	12000	17000	19000	21000	24000	27000	32000	39000	55000
0.0050	0.10	0.050	8200	13000	15000	17000	19000	22000	26000	33000	47000
0.0050	0.20	0.100	5000	84000	9800	12000	14000	17000	20000	26000	39000
0.0010	0.01	0.005	480000	640000	700000	770000	850000	940000	1080000	1290000	1730000
0.0010	0.02	0.010	400000	540000	590000	650000	730000	820000	940000	1140000	1560000
0.0010	0.05	0.025	280000	400000	450000	500000	570000	650000	760000	940000	1330000
0.0010	0.10	0.050	200000	300000	340000	390000	450000	520000	620000	780000	1140000
0.0010	0.20	0.100	120000	210000	240000	280000	330000	390000	480000	620000	940000
0.0005	0.01	0.005	1910000	2520000	2760000	3030000	3350000	3740000	4270000	5110000	6890000
0.0005	0.02	0.010	1560000	2110000	2330000	2590000	2880000	3250000	3730000	4520000	6200000
0.0005	0.05	0.025	1110000	1580000	1770000	1990000	2250000	2580000	3010000	3730000	5270000
0.0005	0.10	0.050	780000	1190000	1350000	1550000	1780000	2060000	2460000	3100000	4520000
0.0005	0.20	0.100	480000	800000	940000	1100000	1300000	1540000	1890000	2460000	3730000
						$\lambda_0 = 0.010$					
0.0100	0.01	0.005	1700	2100	2300	2500	2800	3100	3500	4100	5400
0.0100	0.02	0.010	1400	1800	2000	2200	2400	2700	3000	3600	4800
0.0100	0.05	0.025	950	1400	1500	1700	1900	2100	2400	2900	4100
0.0100	0.10	0.050	670	980	1100	1300	1500	1700	2000	2400	3500
0.0100	0.20	0.100	410	660	760	880	1100	1300	1500	1900	2800
0.0050	0.01	0.005	5300	6900	7500	8200	9000	10000	12000	14000	18000
0.0050	0.02	0.010	4300	5700	6300	7000	7700	8600	9900	12000	17000
0.0050	0.05	0.025	3100	4300	4800	5400	6000	6800	7900	9700	14000
0.0050	0.10	0.050	2200	3200	3700	4100	4700	5400	6400	8100	12000
0.0050	0.20	0.100	1300	2200	2500	2900	3400	4100	4900	6300	9500
0.0010	0.01	0.005	110000	140000	160000	170000	190000	210000	240000	280000	380000
0.0010	0.02	0.010	86000	120000	130000	150000	160000	180000	210000	250000	340000
0.0010	0.05	0.025	61000	87000	98000	110000	130000	150000	170000	210000	290000
0.0010	0.10	0.050	43000	65000	74000	85000	97000	120000	140000	170000	250000
0.0010	0.20	0.100	26000	44000	52000	60000	71000	84000	110000	140000	210000
0.0005	0.01	0.005	410000	540000	590000	650000	720000	800000	910000	1090000	1470000
0.0005	0.02	0.010	340000	460000	500000	560000	620000	700000	800000	970000	1320000
0.0005	0.05	0.025	240000	340000	380000	430000	480000	550000	650000	800000	1120000
0.0005	0.10	0.050	170000	260000	290000	330000	380000	440000	530000	660000	960000
0.0005	0.20	0.100	110000	170000	200000	240000	280000	330000	400000	530000	800000

Table 6.4

Table 6.4 Continued.

λ_1	α 2-sided	1-sided	Power $1-\beta$ 0.50	0.65	0.70	0.75	0.80	0.85	0.90	0.95	0.99
					$\lambda_0 = 0.005$						
0.0050	0.01	0.005	3300	4300	4600	5100	5500	6100	6900	8200	11000
0.0050	0.02	0.010	2700	3600	3900	4300	4800	5300	6000	7200	9700
0.0050	0.05	0.025	2000	2700	3000	3300	3700	4200	4800	5900	8200
0.0050	0.10	0.050	1400	2000	2300	2600	2900	3300	3900	4900	6900
0.0050	0.20	0.100	820	1400	1600	1800	2100	2500	3000	3800	5700
0.0010	0.01	0.005	57000	74000	81000	89000	98000	110000	130000	150000	200000
0.0010	0.02	0.010	46000	62000	68000	76000	84000	94000	110000	140000	180000
0.0010	0.05	0.025	33000	47000	52000	58000	66000	75000	87000	110000	160000
0.0010	0.10	0.050	23000	35000	40000	45000	52000	60000	71000	89000	130000
0.0010	0.20	0.100	14000	24000	28000	32000	38000	45000	54000	70000	110000
0.0005	0.01	0.005	220000	280000	310000	340000	370000	420000	470000	570000	760000
0.0005	0.02	0.010	180000	240000	260000	290000	320000	360000	410000	500000	680000
0.0005	0.05	0.025	130000	180000	200000	220000	250000	290000	340000	410000	580000
0.0005	0.10	0.050	87000	140000	150000	170000	200000	230000	270000	340000	500000
0.0005	0.20	0.100	53000	88000	110000	130000	150000	170000	210000	270000	410000
					$\lambda_0 = 0.001$						
0.0010	0.01	0.005	17000	22000	24000	26000	28000	31000	35000	41000	55000
0.0010	0.02	0.010	14000	18000	20000	22000	24000	27000	31000	36000	49000
0.0010	0.05	0.025	9600	14000	15000	17000	19000	21000	25000	30000	41000
0.0010	0.10	0.050	6800	9900	12000	13000	15000	17000	20000	25000	35000
0.0010	0.20	0.100	4200	6700	7700	8900	11000	13000	15000	19000	29000
0.0005	0.01	0.005	54000	69000	76000	83000	91000	110000	120000	140000	190000
0.0005	0.02	0.010	44000	58000	64000	70000	78000	87000	100000	120000	170000
0.0005	0.05	0.025	31000	44000	48000	54000	61000	69000	80000	98000	140000
0.0005	0.10	0.050	22000	33000	37000	42000	48000	55000	65000	81000	120000
0.0005	0.20	0.100	14000	22000	26000	30000	35000	41000	50000	64000	96000

Table 6.4

Table 6.5 Sample sizes for detection of adverse reactions: control group 5 times treated group.

λ_1	α 2-sided	1-sided	0.50	0.65	0.70	0.75	0.80	0.85	0.90	0.95	0.99
						Power $1-\beta$					
					$\lambda_0 = 0.100$						
0.0500	0.01	0.005	390	500	550	590	650	720	810	960	1300
0.0500	0.02	0.010	320	420	460	510	560	620	710	840	1200
0.0500	0.05	0.025	230	320	350	390	430	490	570	690	950
0.0500	0.10	0.050	160	240	270	300	340	390	460	570	810
0.0500	0.20	0.100	100	160	180	210	250	290	350	450	660
0.0100	0.01	0.005	7700	11000	12000	13000	14000	15000	17000	21000	28000
0.0100	0.02	0.010	6300	8500	9400	11000	12000	13000	15000	18000	25000
0.0100	0.05	0.025	4500	6400	7100	8000	9000	11000	12000	15000	21000
0.0100	0.10	0.050	3200	4800	5400	6200	7100	8200	9700	13000	18000
0.0100	0.20	0.100	2000	3200	3800	4400	5200	6100	7500	9700	15000
0.0050	0.01	0.005	30000	40000	43000	48000	52000	58000	67000	79000	110000
0.0050	0.02	0.010	25000	33000	37000	41000	45000	51000	58000	70000	96000
0.0050	0.05	0.025	18000	25000	28000	31000	35000	40000	47000	58000	82000
0.0050	0.10	0.050	13000	19000	21000	24000	28000	32000	38000	48000	70000
0.0050	0.20	0.100	7400	13000	15000	17000	20000	24000	30000	38000	58000
0.0010	0.01	0.005	730000	960000	1050000	1150000	1270000	1420000	1620000	1940000	2610000
0.0010	0.02	0.010	590000	800000	890000	980000	1100000	1230000	1420000	1720000	2350000
0.0010	0.05	0.025	420000	600000	680000	760000	860000	980000	1150000	1420000	2000000
0.0010	0.10	0.050	300000	450000	520000	590000	680000	790000	930000	1180000	1710000
0.0010	0.20	0.100	180000	310000	360000	420000	490000	590000	720000	930000	1420000
0.0005	0.01	0.005	2880000	3810000	4170000	4580000	5070000	5660000	6450000	7720000	10410000
0.0005	0.02	0.010	2350000	3190000	3530000	3910000	4350000	4900000	5640000	6830000	9380000
0.0005	0.05	0.025	1670000	2390000	2680000	3010000	3410000	3890000	4560000	5630000	7960000
0.0005	0.10	0.050	1180000	1790000	2040000	2340000	2680000	3120000	3710000	4690000	6830000
0.0005	0.20	0.100	720000	1210000	1420000	1660000	1960000	2330000	2850000	3710000	5640000
					$\lambda_0 = 0.050$						
0.0500	0.01	0.005	270	340	370	400	430	470	530	620	800
0.0500	0.02	0.010	220	280	310	340	370	410	460	540	710
0.0500	0.05	0.025	160	210	230	260	280	320	370	440	600
0.0500	0.10	0.050	110	160	180	200	220	250	290	360	500
0.0500	0.20	0.100	70	110	120	140	160	190	220	280	410
0.0100	0.01	0.005	4400	5700	6300	6900	7500	8400	9500	12000	16000
0.0100	0.02	0.010	3600	4800	5300	5800	6500	7300	8300	10000	14000
0.0100	0.05	0.025	2600	3600	4000	4500	5100	5700	6700	8200	12000
0.0100	0.10	0.050	1800	2700	3100	3500	4000	4600	5400	6800	9800
0.0100	0.20	0.100	1100	1800	2100	2500	2900	3400	4100	5300	8000
0.0050	0.01	0.005	17000	22000	24000	26000	29000	32000	36000	43000	58000
0.0050	0.02	0.010	14000	18000	20000	22000	25000	28000	32000	38000	52000
0.0050	0.05	0.025	9500	14000	15000	17000	19000	22000	26000	32000	44000
0.0050	0.10	0.050	6700	11000	12000	14000	15000	18000	21000	26000	38000
0.0050	0.20	0.100	4100	6800	7900	9300	11000	13000	16000	21000	31000
0.0010	0.01	0.005	390000	510000	560000	620000	680000	760000	860000	1030000	1390000
0.0010	0.02	0.010	320000	430000	470000	520000	580000	660000	760000	910000	1250000
0.0010	0.05	0.025	230000	320000	360000	410000	460000	520000	610000	750000	1060000
0.0010	0.10	0.050	160000	240000	280000	320000	360000	420000	500000	630000	910000
0.0010	0.20	0.100	96000	170000	190000	230000	260000	310000	380000	500000	750000
0.0005	0.01	0.005	1530000	2020000	2210000	2430000	2680000	3000000	3420000	4090000	5510000
0.0005	0.02	0.010	1250000	1690000	1870000	2070000	2310000	2600000	2990000	3620000	4960000
0.0005	0.05	0.025	890000	1270000	1420000	1600000	1810000	2060000	2410000	2980000	4210000
0.0005	0.10	0.050	630000	950000	1080000	1240000	1420000	1650000	1970000	2480000	3620000
0.0005	0.20	0.100	380000	640000	750000	880000	1040000	1240000	1510000	1970000	2980000

Table 6.5 Continued.

λ_1	α 2-sided	1-sided	0.50	0.65	0.70	Power $1-\beta$ 0.75	0.80	0.85	0.90	0.95	0.99
						$\lambda_0 = 0.010$					
0.0100	0.01	0.005	1500	1800	2000	2100	2300	2600	2900	3300	4300
0.0100	0.02	0.010	1200	1600	1700	1800	2000	2200	2500	2900	3800
0.0100	0.05	0.025	840	1200	1300	1400	1500	1700	2000	2400	3200
0.0100	0.10	0.050	590	830	930	1100	1200	1400	1600	1900	2700
0.0100	0.20	0.100	360	550	630	720	830	970	1200	1500	2200
0.0050	0.01	0.005	4500	5700	6200	6800	7400	8200	9200	11000	15000
0.0050	0.02	0.010	3700	4800	5300	5800	6300	7100	8000	9600	13000
0.0050	0.05	0.025	2600	3600	4000	4400	4900	5600	6400	7800	11000
0.0050	0.10	0.050	1900	2700	3000	3400	3900	4400	5200	6400	9100
0.0050	0.20	0.100	1200	1800	2100	2400	2800	3300	3900	5000	7400
0.0010	0.01	0.005	86000	120000	130000	140000	150000	170000	190000	230000	310000
0.0010	0.02	0.010	70000	94000	110000	120000	130000	150000	170000	200000	270000
0.0010	0.05	0.025	50000	71000	79000	88000	100000	120000	140000	170000	230000
0.0010	0.10	0.050	35000	53000	60000	68000	78000	91000	110000	140000	200000
0.0010	0.20	0.100	22000	36000	42000	49000	57000	68000	82000	110000	170000
0.0005	0.01	0.005	330000	440000	480000	520000	580000	640000	730000	880000	1180000
0.0005	0.02	0.010	270000	370000	400000	450000	500000	560000	640000	780000	1060000
0.0005	0.05	0.025	200000	280000	310000	350000	390000	440000	520000	640000	900000
0.0005	0.10	0.050	140000	210000	240000	270000	310000	360000	420000	530000	770000
0.0005	0.20	0.100	82000	140000	160000	190000	230000	270000	320000	420000	640000
						$\lambda_0 = 0.005$					
0.0050	0.01	0.005	2900	3700	4000	4300	4700	5100	5700	6600	8600
0.0050	0.02	0.010	2400	3100	3300	3600	4000	4400	4900	5800	7700
0.0050	0.05	0.025	1700	2300	2500	2800	3100	3400	3900	4700	6400
0.0050	0.10	0.050	1200	1700	1900	2100	2400	2700	3200	3900	5400
0.0050	0.20	0.100	720	1200	1300	1500	1700	2000	2400	3000	4300
0.0010	0.01	0.005	47000	61000	66000	72000	79000	88000	100000	120000	160000
0.0010	0.02	0.010	38000	51000	56000	62000	68000	76000	87000	110000	150000
0.0010	0.05	0.025	27000	38000	42000	47000	53000	60000	70000	86000	120000
0.0010	0.10	0.050	19000	29000	32000	37000	42000	48000	57000	71000	110000
0.0010	0.20	0.100	12000	19000	22000	26000	30000	36000	44000	56000	84000
0.0005	0.01	0.005	180000	230000	250000	270000	300000	340000	380000	450000	610000
0.0005	0.02	0.010	140000	190000	210000	230000	260000	290000	330000	400000	550000
0.0005	0.05	0.025	100000	150000	160000	180000	200000	230000	270000	330000	460000
0.0005	0.10	0.050	70000	110000	120000	140000	160000	190000	220000	280000	400000
0.0005	0.20	0.100	43000	71000	83000	97000	120000	140000	170000	220000	330000
						$\lambda_0 = 0.001$					
0.0010	0.01	0.005	15000	19000	20000	22000	24000	26000	29000	34000	44000
0.0010	0.02	0.010	12000	16000	17000	19000	20000	22000	25000	30000	39000
0.0010	0.05	0.025	8500	12000	13000	14000	16000	18000	20000	24000	32000
0.0010	0.10	0.050	6000	8400	9400	11000	12000	14000	16000	20000	27000
0.0010	0.20	0.100	3700	5600	6400	7300	8400	9800	12000	15000	22000
0.0005	0.01	0.005	46000	58000	63000	69000	75000	83000	93000	110000	150000
0.0005	0.02	0.010	37000	49000	53000	58000	64000	71000	81000	97000	130000
0.0005	0.05	0.025	27000	36000	40000	45000	50000	56000	65000	79000	110000
0.0005	0.10	0.050	19000	27000	31000	34000	39000	45000	52000	65000	92000
0.0005	0.20	0.100	12000	18000	21000	24000	28000	33000	40000	51000	75000

Table 6.5   74

Table 6.6 Sample sizes for detection of adverse reactions: background incidence known and allowing for examination of 50 types of reaction.

λ_1	α 2-sided	1-sided	0.50	0.65	0.70	0.75	0.80	0.85	0.90	0.95	0.99
							Power $1-\beta$				
						$\lambda_0 = \mathbf{0.100}$					
0.0500	0.01	0.005	560	710	770	830	910	1000	1200	1400	1800
0.0500	0.02	0.010	510	650	710	770	840	930	1100	1300	1700
0.0500	0.05	0.025	440	570	620	680	750	840	950	1200	1600
0.0500	0.10	0.050	390	510	560	620	690	770	870	1100	1500
0.0500	0.20	0.100	340	450	500	550	620	690	800	960	1400
0.0100	0.01	0.005	14000	18000	19000	20000	22000	24000	26000	30000	38000
0.0100	0.02	0.010	13000	16000	17000	19000	20000	22000	24000	28000	36000
0.0100	0.05	0.025	11000	14000	15000	16000	18000	20000	22000	26000	33000
0.0100	0.10	0.050	9600	13000	14000	15000	16000	18000	20000	24000	31000
0.0100	0.20	0.100	8300	11000	12000	13000	15000	16000	18000	22000	29000
0.0050	0.01	0.005	56000	68000	73000	78000	84000	92000	110000	120000	150000
0.0050	0.02	0.010	51000	62000	67000	72000	78000	85000	95000	110000	150000
0.0050	0.05	0.025	44000	55000	59000	64000	69000	76000	85000	100000	130000
0.0050	0.10	0.050	39000	49000	53000	58000	63000	69000	78000	92000	120000
0.0050	0.20	0.100	34000	43000	47000	51000	56000	63000	71000	84000	120000
0.0010	0.01	0.005	1390000	1690000	1810000	1940000	2090000	2270000	2510000	2890000	3670000
0.0010	0.02	0.010	1260000	1550000	1660000	1780000	1930000	2100000	2340000	2700000	3460000
0.0010	0.05	0.025	1090000	1360000	1460000	1580000	1720000	1880000	2100000	2450000	3170000
0.0010	0.10	0.050	960000	1210000	1310000	1420000	1550000	1710000	1920000	2250000	2950000
0.0010	0.20	0.100	830000	1070000	1160000	1270000	1390000	1540000	1740000	2060000	2730000
0.0005	0.01	0.005	5540000	6750000	7210000	7730000	8330000	9060000	10020000	11530000	14650000
0.0005	0.02	0.010	5020000	6170000	6620000	7120000	7690000	8390000	9320000	10780000	13800000
0.0005	0.05	0.025	4340000	5410000	5830000	6300000	6840000	7500000	8380000	9760000	12650000
0.0005	0.10	0.050	3820000	4840000	5240000	5680000	6200000	6830000	7660000	8990000	11770000
0.0005	0.20	0.100	3320000	4270000	4640000	5060000	5550000	6140000	6940000	8200000	10860000
						$\lambda_0 = \mathbf{0.050}$					
0.0500	0.01	0.005	280	370	400	440	490	540	620	740	990
0.0500	0.02	0.010	260	340	370	410	450	510	580	690	940
0.0500	0.05	0.025	220	300	330	370	410	460	530	640	870
0.0500	0.10	0.050	200	270	300	330	370	420	490	590	820
0.0500	0.20	0.100	170	240	270	300	340	380	450	550	770
0.0100	0.01	0.005	7000	8600	9300	10000	11000	12000	14000	16000	20000
0.0100	0.02	0.010	6300	7900	8500	9200	10000	11000	13000	15000	19000
0.0100	0.05	0.025	5500	6900	7500	8200	8900	9800	12000	13000	18000
0.0100	0.10	0.050	4800	6200	6800	7400	8100	9000	11000	12000	16000
0.0100	0.20	0.100	4200	5500	6000	6600	7300	8100	9200	11000	15000
0.0050	0.01	0.005	28000	35000	37000	40000	43000	47000	52000	60000	76000
0.0050	0.02	0.010	26000	32000	34000	37000	40000	43000	48000	56000	72000
0.0050	0.05	0.025	22000	28000	30000	32000	35000	39000	43000	51000	66000
0.0050	0.10	0.050	20000	25000	27000	29000	32000	35000	40000	47000	62000
0.0050	0.20	0.100	17000	22000	24000	26000	29000	32000	36000	43000	57000
0.0010	0.01	0.005	700000	850000	910000	970000	1050000	1140000	1260000	1450000	1850000
0.0010	0.02	0.010	630000	780000	830000	900000	970000	1060000	1170000	1360000	1740000
0.0010	0.05	0.025	550000	680000	730000	790000	860000	950000	1060000	1230000	1600000
0.0010	0.10	0.050	480000	610000	660000	720000	780000	860000	970000	1130000	1480000
0.0010	0.20	0.100	420000	540000	590000	640000	700000	780000	880000	1040000	1370000
0.0005	0.01	0.005	2770000	3380000	3610000	3870000	4170000	4540000	5020000	5780000	7340000
0.0005	0.02	0.010	2510000	3090000	3310000	3560000	3850000	4200000	4670000	5400000	6920000
0.0005	0.05	0.025	2170000	2710000	2920000	3150000	3430000	3760000	4200000	4890000	6340000
0.0005	0.10	0.050	1910000	2420000	2620000	2840000	3100000	3420000	3840000	4500000	5900000
0.0005	0.20	0.100	1660000	2140000	2320000	2530000	2780000	3080000	3480000	4110000	5450000

Table 6.6

Table 6.6 Continued.

λ_1	α 2-sided	1-sided	0.50	0.65	0.70	Power $1-\beta$ 0.75	0.80	0.85	0.90	0.95	0.99
						$\lambda_0 = 0.010$					
0.0100	0.01	0.005	1400	1900	2000	2200	2500	2700	3100	3700	5000
0.0100	0.02	0.010	1300	1700	1900	2100	2300	2600	2900	3500	4700
0.0100	0.05	0.025	1100	1500	1700	1900	2100	2300	2700	3200	4400
0.0100	0.10	0.050	960	1400	1500	1700	1900	2100	2500	3000	4100
0.0100	0.20	0.100	830	1200	1400	1500	1700	1900	2300	2800	3900
0.0050	0.01	0.005	5600	7100	7700	8300	9100	10000	12000	14000	18000
0.0050	0.02	0.010	5100	6500	7000	7700	8400	9300	11000	13000	17000
0.0050	0.05	0.025	4400	5700	6200	6800	7500	8400	9500	12000	16000
0.0050	0.10	0.050	3900	5100	5600	6200	6800	7700	8700	11000	15000
0.0050	0.20	0.100	3400	4500	5000	5500	6200	6900	8000	9600	14000
0.0010	0.01	0.005	140000	180000	190000	200000	220000	240000	260000	300000	380000
0.0010	0.02	0.010	130000	160000	170000	190000	200000	220000	240000	280000	360000
0.0010	0.05	0.025	110000	140000	150000	160000	180000	200000	220000	260000	330000
0.0010	0.10	0.050	96000	130000	140000	150000	160000	180000	200000	240000	310000
0.0010	0.20	0.100	83000	110000	120000	130000	150000	160000	180000	220000	290000
0.0005	0.01	0.005	560000	680000	730000	780000	840000	920000	1020000	1170000	1490000
0.0005	0.02	0.010	510000	620000	670000	720000	780000	850000	950000	1100000	1410000
0.0005	0.05	0.025	440000	550000	590000	640000	690000	760000	850000	1000000	1290000
0.0005	0.10	0.050	390000	490000	530000	580000	630000	690000	780000	920000	1200000
0.0005	0.20	0.100	340000	430000	470000	510000	560000	630000	710000	840000	1110000
						$\lambda_0 = 0.005$					
0.0050	0.01	0.005	2800	3700	4000	4400	4900	5400	6200	7400	9900
0.0050	0.02	0.010	2600	3400	3700	4100	4500	5100	5800	6900	9400
0.0050	0.05	0.025	2200	3000	3300	3700	4100	4600	5300	6400	8700
0.0050	0.10	0.050	2000	2700	3000	3300	3700	4200	4900	5900	8200
0.0050	0.20	0.100	1700	2400	2700	3000	3400	3800	4500	5500	7700
0.0010	0.01	0.005	70000	86000	93000	100000	110000	120000	140000	160000	200000
0.0010	0.02	0.010	63000	79000	85000	92000	100000	110000	130000	150000	190000
0.0010	0.05	0.025	55000	69000	75000	82000	89000	98000	120000	130000	180000
0.0010	0.10	0.050	48000	62000	68000	74000	81000	90000	110000	120000	160000
0.0010	0.20	0.100	42000	55000	60000	66000	73000	81000	92000	110000	150000
0.0005	0.01	0.005	280000	350000	370000	400000	430000	470000	520000	600000	760000
0.0005	0.02	0.010	260000	320000	340000	370000	400000	430000	480000	560000	720000
0.0005	0.05	0.025	220000	280000	300000	320000	350000	390000	430000	510000	660000
0.0005	0.10	0.050	200000	250000	270000	290000	320000	350000	400000	470000	620000
0.0005	0.20	0.100	170000	220000	240000	260000	290000	320000	360000	430000	570000
						$\lambda_0 = 0.001$					
0.0010	0.01	0.005	14000	19000	20000	22000	25000	27000	31000	37000	50000
0.0010	0.02	0.010	13000	17000	19000	21000	23000	26000	29000	35000	47000
0.0010	0.05	0.025	11000	15000	17000	19000	21000	23000	27000	32000	44000
0.0010	0.10	0.050	9600	14000	15000	17000	19000	21000	25000	30000	41000
0.0010	0.20	0.100	8300	12000	14000	15000	17000	19000	23000	28000	39000
0.0005	0.01	0.005	56000	71000	77000	83000	91000	100000	120000	140000	180000
0.0005	0.02	0.010	51000	65000	70000	77000	84000	93000	110000	130000	170000
0.0005	0.05	0.025	44000	57000	62000	68000	75000	84000	95000	120000	160000
0.0005	0.10	0.050	39000	51000	56000	62000	68000	77000	87000	110000	150000
0.0005	0.20	0.100	34000	45000	50000	55000	62000	69000	80000	96000	140000

Table 6.6

76

Table 6.7 Sample sizes for detection of adverse reactions: background incidence known and allowing for examination of 100 types of reaction.

λ_1	α 2-sided	1-sided	0.50	0.65	0.70	0.75	0.80	0.85	0.90	0.95	0.99
							Power $1-\beta$				

$\lambda_0 = 0.100$

λ_1	2-sided	1-sided	0.50	0.65	0.70	0.75	0.80	0.85	0.90	0.95	0.99
0.0500	0.01	0.005	610	770	830	900	970	1100	1200	1400	1900
0.0500	0.02	0.010	560	710	770	830	910	1000	1200	1400	1800
0.0500	0.05	0.025	490	630	690	750	820	910	1100	1300	1700
0.0500	0.10	0.050	440	570	620	680	750	840	950	1200	1600
0.0500	0.20	0.100	390	510	560	620	690	770	870	1100	1500
0.0100	0.01	0.005	16000	19000	20000	22000	23000	25000	28000	32000	41000
0.0100	0.02	0.010	14000	18000	19000	20000	22000	24000	26000	30000	38000
0.0100	0.05	0.025	13000	16000	17000	18000	20000	21000	24000	28000	36000
0.0100	0.10	0.050	11000	14000	15000	16000	18000	20000	22000	26000	33000
0.0100	0.20	0.100	9600	13000	14000	15000	16000	18000	20000	24000	31000
0.0050	0.01	0.005	61000	74000	79000	84000	91000	99000	110000	130000	160000
0.0050	0.02	0.010	56000	68000	73000	78000	84000	92000	110000	120000	150000
0.0050	0.05	0.025	49000	61000	65000	70000	76000	83000	92000	110000	140000
0.0050	0.10	0.050	44000	55000	59000	64000	69000	76000	85000	100000	130000
0.0050	0.20	0.100	39000	49000	53000	58000	63000	69000	78000	92000	120000
0.0010	0.01	0.005	1520000	1830000	1960000	2090000	2250000	2440000	2690000	3080000	3880000
0.0010	0.02	0.010	1390000	1690000	1810000	1940000	2090000	2270000	2510000	2890000	3670000
0.0010	0.05	0.025	1220000	1500000	1610000	1730000	1880000	2050000	2280000	2640000	3390000
0.0010	0.10	0.050	1090000	1360000	1460000	1580000	1720000	1880000	2100000	2450000	3170000
0.0010	0.20	0.100	960000	1210000	1310000	1420000	1550000	1710000	1920000	2250000	2950000
0.0005	0.01	0.005	6060000	7320000	7810000	8350000	8970000	9730000	10720000	12280000	15490000
0.0005	0.02	0.010	5540000	6750000	7210000	7730000	8330000	9060000	10020000	11530000	14650000
0.0005	0.05	0.025	4850000	5990000	6430000	6920000	7490000	8180000	9090000	10530000	13520000
0.0005	0.10	0.050	4340000	5410000	5830000	6300000	6840000	7500000	8380000	9760000	12650000
0.0005	0.20	0.100	3820000	4840000	5240000	5680000	6200000	6830000	7660000	8990000	11770000

$\lambda_0 = 0.050$

λ_1	2-sided	1-sided	0.50	0.65	0.70	0.75	0.80	0.85	0.90	0.95	0.99
0.0500	0.01	0.005	310	400	440	480	520	580	660	780	1100
0.0500	0.02	0.010	280	370	400	440	490	540	620	740	990
0.0500	0.05	0.025	250	330	360	400	440	500	570	680	920
0.0500	0.10	0.050	220	300	330	370	410	460	530	640	870
0.0500	0.20	0.100	200	270	300	330	370	420	490	590	820
0.0100	0.01	0.005	7600	9400	10000	11000	12000	13000	15000	17000	21000
0.0100	0.02	0.010	7000	8600	9300	10000	11000	12000	14000	16000	20000
0.0100	0.05	0.025	6100	7700	8300	9000	9700	11000	12000	14000	19000
0.0100	0.10	0.050	5500	6900	7500	8200	8900	9800	12000	13000	18000
0.0100	0.20	0.100	4800	6200	6800	7400	8100	9000	11000	12000	16000
0.0050	0.01	0.005	31000	37000	40000	43000	46000	50000	55000	64000	81000
0.0050	0.02	0.010	28000	35000	37000	40000	43000	47000	52000	60000	76000
0.0050	0.05	0.025	25000	31000	33000	36000	39000	42000	47000	55000	71000
0.0050	0.10	0.050	22000	28000	30000	32000	35000	39000	43000	51000	66000
0.0050	0.20	0.100	20000	25000	27000	29000	32000	35000	40000	47000	62000
0.0010	0.01	0.005	760000	920000	980000	1050000	1130000	1220000	1350000	1550000	1950000
0.0010	0.02	0.010	700000	850000	910000	970000	1050000	1140000	1260000	1450000	1850000
0.0010	0.05	0.025	610000	750000	810000	870000	940000	1030000	1150000	1330000	1700000
0.0010	0.10	0.050	550000	680000	730000	790000	860000	950000	1060000	1230000	1600000
0.0010	0.20	0.100	480000	610000	660000	720000	780000	860000	970000	1130000	1480000
0.0005	0.01	0.005	3030000	3660000	3910000	4180000	4490000	4870000	5370000	6150000	7760000
0.0005	0.02	0.010	2770000	3380000	3610000	3870000	4170000	4540000	5020000	5780000	7340000
0.0005	0.05	0.025	2430000	3000000	3220000	3460000	3750000	4100000	4550000	5280000	6780000
0.0005	0.10	0.050	2170000	2710000	2920000	3150000	3430000	3760000	4200000	4890000	6340000
0.0005	0.20	0.100	1910000	2420000	2620000	2840000	3100000	3420000	3840000	4500000	5900000

Table 6.7

Table 6.7 Continued.

λ_1	α 2-sided	1-sided	0.50	0.65	0.70	0.75	Power $1-\beta$ 0.80	0.85	0.90	0.95	0.99
						$\lambda_0 = 0.010$					
0.0100	0.01	0.005	1600	2000	2200	2400	2600	2900	3300	3900	5200
0.0100	0.02	0.010	1400	1900	2000	2200	2500	2700	3100	3700	5000
0.0100	0.05	0.025	1300	1700	1800	2000	2200	2500	2900	3400	4600
0.0100	0.10	0.050	1100	1500	1700	1900	2100	2300	2700	3200	4400
0.0100	0.20	0.100	960	1400	1500	1700	1900	2100	2500	3000	4100
0.0050	0.01	0.005	6100	7700	8300	8900	9700	11000	12000	14000	19000
0.0050	0.02	0.010	5600	7100	7700	8300	9100	10000	12000	14000	18000
0.0050	0.05	0.025	4900	6300	6900	7500	8200	9100	11000	13000	17000
0.0050	0.10	0.050	4400	5700	6200	6800	7500	8400	9500	12000	16000
0.0050	0.20	0.100	3900	5100	5600	6200	6800	7700	8700	11000	15000
0.0010	0.01	0.005	160000	190000	200000	220000	230000	250000	280000	320000	410000
0.0010	0.02	0.010	140000	180000	190000	200000	220000	240000	260000	300000	380000
0.0010	0.05	0.025	130000	160000	170000	180000	200000	210000	240000	280000	360000
0.0010	0.10	0.050	110000	140000	150000	160000	180000	200000	220000	260000	330000
0.0010	0.20	0.100	96000	130000	140000	150000	160000	180000	200000	240000	310000
0.0005	0.01	0.005	610000	740000	790000	840000	910000	990000	1090000	1250000	1580000
0.0005	0.02	0.010	560000	680000	730000	780000	840000	920000	1020000	1170000	1490000
0.0005	0.05	0.025	490000	610000	650000	700000	760000	830000	920000	1070000	1380000
0.0005	0.10	0.050	440000	550000	590000	640000	690000	760000	850000	1000000	1290000
0.0005	0.20	0.100	390000	490000	530000	580000	630000	690000	780000	920000	1200000
						$\lambda_0 = 0.005$					
0.0050	0.01	0.005	3100	4000	4300	4700	5200	5800	6600	7800	11000
0.0050	0.02	0.010	2800	3700	4000	4400	4900	5400	6200	7400	9900
0.0050	0.05	0.025	2500	3300	3600	4000	4400	4900	5700	6800	9200
0.0050	0.10	0.050	2200	3000	3300	3700	4100	4600	5300	6400	8700
0.0050	0.20	0.100	2000	2700	3000	3300	3700	4200	4900	5900	8200
0.0010	0.01	0.005	76000	93000	100000	110000	120000	130000	150000	170000	210000
0.0010	0.02	0.010	70000	86000	93000	100000	110000	120000	140000	160000	200000
0.0010	0.05	0.025	61000	77000	83000	90000	97000	110000	120000	140000	190000
0.0010	0.10	0.050	55000	69000	75000	82000	89000	98000	120000	130000	180000
0.0010	0.20	0.100	48000	62000	68000	74000	81000	90000	110000	120000	160000
0.0005	0.01	0.005	310000	370000	400000	430000	460000	500000	550000	640000	810000
0.0005	0.02	0.010	280000	350000	370000	400000	430000	470000	520000	600000	760000
0.0005	0.05	0.025	250000	310000	330000	360000	390000	420000	470000	550000	710000
0.0005	0.10	0.050	220000	280000	300000	320000	350000	390000	430000	510000	660000
0.0005	0.20	0.100	200000	250000	270000	290000	320000	350000	400000	470000	620000
						$\lambda_0 = 0.001$					
0.0010	0.01	0.005	16000	20000	22000	24000	26000	29000	33000	39000	52000
0.0010	0.02	0.010	14000	19000	20000	22000	25000	27000	31000	37000	50000
0.0010	0.05	0.025	13000	17000	18000	20000	22000	25000	29000	34000	46000
0.0010	0.10	0.050	11000	15000	17000	19000	21000	23000	27000	32000	44000
0.0010	0.20	0.100	9600	14000	15000	17000	19000	21000	25000	30000	41000
0.0005	0.01	0.005	61000	77000	83000	89000	97000	110000	120000	140000	190000
0.0005	0.02	0.010	56000	71000	77000	83000	91000	100000	120000	140000	180000
0.0005	0.05	0.025	49000	63000	68000	75000	82000	91000	110000	130000	170000
0.0005	0.10	0.050	44000	57000	62000	68000	75000	84000	95000	120000	160000
0.0005	0.20	0.100	39000	51000	56000	62000	68000	77000	87000	110000	150000

Table 6.7

Chapter 7
Comparing Two Means

7.1 INTRODUCTION

Often a clinical trial will result in quantitative measurements. For example, in a trial of an anti-hypertensive drug one might have two groups of patients, one taking the drug and the other a placebo, and the outcome is the level of blood pressure in the two groups. Alternatively one might compare the change in blood pressure before and after the trial for the two groups. If the observations are plausibly sampled from a normal distribution, then the best summary of the data is the mean, and the usual test is the one-sample or two-sample t-test. If the observations are not normally distributed, then a popular test for a shift of location is the Wilcoxon rank sum, or equivalently the Mann-Whitney U, test. This is nearly as efficient as the t-test in the normal situation and can be much more efficient for non-normal situations.

7.2 THEORY AND FORMULAE

In order to produce a set of general tables we need an index which is dimensionless, that is, one that is free of the original measurement units. Given two samples, we might postulate that they have different means, but the same standard deviation, σ. Thus the alternative hypothesis might be that the two means are d_t standard deviations apart. Denoting the two alternative populations means by μ_0 and μ_1 we have

$$d_t = |\mu_0 - \mu_1|/\delta \qquad (7.1)$$

where σ is the postulated population standard deviation.

In practical situations, one is more likely to have a feel for d_t than for the individual values of μ_0, μ_1 and σ. Cohen (1977) suggests a range of d_t from 0.1 to 1.0, where a 'small' effect, i.e. a stringent criterion, might be $d_t = 0.2$, and a 'large' effect, i.e. a liberal criterion, might be $d_t = 0.8$.

In other situations the investigator may know the likely range of the measurements, even though he does not know the standard deviation. On the assumption of a normal distribution he can find an approximation for the standard deviation by dividing the range by 4.

One–Sample t–Test

In this test we require the number of subjects necessary to show that a given mean, μ_1, differs from a target value μ_0. Given a significance level α, and a power $1 - \beta$ against the specified alternative μ_1, then the number of subjects in the group should satisfy

$$m = \frac{(z_{1-\alpha} + z_{1-\beta})^2}{d_t^2} + \tfrac{1}{2}z_{1-\alpha}^2 \cdot \qquad (7.2)$$

In the paired t-test situation μ_0 is often chosen to be zero. Observation from cross-over trials, in which subjects receive each of two treatments at different times, can also be considered as paired data.

Two-Sample t-Test

In this case μ_0 is the specified mean for one group and μ_1 the specified mean for the other. The number of subjects in each group should satisfy

$$m = \frac{2(z_{1-\alpha} + z_{1-\beta})^2}{d_t^2} + \tfrac{1}{4}z_{1-\alpha}^2 \tag{7.3}$$

Power

In contrast if one were asked for the power of a trial for a given m, α and d_t, then either equations (7.2) or (7.3) can be inverted to give

$$z_{1-\beta} = d_t\sqrt{\{\lambda(m - vz_{1-\alpha}^2)\}} - z_{1-\alpha} \tag{7.4}$$

where $\lambda = 1$, $v = \tfrac{1}{2}$ for the one-sample t-test and $\lambda = \tfrac{1}{2}$, $v = \tfrac{1}{4}$ for the two-sample t-test.

Wilcoxon Rank-Sum or Mann-Whitney U Test

Let $F_1(x)$ and $F_2(x)$ be the probability of a response being less than a particular value x under treatments 1 and 2 respectively. Then a non-parametric test has, as null hypothesis, $F_1(x) = F_2(x)$ and as alternative $F_1(x) \leq F_2(x)$ with strict inequality for at least one x. A simple alternative hypothesis is one which assumes that $F_1(x) = F_2(x - \triangle)$ for some $\triangle > 0$. $\triangle$ represents a shift in location between the two alternative treatments. It can be shown that an approximate formula for the sample size to give a significance level α and power $1-\beta$ is given by

$$m = \frac{(z_{1-\alpha} + z_{1-\beta})^2}{6\triangle f^2} \tag{7.5}$$

where f depends on the particular form of the distribution of F_1 and F_2.

To calculate f it is necessary to derive the cumulative distribution of the difference of two independent random variables under the null hypothesis assumption $F_1 = F_2$. The parameter f is the first derivative of this derived distribution evaluated at $x = 0$.

In the two sample case, if we assume the underlying distributions are normal then $f = 1/(2\sigma\sqrt{\pi})$.

Equations (7.2), (7.3) and (7.5) can be evaluated by using Table 2.3 which gives $(z_{1-\alpha} + z_{1-\beta})^2$ for different values of α and β.

7.3 BIBLIOGRAPHY

Guenther (1981) gave formulae (7.2) and (7.3) and Cohen (1977, Chapter 2) discussed sample size formulae for the t-test. Lehman (1975, p74) gives the derivation of formula (7.5). Tables for the sample size requirements for the t-test have been given by Lesser (1982).

7.4 DESCRIPTION OF THE TABLES

Table 7.1

Table 7.1 gives the sample sizes required for the one sample t-test for a range of two-sided test sizes of α from 0.01 to 0.20, one-sided values from 0.005 to 0.10, power ranging from 0.50 to 0.99 for d_t from 0.05 to 1.50. The entries in the table are calculated using equation (7.1) and are rounded upwards to the nearest whole number.

Table 7.2

Table 7.2 gives the sample sizes required for the two sample t-test for a range of two-sided test sizes of α from 0.01 to 0.02, one sided from 0.005 to 0.10, power ranging from 0.50 to 0.99 for d_t from 0.05 to 1.50. The entries in the table are calculated using equation (7.2) and are rounded upwards to the nearest whole number.

7.5 USE OF THE TABLES

Table 7.1

Example 7.1 One-sample t-test

A psychologist wishes to test the IQ of a certain population. His null hypothesis is that the IQ is 100, and he has no preconceived notion of whether the group are likely to be above or below it. He wishes to be able to detect a fairly small difference from 100, e.g. 0.2 standard deviations from it, so that if he gets a non-significant result from his analysis, he can be sure that the average IQ from his population lies very close to 100. How many subjects should he recruit?

The test to use is the one-sample t-test. If the psychologist specifies $\alpha = 0.01$ for a two-sided test, and a power $1 - \beta = 0.95$, then, with $d_t = 0.2$ from Table 7.1, he will require 447 subjects in his sample.

Example 7.2 Paired t-tests

An investigator wishes to test whether a particular drug reduces blood pressure. He proposes to measure blood pressure on a group of patients, administer the drug, and then measure the blood pressure again one hour later. Previous studies have shown that, in the absence of any drug effect, the standard deviation of hourly, within-subjects blood pressure measurements is about 10mm.Hg. The investigator decides that he is looking for a fall in blood pressure of more than 10 mm.Hg to be significant at the 5% level, and with a power of 90%. How many patients should he recruit?

Using Table 7.1, $d_t = 1.0$, one-sided $\alpha = 0.05$ and $1 - \beta = 0.90$, the investigator would require at least 10 subjects.

Table 7.2

Example 7.3: Two sample t-test

The investigator of Example 7.2 compares the change in blood pressure due to a placebo with that due to the drug. In this case a two-sided test is probably more appropriate, because he may not know if the placebo effect is going to be less than or greater than the drug effect. If the investigator is now looking for a difference between groups of 5 mm.Hg, then, with the standard deviation as 10 mm.Hg, how many patients should he recruit?

Now $d_t = 0.5$ with, $\alpha = 0.05$ (two-sided) and $1 - \beta = 0.9$. Table 7.2 gives 85 patients per group. If, on the other hand, he had retained $d_t = 1.0$ he would have required only 22 patients per group.

The increased sample size from Example 7.2 is due to the presence of uncertainty in two means, as opposed to one, and the change from a one-sided to a two-sided hypothesis.

Example 7.4

Consider a clinical trial of the use of a drug in twin pregnancies. An obstetrician wishes to show a significant prolongation of pregnancy by use of the drug when compared to placebo. In the absence of any data on the standard deviation of pregnancy duration he argues that normal pregnancies range from 33 to 40 weeks, and so a rough guess for the standard deviation is $(40 - 33)/4 = 1.75$ weeks. How many pregnancies must he observe if he decides that one week is a clinically significant increase in the length of a pregnancy?

Here, $d_t = 1/1.75 = 0.57$. For a two-sided test with $\alpha = 0.05$ and power, $1 - \beta = 0.80$, Table 7.2 gives $m = 53$ for $d_t = 0.55$ and $m = 45$ for $d_t = 0.60$. Thus approximately 49 mothers would correspond to $d_t = 0.57$. The obstetrician would need to observe 100 pregnancies for comparison between the drug and a placebo.

Example 7.5 Power calculations

Woollard & Couper (1983) describe a clinical trial for comparing Moducren with Propranolol as initial therapies in essential hypertension. They proposed to compare the change in blood pressure due to the two drugs. Given that they can recruit only about 50 patients for each drug, and that they are looking for a 'medium' effect size of about $d_t = 0.5$ what is the power of the test, given a two-sided significance level of $\alpha = 0.05$?

From Table 7.2 with $\alpha = 0.05$ (two-sided), $d_t = 0.5$, the corresponding power is about 0.70. Using equation (7.4) directly with $\lambda = \frac{1}{2}$ and $v = \frac{1}{4}$ gives $z_{1-\beta} = 0.52$ and Table 2.1 confirms the power $1 - \beta = 0.7$.

Example 7.6

What size of effect could Woollard & Couper (*loc cit*), of Example 7.5, reasonably expect to detect with their trial?

The answer depends on the definition of reasonable, but if we say a power of $1 - \beta = 0.8$, then with $\alpha = 0.05$ (two-sided), $m = 50$ then equations (7.2) and (7.3) can be inverted to give

$$d_t = \frac{(z_{1-\alpha} + z_{1-\beta})}{\sqrt{\{\lambda(m - vz_{1-\alpha}^2)\}}}$$

where $\lambda = 1$, $v = \frac{1}{2}$ for a one-sample t-test, $\lambda = \frac{1}{2}$, $v = \frac{1}{4}$ for a two-sample t-test.

Substituting $\lambda = \frac{1}{2}$, $v = \frac{1}{4}$, $m = 50$ and from Table 2.2, $z_{1-\alpha} = 1.96$, $z_{1-\beta} = 0.8416$ we obtain $d_t = 0.565$. An approximation to this value of d_t can be obtained by searching in Table 7.2 in the column corresponding to $1 - \beta = 0.8$ and the rows corresponding to $\alpha = 0.05$ (two-sided), for a sample size m as close to 50 as possible. A value of $m = 53$ gives $d_t = 0.55$, and $m = 45$ gives $d_t = 0.60$.

Given that diastolic blood pressure change has a standard deviation of about 7mm.Hg, this implies we can detect a difference in effect between the two drugs of about $7d_t$ or 4 mm.Hg. In fact, the authors did obtain a statistically significant result with an estimated difference between the drugs of only 3 mm.Hg so the true difference is probably greater than 3 mm.Hg.

Example 7.7 Non-parametric test

In the two-sample case described in Example 7.3 above, how is the sample size affected if a non-parametric Wilcoxon test is used for analysis?

If we assume that in fact the underlying distribution is normal, then $f = 1/(2\sigma\sqrt{\pi})$ and (7.5) becomes

$$m = \frac{2\pi(z_{1-\alpha} + z_{1-\beta})^2}{3d_t^2}$$

Thus, if we assume the same parameters as Example 7.3, that is $d_t = 0.5$, $\alpha = 0.05$, power $1 - \beta = 0.90$, we find that $m = 88$. In this case the Wilcoxon test, being slightly less powerful than the t-test in the normal situation, requires rather more subjects than the t-test.

7.6 REFERENCES

Cohen J. (1977) *Statistical Power Analysis for the Behavioral Sciences,* revised edn. Academic Press, New York.

Guenther W. C. (1981) Sample size formulas for normal theory t-tests. *The American Statistician,* **35,** 243-244.

Lehman E. L. (1975) *Nonparametrics. Statistical Methods Based on Ranks.* Holden-Day Inc, San Fransisco.

Lesser M.L. (1982) *In* Mike V. & Stanley K.E. (eds) *Statistics in Medical Research Methods and Issues, with Applications in Cancer Research.* Wiley, New York.

Woollard M. L. and Couper W. D. (1981) Moducren and Propranolol as initial therapies in essential hypertension. *Clinical Trials Journal,* **20,** 89-97.

Table 7.1 Sample sizes for the one-sample *t*-test.

d_t	α 2-sided	1-sided	Power $1 - \beta$ 0.50	0.65	0.70	0.75	0.80	0.85	0.90	0.95	0.99
0.05	0.01	0.005	2656	3509	3846	4228	4673	5221	5954	7127	9614
0.05	0.02	0.010	2166	2943	3252	3604	4016	4525	5208	6310	8661
0.05	0.05	0.025	1538	2202	2470	2778	3141	3593	4204	5199	7350
0.05	0.10	0.050	1084	1650	1884	2153	2474	2877	3427	4330	6309
0.05	0.20	0.100	658	1113	1306	1532	1804	2150	2629	3427	5208
0.10	0.01	0.005	665	879	963	1058	1170	1307	1490	1783	2405
0.10	0.02	0.010	543	737	814	902	1005	1132	1303	1579	2166
0.10	0.05	0.025	386	552	619	696	786	899	1052	1301	1839
0.10	0.10	0.050	272	413	472	539	620	720	858	1084	1578
0.10	0.20	0.100	165	279	327	384	452	538	658	858	1303
0.15	0.01	0.005	297	391	429	471	521	582	663	794	1070
0.15	0.02	0.010	242	328	363	402	448	504	580	703	964
0.15	0.05	0.025	172	246	276	310	350	401	468	579	818
0.15	0.10	0.050	122	185	210	240	276	321	382	482	702
0.15	0.20	0.100	74	125	146	171	201	240	293	382	580
0.20	0.01	0.005	168	221	242	266	294	328	374	447	603
0.20	0.02	0.010	137	185	205	227	253	284	327	396	543
0.20	0.05	0.025	98	139	156	175	198	226	264	326	461
0.20	0.10	0.050	69	104	119	136	156	181	215	272	396
0.20	0.20	0.100	42	71	83	97	114	135	165	215	327
0.25	0.01	0.005	108	142	156	171	189	211	240	287	386
0.25	0.02	0.010	88	119	132	146	162	183	210	254	348
0.25	0.05	0.025	63	89	100	113	127	145	170	209	295
0.25	0.10	0.050	45	67	77	87	100	116	138	174	254
0.25	0.20	0.100	27	46	53	62	73	87	106	138	209
0.30	0.01	0.005	76	99	109	119	132	147	167	200	269
0.30	0.02	0.010	62	83	92	102	113	127	146	177	242
0.30	0.05	0.025	44	63	70	79	89	101	118	146	206
0.30	0.10	0.050	31	47	54	61	70	81	96	122	177
0.30	0.20	0.100	19	32	37	44	51	61	74	96	146
0.35	0.01	0.005	56	73	80	88	97	108	123	147	198
0.35	0.02	0.010	46	62	68	75	84	94	108	130	178
0.35	0.05	0.025	33	46	52	58	66	75	87	108	151
0.35	0.10	0.050	23	35	40	45	52	60	71	90	130
0.35	0.20	0.100	15	24	28	32	38	45	55	71	107
0.40	0.01	0.005	43	57	62	68	75	83	95	113	152
0.40	0.02	0.010	35	48	52	58	64	72	83	100	137
0.40	0.05	0.025	25	36	40	45	51	58	67	83	116
0.40	0.10	0.050	18	27	31	35	40	46	55	69	100
0.40	0.20	0.100	11	19	22	25	29	35	42	55	82

Table 7.1 Continued.

d_t	α 2-sided	1-sided	Power $1-\beta$ 0.50	0.65	0.70	0.75	0.80	0.85	0.90	0.95	0.99
0.45	0.01	0.005	35	45	49	54	59	66	75	90	120
0.45	0.02	0.010	28	38	42	46	51	58	66	80	109
0.45	0.05	0.025	20	29	32	36	40	46	53	66	92
0.45	0.10	0.050	15	22	25	28	32	37	44	55	79
0.45	0.20	0.100	9	15	17	20	23	28	34	43	65
0.50	0.01	0.005	28	37	40	44	49	54	61	73	98
0.50	0.02	0.010	23	31	34	38	42	47	54	65	88
0.50	0.05	0.025	17	23	26	29	33	37	44	53	75
0.50	0.10	0.050	12	18	20	23	26	30	36	45	64
0.50	0.20	0.100	8	12	14	16	19	23	27	35	53
0.55	0.01	0.005	24	31	34	37	40	45	51	61	81
0.55	0.02	0.010	20	26	29	31	35	39	45	54	73
0.55	0.05	0.025	14	20	22	24	27	31	36	44	62
0.55	0.10	0.050	10	15	17	19	22	25	30	37	53
0.55	0.20	0.100	7	10	12	14	16	19	23	29	44
0.60	0.01	0.005	20	26	28	31	34	38	43	51	69
0.60	0.02	0.010	17	22	24	27	30	33	38	45	62
0.60	0.05	0.025	12	17	19	21	23	26	31	38	53
0.60	0.10	0.050	9	13	14	16	18	21	25	31	45
0.60	0.20	0.100	6	9	10	12	14	16	19	25	37
0.65	0.01	0.005	17	23	25	27	29	33	37	44	59
0.65	0.02	0.010	14	19	21	23	25	28	32	39	53
0.65	0.05	0.025	11	14	16	18	20	23	26	32	45
0.65	0.10	0.050	8	11	12	14	16	18	22	27	39
0.65	0.20	0.100	5	8	9	10	12	14	17	21	32
0.70	0.01	0.005	15	20	21	23	26	28	32	38	51
0.70	0.02	0.010	13	17	18	20	22	25	28	34	46
0.70	0.05	0.025	9	13	14	16	17	20	23	28	39
0.70	0.10	0.050	7	10	11	12	14	16	19	23	34
0.70	0.20	0.100	4	7	8	9	10	12	15	19	28
0.75	0.01	0.005	14	17	19	21	23	25	28	33	45
0.75	0.02	0.010	11	15	16	18	20	22	25	30	40
0.75	0.05	0.025	8	11	12	14	15	17	20	25	34
0.75	0.10	0.050	6	9	10	11	12	14	17	21	29
0.75	0.20	0.100	4	6	7	8	9	11	13	16	24
0.80	0.01	0.005	12	15	17	18	20	22	25	30	39
0.80	0.02	0.010	10	13	14	16	17	19	22	26	35
0.80	0.05	0.025	7	10	11	12	14	16	18	22	30
0.80	0.10	0.050	6	8	9	10	11	13	15	18	26
0.80	0.20	0.100	4	5	6	7	8	10	11	15	21
0.85	0.01	0.005	11	14	15	16	18	20	22	26	35
0.85	0.02	0.010	9	12	13	14	16	17	20	23	32
0.85	0.05	0.025	7	9	10	11	12	14	16	19	27
0.85	0.10	0.050	5	7	8	9	10	11	13	16	23
0.85	0.20	0.100	3	5	6	6	7	9	10	13	19
0.90	0.01	0.005	10	13	14	15	16	18	20	24	31
0.90	0.02	0.010	8	11	12	13	14	16	18	21	28
0.90	0.05	0.025	6	8	9	10	11	13	14	18	24
0.90	0.10	0.050	5	6	7	8	9	10	12	15	21
0.90	0.20	0.100	3	5	5	6	7	8	9	12	17
0.95	0.01	0.005	9	12	12	13	15	16	18	22	28
0.95	0.02	0.010	8	10	11	12	13	14	16	19	26
0.95	0.05	0.025	6	8	8	9	10	11	13	16	22
0.95	0.10	0.050	4	6	7	7	8	9	11	13	19
0.95	0.20	0.100	3	4	5	5	6	7	8	11	16

Table 7.1

Table 7.1 Continued.

d_t	α 2-sided	1-sided	Power $1-\beta$ 0.50	0.65	0.70	0.75	0.80	0.85	0.90	0.95	0.99
1.00	0.01	0.005	8	11	11	12	13	15	17	20	26
1.00	0.02	0.010	7	9	10	11	12	13	15	17	23
1.00	0.05	0.025	5	7	8	8	9	10	12	14	20
1.00	0.10	0.050	4	5	6	7	8	9	10	12	17
1.00	0.20	0.100	3	4	4	5	6	7	8	10	14
1.05	0.01	0.005	8	10	11	11	12	14	15	18	24
1.05	0.02	0.010	7	8	9	10	11	12	13	16	21
1.05	0.05	0.025	5	6	7	8	9	10	11	13	18
1.05	0.10	0.050	4	5	6	6	7	8	9	11	16
1.05	0.20	0.100	3	4	4	5	5	6	7	9	13
1.10	0.01	0.005	7	9	10	11	11	13	14	17	22
1.10	0.02	0.010	6	8	8	9	10	11	12	15	20
1.10	0.05	0.025	5	6	7	7	8	9	10	12	17
1.10	0.10	0.050	4	5	5	6	6	7	8	10	14
1.10	0.20	0.100	2	3	4	4	5	6	7	8	12
1.15	0.01	0.005	7	8	9	10	11	12	13	15	20
1.15	0.02	0.010	6	7	8	8	9	10	12	14	18
1.15	0.05	0.025	4	6	6	7	7	8	9	11	15
1.15	0.10	0.050	3	4	5	5	6	7	8	10	13
1.15	0.20	0.100	2	3	4	4	5	5	6	8	11
1.20	0.01	0.005	6	8	8	9	10	11	12	14	18
1.20	0.02	0.010	5	7	7	8	9	10	11	13	17
1.20	0.05	0.025	4	5	6	6	7	8	9	11	14
1.20	0.10	0.050	3	4	5	5	6	6	7	9	12
1.20	0.20	0.100	2	3	3	4	4	5	6	7	10
1.25	0.01	0.005	6	7	8	9	9	10	11	13	17
1.25	0.02	0.010	5	6	7	7	8	9	10	12	16
1.25	0.05	0.025	4	5	5	6	7	7	8	10	13
1.25	0.10	0.050	3	4	4	5	5	6	7	8	11
1.25	0.20	0.100	2	3	3	4	4	5	5	7	9
1.30	0.01	0.005	6	7	7	8	9	10	11	12	16
1.30	0.02	0.010	5	6	6	7	8	8	9	11	14
1.30	0.05	0.025	4	5	5	6	6	7	8	9	12
1.30	0.10	0.050	3	4	4	5	5	6	6	8	11
1.30	0.20	0.100	2	3	3	3	4	4	5	6	9
1.35	0.01	0.005	5	7	7	8	8	9	10	12	15
1.35	0.02	0.010	5	6	6	7	7	8	9	10	14
1.35	0.05	0.025	4	4	5	5	6	6	7	9	12
1.35	0.10	0.050	3	4	4	4	5	5	6	7	10
1.35	0.20	0.100	2	3	3	3	4	4	5	6	8
1.40	0.01	0.005	5	6	7	7	8	8	9	11	14
1.40	0.02	0.010	4	5	6	6	7	7	8	10	13
1.40	0.05	0.025	3	4	5	5	5	6	7	8	11
1.40	0.10	0.050	3	3	4	4	4	5	6	7	9
1.40	0.20	0.100	2	3	3	3	3	4	4	6	8
1.45	0.01	0.005	5	6	6	7	7	8	9	10	13
1.45	0.02	0.010	4	5	6	6	6	7	8	9	12
1.45	0.05	0.025	3	4	4	5	5	6	6	8	10
1.45	0.10	0.050	3	3	4	4	4	5	5	6	9
1.45	0.20	0.100	2	2	3	3	3	4	4	5	7
1.50	0.01	0.005	5	6	6	6	7	8	8	10	12
1.50	0.02	0.010	4	5	5	6	6	7	7	9	11
1.50	0.05	0.025	3	4	4	5	5	5	6	7	10
1.50	0.10	0.050	3	3	3	4	4	5	5	6	8
1.50	0.20	0.100	2	2	3	3	3	4	4	5	7

Table 7.1

Table 7.2 Sample sizes for 2-sample *t*-test.

d_t	α 2-sided	α 1-sided	Power $1-\beta$ 0.50	0.65	0.70	0.75	0.80	0.85	0.90	0.95	0.99
0.05	0.01	0.005	5309	7016	7690	8453	9344	10440	11905	14252	19226
0.05	0.02	0.010	4331	5884	6502	7205	8030	9048	10415	12617	17319
0.05	0.05	0.025	3074	4401	4939	5553	6280	7184	8407	10397	14699
0.05	0.10	0.050	2165	3298	3765	4304	4947	5752	6852	8659	12617
0.05	0.20	0.100	1315	2224	2610	3062	3607	4299	5256	6852	10414
0.10	0.01	0.005	1328	1755	1923	2114	2337	2611	2977	3564	4807
0.10	0.02	0.010	1083	1472	1626	1802	2008	2263	2604	3155	4331
0.10	0.05	0.025	769	1101	1235	1389	1571	1797	2102	2600	3675
0.10	0.10	0.050	542	825	942	1077	1237	1439	1714	2165	3155
0.10	0.20	0.100	329	557	653	766	902	1075	1315	1714	2604
0.15	0.01	0.005	591	781	855	940	1039	1161	1324	1585	2137
0.15	0.02	0.010	482	655	723	802	893	1006	1158	1403	1925
0.15	0.05	0.025	342	490	550	618	699	799	935	1156	1634
0.15	0.10	0.050	241	367	419	479	550	640	762	963	1403
0.15	0.20	0.100	147	248	291	341	402	478	585	762	1158
0.20	0.01	0.005	333	440	482	529	585	654	745	892	1203
0.20	0.02	0.010	272	369	407	451	503	566	652	790	1083
0.20	0.05	0.025	193	276	310	348	393	450	526	651	920
0.20	0.10	0.050	136	207	236	270	310	360	429	542	789
0.20	0.20	0.100	83	140	164	192	226	269	329	429	652
0.25	0.01	0.005	213	282	309	339	375	419	477	571	770
0.25	0.02	0.010	174	236	261	289	322	363	418	506	694
0.25	0.05	0.025	124	177	198	223	252	288	337	417	589
0.25	0.10	0.050	87	133	151	173	199	231	275	347	506
0.25	0.20	0.100	53	90	105	123	145	173	211	275	417
0.30	0.01	0.005	149	196	215	236	261	291	332	397	535
0.30	0.02	0.010	121	164	182	201	224	252	290	352	482
0.30	0.05	0.025	86	123	138	155	175	201	234	290	409
0.30	0.10	0.050	61	93	105	120	138	161	191	241	351
0.30	0.20	0.100	37	63	73	86	101	120	147	191	290
0.35	0.01	0.005	109	144	158	174	192	214	244	292	393
0.35	0.02	0.010	89	121	134	148	165	186	214	259	355
0.35	0.05	0.025	64	91	102	114	129	148	173	213	301
0.35	0.10	0.050	45	68	78	89	102	118	141	178	258
0.35	0.20	0.100	28	46	54	63	74	89	108	141	213
0.40	0.01	0.005	84	111	121	133	147	164	187	224	302
0.40	0.02	0.010	69	93	103	114	127	142	164	198	272
0.40	0.05	0.025	49	70	78	88	99	113	132	163	231
0.40	0.10	0.050	35	52	60	68	78	91	108	136	198
0.40	0.20	0.100	21	36	42	49	57	68	83	108	164
0.45	0.01	0.005	67	88	96	105	116	130	148	177	238
0.45	0.02	0.010	55	74	81	90	100	113	130	157	215
0.45	0.05	0.025	39	55	62	70	79	90	105	129	182
0.45	0.10	0.050	28	42	47	54	62	72	85	108	157
0.45	0.20	0.100	17	28	33	39	45	54	66	85	129
0.50	0.01	0.005	54	71	78	86	95	106	120	144	193
0.50	0.02	0.010	44	60	66	73	81	92	105	127	174
0.50	0.05	0.025	32	45	50	57	64	73	85	105	148
0.50	0.10	0.050	23	34	39	44	50	58	69	87	127
0.50	0.20	0.100	14	23	27	31	37	44	53	69	105
0.55	0.01	0.005	45	59	65	71	78	87	100	119	160
0.55	0.02	0.010	37	50	55	61	67	76	87	105	144
0.55	0.05	0.025	26	37	42	47	53	60	70	87	122
0.55	0.10	0.050	19	28	32	36	42	48	58	72	105
0.55	0.20	0.100	12	19	22	26	31	36	44	57	87

Table 7.2

Table 7.2 Continued.

d_t	α 2-sided	1-sided	Power $1-\beta$ 0.50	0.65	0.70	0.75	0.80	0.85	0.90	0.95	0.99
0.60	0.01	0.005	38	50	55	60	66	74	84	100	135
0.60	0.02	0.010	31	42	46	51	57	64	73	89	121
0.60	0.05	0.025	22	32	35	40	45	51	59	73	103
0.60	0.10	0.050	16	24	27	31	35	41	48	61	89
0.60	0.20	0.100	10	16	19	22	26	31	37	48	73
0.65	0.01	0.005	33	43	47	51	56	63	72	85	115
0.65	0.02	0.010	27	36	40	44	49	55	63	76	104
0.65	0.05	0.025	19	27	30	34	38	43	51	63	88
0.65	0.10	0.050	14	20	23	26	30	35	41	52	76
0.65	0.20	0.100	9	14	16	19	22	26	32	41	62
0.70	0.01	0.005	28	37	40	44	49	54	62	74	99
0.70	0.02	0.010	23	31	34	38	42	47	54	65	89
0.70	0.05	0.025	17	23	26	29	33	38	44	54	76
0.70	0.10	0.050	12	18	20	23	26	30	36	45	65
0.70	0.20	0.100	8	12	14	16	19	23	28	36	54
0.75	0.01	0.005	25	32	35	39	43	48	54	64	87
0.75	0.02	0.010	20	27	30	33	37	41	47	57	78
0.75	0.05	0.025	15	21	23	26	29	33	38	47	66
0.75	0.10	0.050	11	16	18	20	23	26	31	39	57
0.75	0.20	0.100	7	11	12	14	17	20	24	31	47
0.80	0.01	0.005	22	29	31	34	38	42	48	57	76
0.80	0.02	0.010	18	24	26	29	32	36	42	50	69
0.80	0.05	0.025	13	18	20	23	26	29	34	42	58
0.80	0.10	0.050	9	14	16	18	20	23	28	35	50
0.80	0.20	0.100	6	10	11	13	15	18	21	28	41
0.85	0.01	0.005	20	25	28	30	33	37	42	50	68
0.85	0.02	0.010	16	21	24	26	29	32	37	45	61
0.85	0.05	0.025	12	16	18	20	23	26	30	37	52
0.85	0.10	0.050	8	12	14	16	18	21	25	31	45
0.85	0.20	0.100	5	9	10	11	13	16	19	25	37
0.90	0.01	0.005	18	23	25	27	30	33	38	45	60
0.90	0.02	0.010	14	19	21	23	26	29	33	40	55
0.90	0.05	0.025	10	15	16	18	20	23	27	33	46
0.90	0.10	0.050	8	11	13	14	16	19	22	28	40
0.90	0.20	0.100	5	8	9	10	12	14	17	22	33
0.95	0.01	0.005	16	21	22	25	27	30	34	41	54
0.95	0.02	0.010	13	17	19	21	23	26	30	36	49
0.95	0.05	0.025	10	13	15	16	18	21	24	30	42
0.95	0.10	0.050	7	10	11	13	15	17	20	25	36
0.95	0.20	0.100	4	7	8	9	11	13	15	20	30
1.00	0.01	0.005	14	19	20	22	25	27	31	37	49
1.00	0.02	0.010	12	16	17	19	21	24	27	33	44
1.00	0.05	0.025	9	12	13	15	17	19	22	27	38
1.00	0.10	0.050	6	9	10	12	13	15	18	23	32
1.00	0.20	0.100	4	6	7	8	10	12	14	18	27
1.05	0.01	0.005	13	17	19	20	22	25	28	33	45
1.05	0.02	0.010	11	14	16	17	19	22	25	30	40
1.05	0.05	0.025	8	11	12	14	15	17	20	25	34
1.05	0.10	0.050	6	8	9	11	12	14	16	21	30
1.05	0.20	0.100	4	6	7	8	9	11	13	16	24
1.10	0.01	0.005	12	16	17	19	20	23	26	31	41
1.10	0.02	0.010	10	13	15	16	18	20	23	27	37
1.10	0.05	0.025	7	10	11	12	14	16	18	22	31
1.10	0.10	0.050	5	8	9	10	11	13	15	19	27
1.10	0.20	0.100	4	5	6	7	8	10	12	15	22

Table 7.2

Table 7.2 Continued.

d_t	α 2-sided	1-sided	Power $1-\beta$ 0.50	0.65	0.70	0.75	0.80	0.85	0.90	0.95	0.99
1.15	0.01	0.005	11	14	16	17	19	21	24	28	37
1.15	0.02	0.010	9	12	13	15	16	18	21	25	34
1.15	0.05	0.025	7	9	10	11	13	15	17	21	29
1.15	0.10	0.050	5	7	8	9	10	12	14	17	25
1.15	0.20	0.100	3	5	6	7	8	9	11	14	21
1.20	0.01	0.005	10	13	14	16	17	19	22	26	35
1.20	0.02	0.010	9	11	12	14	15	17	19	23	31
1.20	0.05	0.025	6	9	10	11	12	13	16	19	27
1.20	0.10	0.050	5	7	7	8	9	11	13	16	23
1.20	0.20	0.100	3	5	5	6	7	8	10	13	19
1.25	0.01	0.005	10	12	13	15	16	18	20	24	32
1.25	0.02	0.010	8	10	11	13	14	16	18	21	29
1.25	0.05	0.025	6	8	9	10	11	12	14	18	25
1.25	0.10	0.050	4	6	7	8	9	10	12	15	21
1.25	0.20	0.100	3	4	5	6	7	8	9	12	17
1.30	0.01	0.005	9	12	13	14	15	17	19	22	30
1.30	0.02	0.010	7	10	11	12	13	14	16	20	27
1.30	0.05	0.025	6	7	8	9	10	12	13	16	23
1.30	0.10	0.050	4	6	6	7	8	9	11	14	20
1.30	0.20	0.100	3	4	5	5	6	7	9	11	16
1.35	0.01	0.005	8	11	12	13	14	15	17	21	28
1.35	0.02	0.010	7	9	10	11	12	13	15	18	25
1.35	0.05	0.025	5	7	8	9	10	11	13	15	21
1.35	0.10	0.050	4	5	6	7	8	9	10	13	18
1.35	0.20	0.100	3	4	4	5	6	7	8	10	15
1.40	0.01	0.005	8	10	11	12	13	14	16	19	26
1.40	0.02	0.010	7	9	9	10	11	13	14	17	23
1.40	0.05	0.025	5	7	7	8	9	10	12	14	20
1.40	0.10	0.050	4	5	6	6	7	8	10	12	17
1.40	0.20	0.100	2	4	4	5	5	6	8	10	14
1.45	0.01	0.005	7	9	10	11	12	14	15	18	24
1.45	0.02	0.010	6	8	9	10	11	12	13	16	22
1.45	0.05	0.025	5	6	7	8	8	10	11	13	18
1.45	0.10	0.050	3	5	5	6	7	8	9	11	16
1.45	0.20	0.100	2	3	4	4	5	6	7	9	13
1.50	0.01	0.005	7	9	10	11	12	13	14	17	23
1.50	0.02	0.010	6	8	8	9	10	11	13	15	20
1.50	0.05	0.025	4	6	6	7	8	9	10	13	17
1.50	0.10	0.050	3	5	5	6	6	7	9	11	15
1.50	0.20	0.100	2	3	4	4	5	6	7	8	12

Table 7.2

Chapter 8
Significance of The Correlation Coefficient

8.1 INTRODUCTION

A doctor may wish to show that two measurements are associated. For example, in patients with a particular disease the number of monocytes in the blood may be correlated with the estrone to estradiol conversion rate. One method of measuring association is to compute Pearson's product-moment correlation coefficient, r, and test it for statistical significance, the corresponding null hypothesis being that the true correlation coefficient $\rho = 0$. The correlation coefficient measures the degree of linear relationship between two variables, but is inappropriate in the case of non-linear relationships. Given that the number of monocytes in the blood and the estrone to estradiol conversion rates differ between patients, then the doctor can calculate the required number of patients to be observed by specifying the magnitude of the correlation coefficient that he wishes to detect. Specifying the correlation coefficient avoids the problem of defining the ranges of the two variables under consideration; however the wider their ranges, the more sensitive the study.

8.2 THEORY AND FORMULAE

Sample Size

The correlation coefficient is dimensionless, and so can act as an effect size index. Before the trial we need to specify the size of the correlation coefficient, ρ, that we think might represent the true correlation between the two variables. Clearly the smaller the true correlation coefficient the larger the study has to be to detect it. To get a feel for the correlation coefficient between the two variables, we note that ρ^2 is the proportion of variance in either of the variables which may be accounted for by the other, using a linear relationship. Cohen (1977) suggested values of ρ of 0.10, 0.30 and 0.50 as 'small', 'medium' and 'large' effects and pointed out that correlations in many sociological studies are generally less than 0.50.

Usually the direction of the relationship will be specified in advance. For example systolic and diastolic blood pressure are certain to be positively associated and so one would use a one-sided test. However, if one is looking for an association with no sign specified, then a two-sided test is warranted.

To calculate appropriate sample sizes, we assume two normally distributed variables with correlation coefficient ρ. It can be shown that

$$u_\rho = \tfrac{1}{2} log \left[\frac{1+\rho}{1-\rho} \right] + \frac{\rho}{2(m-1)} \tag{8.1}$$

is approximately normally distributed with standard deviation $1/(m-3)$. This leads to the appropriate sample size to detect a correlation ρ, for significance level α and power $1-\beta$ of

$$m = \frac{(z_{1-\alpha} + z_{1-\beta})^2}{u_\rho^2} + 3. \tag{8.2}$$

It should be emphasised that to calculate u_ρ for equation (8.2) we require some value for m, the number we are trying to estimate! To get round this problem an initial value for u_ρ labelled u'_ρ is calculated where

$$u'_\rho = \tfrac{1}{2} log \left[\frac{1+\rho}{1-\rho} \right] \tag{8.3}$$

which is the first term on the right hand side of equation (8.1). This value is then used in (8.2) to give an initial value for m labelled m'. This m' is then used in (8.1) to obtain a new value for u_p which is then used in equation (8.2) to obtain a new value for m and the whole process repeated again. Such a procedure is called an iteration. To calculate the tables the iteration was repeated until two consecutive values of m within unity of each other were found. It should be noted that Table 2.3 which gives $(z_{1-\alpha} + z_{1-\beta})^2$ for different values of α and β can be used to calculate (8.2).

Power

In contrast, if one were asked for the power of a study for given ρ, m and α then equation (8.2) can be inverted to give

$$z_{1-\beta} = u_\rho\sqrt{(m-3)} - z_{1-\alpha} \tag{8.4}$$

Lack of Association

Cohen (1977) cites an example of a social psychologist planning an experiment in which college students are subject to two questionnaries, one on personality (y) and one on social desirability (x). He wishes to show that y and x are *not* associated. This is a similar situation to that described in Chapter 5, where we are trying to do the impossible and prove the null hypothesis. We avoid this problem by attempting to demonstrate that ρ is small, for example no greater in absolute value than 0.10.

8.3 BIBLIOGRAPHY

Problems connected with the use of and interpretation of the correlation coefficient in medical studies have been reviewed by Altman & Bland (1983). Cohen (1977, Chapter 3) described sample sizes for tests of significance of the correlation coefficient. Tables for testing significance are given by Fisher & Yates (1963).

8.4 DESCRIPTION OF THE TABLE

Table 8.1

Table 8.1 gives the sample sizes required to detect a correlation for a range of two-sided test sizes of α from 0.01 to 0.20, one sided from 0.005 to 0.10, power $1-\beta$ ranging from 0.50 to 0.99 for postulated ρ between 0.05 and 0.95. It should be emphasised that the values in the tables are for the total number of subjects required, with both variables measured on each subject.

8.5 USE OF THE TABLE

Table 8.1

Example 8.1

In respiratory physiology, suppose the correlation between the forced expiratory volume in 1 second (FEV$_1$), and the forced vital capacity (FVC) in healthy subjects is thought to be about 0.60. Also suppose patients with a certain lung disease are available at a clinic and one wishes to test if there is a significance correlation between the FEV_1 and the FVC in these patients. We do not expect a negative correlation, and so the test is a one-sided one, say at the 5% level, and power 80%. How many subjects are required?

Here, one-sided $\alpha = 0.05$, $1 - \beta = 0.80$, $\rho = 0.60$, and then from Table 8.1 we require at least 16 subjects.

Example 8.2

If subjects suffering from mild hypertension are given a certain dose of a drug, the subsequent fall in blood pressure is associated with a decrease in blood viscosity, with a correlation coefficient of about 0.3. Suppose we conduct the same experiment on patients with severe hypertension and observe their fall in blood pressure and decrease in viscosity following the drug. How many patients do we need to recruit to obtain a significant correlation, given that its magnitude is expected to be 0.3?

In this case, suppose we do not know which way the correlation may go, so that we are using a two-sided test, with significance level 5%, and power 90%. Thus two-sided $\alpha = 0.05$, $1 - \beta = 0.90$, $\rho = 0.3$, and from Table 8.1 we would require 112 patients.

Example 8.3

If in fact the investigator in Example 8.1 managed to recruit 40 patients but only finds a correlation of $r = 0.30$, what is the power of a test of significance at one sided 5% level, if 0.30 is in fact the true correlation?

Again, one sided $\alpha = 0.05$, but $\rho = 0.30$, and searching along the corresponding row of Table 8.1, it can be seen that with $m = 31$ subjects the power $1 - \beta = 0.50$, whereas for $m = 46$ subjects $1 - \beta = 0.65$. Thus $m = 40$ subjects corresponds to a power of between 0.50 and 0.65. More exactly use can be made of equations (8.1) and (8.4) with $m = 40$, $\rho = 0.3$ and $z_{1-\alpha} = 1.6449$. These give $u_\rho = u_{0.3} = 0.3134$, $z_{1-\beta} = 0.261$ and making use of Table 2.1, the power $1 - \beta = 0.60$.

Example 8.4

A psychologist wishes to plan an experiment in which students are asked questions from which measures of personality and social desirability can be obtained. The investigator wishes to show that these two variables are not associated. How many students must he recruit to his study?

It is first necessary to specify the size of the type I error α. In such a situation, as described here, the investigator may be willing to effectively assume $\rho = 0$ when it is small and so choose a large type I error $\alpha = 0.2$. However he will require a relatively small type II error and perhaps choose $1 - \beta = 0.95$. In addition he must set a value of ρ below which he would regard the association as effectively zero, he chooses $\rho = 0.1$. Table 8.1 with two sided $\alpha = 0.2$, $1 - \beta = 0.95$ and $\rho = 0.1$ gives the number of students to be recruited $m = 853$.

REFERENCES

Altman D. G. & Bland J. M. (1983). Measurement in medicine: the analysis of method comparison studies. *The Statistician*, **32**, 307-317

Cohen J. (1977). *Statistical Power Analysis for the Behavioral Sciences*, revised edn. Academic Press, New York.

Fisher R. A. & Yates F. (1963). *Statistical Tables for Use in Biological, Agricultural and Medical Research*, 6th edn. Oliver and Boyd, Edinburgh.

Table 8.1 Sample sizes for detecting a statistically significant correlation coefficient.

ρ	α 2-sided	1-sided	Power 1−β 0.50	0.65	0.70	0.75	0.80	0.85	0.90	0.95	0.99
0.05	0.01	0.005	2652	3504	3841	4221	4666	5213	5944	7116	9599
0.05	0.02	0.010	2164	2939	3248	3599	4010	4518	5201	6300	8647
0.05	0.05	0.025	1537	2199	2467	2774	3137	3588	4198	5192	7339
0.05	0.10	0.050	1083	1648	1882	2151	2471	2873	3422	4324	6300
0.05	0.20	0.100	658	1112	1305	1530	1803	2148	2626	3422	5201
0.10	0.01	0.005	662	873	957	1052	1163	1299	1481	1772	2390
0.10	0.02	0.010	540	733	810	897	999	1126	1296	1569	2153
0.10	0.05	0.025	384	549	616	692	782	894	1046	1293	1827
0.10	0.10	0.050	271	412	470	537	617	717	853	1078	1569
0.10	0.20	0.100	166	279	326	383	450	536	655	853	1296
0.15	0.01	0.005	293	386	423	465	514	574	654	782	1055
0.15	0.02	0.010	239	324	358	397	442	498	572	693	950
0.15	0.05	0.025	171	243	273	306	346	396	462	571	807
0.15	0.10	0.050	121	183	209	238	273	317	377	476	693
0.15	0.20	0.100	74	124	145	170	200	238	290	377	572
0.20	0.01	0.005	164	216	236	260	287	320	365	436	587
0.20	0.02	0.010	134	181	200	222	247	278	319	386	529
0.20	0.05	0.025	96	136	153	171	193	221	258	319	450
0.20	0.10	0.050	68	103	117	133	153	177	211	266	386
0.20	0.20	0.100	43	70	82	96	112	133	162	211	319
0.25	0.01	0.005	104	137	150	164	182	203	231	276	371
0.25	0.02	0.010	86	115	127	141	156	176	202	244	334
0.25	0.05	0.025	61	87	97	109	123	140	164	202	284
0.25	0.10	0.050	44	66	75	85	97	113	134	168	244
0.25	0.20	0.100	28	45	53	61	72	85	103	134	202
0.30	0.01	0.005	72	94	103	113	124	139	158	188	253
0.30	0.02	0.010	59	79	87	97	107	121	138	167	228
0.30	0.05	0.025	43	60	67	75	84	96	112	138	194
0.30	0.10	0.050	31	46	52	59	67	78	92	116	167
0.30	0.20	0.100	20	32	37	43	50	59	71	92	138
0.35	0.01	0.005	52	68	75	82	90	100	114	136	182
0.35	0.02	0.010	43	58	63	70	78	87	100	121	165
0.35	0.05	0.025	31	44	49	55	61	70	81	100	140
0.35	0.10	0.050	23	33	38	43	49	56	67	84	121
0.35	0.20	0.100	15	23	27	31	36	43	52	67	100
0.40	0.01	0.005	40	51	56	61	68	75	85	102	136
0.40	0.02	0.010	33	44	48	53	59	66	75	90	123
0.40	0.05	0.025	24	33	37	41	46	53	61	75	105
0.40	0.10	0.050	18	26	29	33	37	43	50	63	90
0.40	0.20	0.100	12	18	21	24	28	33	39	50	75
0.45	0.01	0.005	31	40	44	48	52	58	66	78	105
0.45	0.02	0.010	26	34	37	41	45	51	58	70	95
0.45	0.05	0.025	19	26	29	32	36	41	47	58	81
0.45	0.10	0.050	14	20	23	26	29	33	39	49	70
0.45	0.20	0.100	10	15	17	19	22	26	31	39	58
0.50	0.01	0.005	25	32	34	38	41	46	52	62	82
0.50	0.02	0.010	21	27	30	32	36	40	46	55	74
0.50	0.05	0.025	15	21	23	26	29	32	37	46	64
0.50	0.10	0.050	12	16	18	20	23	26	31	38	55
0.50	0.20	0.100	8	12	14	15	18	20	24	31	46
0.55	0.01	0.005	20	26	28	30	33	37	42	49	65
0.55	0.02	0.010	17	22	24	26	29	32	37	44	59
0.55	0.05	0.025	13	17	19	21	23	26	30	37	51
0.55	0.10	0.050	10	14	15	17	19	21	25	31	44
0.55	0.20	0.100	7	10	11	13	15	17	20	25	37

Table 8.1

Table 8.1 Continued.

ρ	α 2-sided	1-sided	Power $1-\beta$ 0.50	0.65	0.70	0.75	0.80	0.85	0.90	0.95	0.99
0.60	0.01	0.005	17	21	23	25	27	30	34	40	53
0.60	0.02	0.010	14	18	20	21	24	26	30	36	48
0.60	0.05	0.025	11	14	16	17	19	21	25	30	41
0.60	0.10	0.050	8	11	13	14	16	18	21	25	36
0.60	0.20	0.100	6	9	10	11	12	14	16	21	30
0.65	0.01	0.005	14	17	19	20	22	24	27	32	43
0.65	0.02	0.010	12	15	16	18	19	22	24	29	39
0.65	0.05	0.025	9	12	13	14	16	18	20	24	33
0.65	0.10	0.050	7	10	11	12	13	15	17	21	29
0.65	0.20	0.100	6	8	8	9	10	12	14	17	24
0.70	0.01	0.005	12	14	16	17	18	20	23	26	35
0.70	0.02	0.010	10	13	14	15	16	18	20	24	32
0.70	0.05	0.025	8	10	11	12	13	15	17	20	27
0.70	0.10	0.050	7	8	9	10	11	12	14	17	24
0.70	0.20	0.100	5	7	7	8	9	10	12	14	20
0.75	0.01	0.005	10	12	13	14	15	17	19	22	28
0.75	0.02	0.010	9	11	11	12	13	15	17	19	26
0.75	0.05	0.025	7	9	9	10	11	12	14	17	22
0.75	0.10	0.050	6	7	8	9	9	10	12	14	19
0.75	0.20	0.100	5	6	6	7	8	9	10	12	17
0.80	0.01	0.005	8	10	11	12	13	14	15	18	23
0.80	0.02	0.010	7	9	10	10	11	12	14	16	21
0.80	0.05	0.025	6	8	8	9	9	10	12	14	18
0.80	0.10	0.050	5	6	7	7	8	9	10	12	16
0.80	0.20	0.100	5	5	6	6	7	7	8	10	14
0.85	0.01	0.005	7	9	9	10	10	11	12	14	18
0.85	0.02	0.010	6	8	8	9	9	10	11	13	17
0.85	0.05	0.025	6	7	7	7	8	9	10	11	15
0.85	0.10	0.050	5	6	6	6	7	8	8	10	13
0.85	0.20	0.100	4	5	5	6	6	6	7	8	11
0.90	0.01	0.005	6	7	7	8	8	9	10	11	14
0.90	0.02	0.010	6	6	7	7	8	8	9	10	13
0.90	0.05	0.025	5	6	6	6	7	7	8	9	11
0.90	0.10	0.050	4	5	5	6	6	6	7	8	10
0.90	0.20	0.100	4	4	5	5	5	6	6	7	9
0.95	0.01	0.005	5	6	6	6	7	7	8	8	10
0.95	0.02	0.010	5	5	6	6	6	7	7	8	10
0.95	0.05	0.025	4	5	5	5	6	6	6	7	9
0.95	0.10	0.050	4	4	5	5	5	5	6	6	8
0.95	0.20	0.100	4	4	4	4	5	5	5	6	7

Table 8.1

Chapter 9
Comparing Two Survival Curves (Logrank Comparisons)

9.1 INTRODUCTION

There are many clinical trials in which patients are recruited and randomized to receive a particular treatment, then followed up until some critical event occurs. The length of follow up for each patient is then used in comparing the efficacy of the two treatments by use of, so called, survival techniques, in particular the logrank test. The basic difference between this type of study and those described in Chapter 3 is that, in the latter, success or failure of treatment is determined at some 'fixed' time after randomization. For example, if patients with a particular cancer are recruited to a trial one might record the individual survival experiences of all the patients rather than merely record how many are dead at 5 years from randomization. It should be emphasized it is not essential that all patients must be followed until the critical event occurs, indeed in many cancer trials some patients will survive many years after randomization. It is usual, in such circumstances, to fix a date beyond which no further information is to be collected on any patient. Any patient still alive at that date will nevertheless be used in the treatment comparison. Such an observation is called 'censored'. For this reason it is the number of observed events rather than the number of subjects recruited to a trial that is important.

Although we are concerned with survival time studies and are using the logrank test for comparison, one often uses survival rates at particular (fixed) times for planning purposes.

9.2 THEORY AND FORMULAE

Sample Size

Suppose the critical event of interest is death and that two treatments give rise to survival proportions π_1 and π_2, where $\pi_2 > \pi_1$, at some chosen time point, for example 1 year after the date of randomization. The ratio of the risks of death in the two groups is called the relative hazard h and if this does not change with time, it is estimated by

$$h = \frac{\log \pi_1}{\log \pi_2} .$$
(9.1)

The total number of events, e, required in each patient group to give a test with significance level α and power $1 - \beta$ is approximately

$$e = \frac{(z_{1-\alpha} + z_{1-\beta})^2 (h + 1)^2}{2(h - 1)^2},$$
(9.2)

and the number of patients per group as

$$m = 2e/(2 - \pi_1 - \pi_2).$$
(9.3)

At the design stage we have the option either to randomize equally to the two alternative treatments or to randomize with unequal allocation. If the allocation ratio in the two groups is to be $r\,(> 1)$ then the number of events required in the smaller group is

$$e = \frac{(z_{1-\alpha} + z_{1-\beta})^2 (rh + 1)^2}{r(1 + r)(h - 1)^2} \cdot \tag{9.4}$$

The corresponding number of patients for the smaller group is

$$m = \frac{(1 + r)\,e}{r(1 - \pi_1) + (1 - \pi_2)} \cdot \tag{9.5}$$

Equations (9.2) and (9.4) can be evaluated using Table 2.3 which gives $(z_{1-\alpha} + z_{1-\beta})^2$ for different values of α and β.

The derivation of these formulae assumes that analysis occurs at a fixed time (date) T after the *last* patient has entered the study and that patient follow-up extending beyond T is excluded. This leads to the required number of patients being over-estimated. Thus T should be chosen in such a way that the rate of occurence of events beyond T is low so that numbers are only slightly over-estimated. It should be emphasized that only when the great majority of the relevant information has been collected is it appropriate to conduct a definitive analysis of the trial.

Withdrawals

One aspect of a trial which can effect the numbers is the proportion of patients who are lost to follow up during the trial. Such patients have censored observations as do those for whom the event of interest has not occurred at the close of the trial. If this anticipated withdrawal rate is $x\%$ then the required number of patients given in (9.3) or (9.5) should be modified to

$$m' = \frac{100m}{(100 - x)} \cdot \tag{9.6}$$

Median Survival

In many survival type studies it is more natural to think of treatment differences in terms of median survival times rather than survival rates. Defining τ_1 and τ_2 as the median survival times for the two groups then, on the assumption that survival times are exponentially distributed, the relation between π_1, π_2 and τ_1, τ_2 can be expressed by

$$\pi_2 = \exp(\log \pi_1 / h) \tag{9.7}$$

where $h = \tau_2 / \tau_1$ is the ratio of the two medians.

Power

In contrast, if one were asked for the power of a study for given π_1, π_2, m and α then equations (9.4) and (9.5) can be combined and then inverted to give

$$z_{1-\beta} = \frac{(h - 1)\sqrt{[mr(1+r)\{r(1-\pi_1) + (1-\pi_2)\}]}}{(rh + 1)} - z_{1-\alpha} \cdot \tag{9.8}$$

9.3 BIBLIOGRAPHY

Basic references for survival time studies and the log rank test are Peto, Pike, Armitage *et al.* (1976, 1977) who give examples from leukaemia trials. Theory, formulae and tables for the calculation of sample sizes are given by Freedman (1982) who also give guidelines for their use and comments on the accuracy of equations (9.2) and (9.4). It should be noted, however, that Freedman's tables give the *total* number of events and the *total* number of patients required for the trial, and not the numbers per patient group.

9.4 DESCRIPTION OF THE TABLES

Table 9.1

This table gives for $r = 1$ the number of critical events e to be observed in one group, for two-sided test size α ranging from 0.01 to 0.20, one-sided from 0.005 to 0.10, power $1 - \beta$ ranging from 0.50 to 0.99 for values of π_1 from 0.05 to 0.9 with $\pi_2 = (\pi_1 + 0.05)$ by steps of 0.05 to 0.95.

Table 9.2

Is tabulated as for Table 9.1 but gives the required numbers of patients to be recruited.

It should also be noted that, although these tables refer to π_1 and π_2 as do those of Chapter 3, the resulting sample sizes in general are not the same. In certain circumstances they are considerably less, since the logrank test is more powerful than the corresponding chi-squared test in this context.

Table 9.3

This table gives the corresponding 'survival' rate π_2 of the patients receiving the second treatment corresponding to a postulated ratio of medians h ranging from 1.1 to 10.0 and survival rate π_1 ranging from 0.05 to 0.95.

9.5 USE OF THE TABLES

Table 9.1

Example 9.1

An adjuvant study of the drug Levamisole is proposed for patients with resectable cancer of the colon (Duke's C) in which the primary objective of the study is to compare the efficacy of Levamisole against a placebo control with respect to relapse free survival. How many relapses need to be observed in the trial if a decrease in relapse rates at one year, from 50% to 40% is anticipated, and a power of 80% is required?

Here we wish to increase the success rate, i.e. failure to relapse, from 50% to 60%, so $\pi_1 = 0.5$, $\pi_2 = 0.6$ and $1 - \beta = 0.8$. Assuming a one-sided test with $\alpha = 0.05$ and $r = 1$, then Table 9.1 gives $e = 135$. Thus a total of 270 relapses would be required.

Table 9.2

Example 9.2

How many subjects need to be recruited to the study described in Example 9.1 if it is assumed there will be a 10%

withdrawal of patients beyond the control of the investigator?

Using Table 9.2 with $\pi_1 = 0.5$, $\pi_2 = 0.6$, one sided $\alpha = 0.05$ and $1 - \beta = 0.8$ we have $m = 300$. Allowing for a withdrawal rate $x = 10\%$ and using equation (9.6) we obtain $m' = 334$. Thus a total of approximately 670 subjects should be recruited to the trial.

Table 9.3

Example 9.3

The Multicenter Study Group (1980) describe a double blind controlled study of long-term oral acetylcysteine against placebo in chronic bronchitis. Their results gave the percentage of exacerbation-free subjects with placebo at 6 months as 25%. They also observed a doubling of median exacerbation-free times with the active treatment. A repeat trial is planned. How many subjects should be recruited if the power is to be set at 90% and the one sided test size at 5%.

Here $\pi_1 = 0.25$ and $h = 2$, using Table 9.3 we have $\pi_2 = 0.50$. In addition one sided $\alpha = 0.05$, power $1 - \beta = 0.9$ and if we assume $r = 1$, then from Table 9.1 the number of events e to be observed per treatment group is 39. Table 9.2 indicates that $m = 62$ patients should be recruited to each treatment in order that the necessary number of events is observed. A total of approximately 120 patients should therefore be recruited to the study.

Example 9.4

A trial is planned involving patients with bed sores. It is postulated that ultrasound treatment will halve the healing time of such sores and a double blind trial is proposed using ultrasound and a placebo, the placebo being a non functioning ultrasound machine. The investigators would like a power of 90% and a two-sided test at 5%. The proportion healed without treatment at 21 days is approximately 70%. How many patients should be recruited to the study?

Here the ratio of the median healing times is $h = 2$. The proportion healed at 21 days without treatment is $\pi_1 = 0.7$ and the corresponding value of π_2 is given from Table 9.3 as 0.837. Thus halving the median healing time increases the proportion healed at 21 days from 0.7 to 0.837. The closest values in Table 9.2 for π_2 are 0.80 and 0.85. Thus for two-sided $\alpha = 0.05$, $1 - \beta = 0.90$, $\pi_1 = 0.70$ and $\pi_2 = 0.80$ the table gives $m = 397$, whereas for $\pi_1 = 0.70$, $\pi_2 = 0.85$ the corresponding $m = 167$. These figures suggest that between 340 and 800 patients should be recruited to the trial.

Direct calculations using equation (9.2) and (9.3), give $m = 205$ or approximately 410 patients should be recruited. This example emphasises the need for care when using tables to estimate patient numbers corresponding to non-tabular values of π_1 and π_2.

Example 9.5

Suppose the investigators in Example 9.4 could only recruit 100 patients to their proposed study. What would be the corresponding power?

Table 9.2 with $\pi_1 = 0.70$, $\pi_2 = 0.80$ and two-sided $\alpha = 0.05$ suggests for $m = 50$ patients per group a power $1 - \beta < 0.50$, as does the entry corresponding to $\pi_1 = 0.70$, $\pi_2 = 0.88$. More exact calculation, using equation (9.8) with allocation ratio $r = 1$, $\pi_1 = 0.70$, $\pi_2 = 0.837$, $h = 2$ and, from Table 2.2, $z_{1-\alpha} = 1.96$ gives $z_{1-\beta} = -0.3562$. Using the symmetry of Table 2.1 for negative values of z_γ gives the power $1 - \beta = 0.36$. Thus the power of the proposed trial, assuming no patient losses, is approximately 36%.

9.6 REFERENCES

Freedman L. S. (1982) Tables of the number of patients required in clinical trials using the logrank test. *Statistics in Medicine,* **1,** 121-129.

Multicenter Study Group (1980). Long-term oral acetylcysteine in chronic bronchitis. A double-blind controlled study. *Europ. J. Resp. Dis.,* **61,** Suppl 111, 93-108.

Peto R., Pike M. C., Armitage P., Breslow N. E., Cox D. R., Howard S. V. Mantel N., MacPherson K., Peto J. & Smith P. G. (1976). Design and analysis of randomized clinical trials requiring prolonged observation of each patient (I) Introduction and design. *Br. J. Cancer,* **34,** 585-612.

Peto R., Pike M. C. Armitage P., Breslow N. E., Cox D. R., Howard S. V., Mantel N., MacPherson K., Peto J. & Smith P. G. (1977). Design and analysis of randomized clinical trials requiring prolonged observation of each patient (II) Analysis and examples. *Br. J. Cancer,* **35,** 1-39.

Table 9.1 Number of critical events for comparison of survival rates (logrank test).

$\pi_1 = 0.05$

π_2	α 2-Sided	1-Sided	0.50	0.65	0.70	0.75	0.80	0.85	0.90	0.95	0.99
						Power $1-\beta$					
					$\pi_1 = 0.05$						
0.10	0.01	0.005	194	257	281	309	342	382	435	521	703
0.10	0.02	0.010	159	215	238	264	294	331	381	461	633
0.10	0.05	0.025	113	161	181	203	230	263	307	380	537
0.10	0.10	0.050	80	121	138	158	181	211	251	317	461
0.10	0.20	0.100	48	82	96	112	132	157	192	251	381
0.15	0.01	0.005	66	87	96	105	116	130	148	177	239
0.15	0.02	0.010	54	73	81	90	100	113	130	157	215
0.15	0.05	0.025	39	55	62	69	78	90	105	129	183
0.15	0.10	0.050	27	41	47	54	62	72	85	108	157
0.15	0.20	0.100	17	28	33	38	45	54	66	85	130
0.20	0.01	0.005	37	49	54	59	65	72	83	99	133
0.20	0.02	0.010	30	41	45	50	56	63	72	88	120
0.20	0.05	0.025	22	31	35	39	44	50	58	72	102
0.20	0.10	0.050	15	23	26	30	35	40	48	60	88
0.20	0.20	0.100	10	16	18	22	25	30	37	48	72
0.25	0.01	0.005	25	33	36	40	44	49	56	67	90
0.25	0.02	0.010	21	28	31	34	38	42	49	59	81
0.25	0.05	0.025	15	21	23	26	30	34	39	49	69
0.25	0.10	0.050	11	16	18	20	23	27	32	41	59
0.25	0.20	0.100	7	11	13	15	17	20	25	32	49
0.30	0.01	0.005	19	25	27	30	33	36	41	49	67
0.30	0.02	0.010	15	21	23	25	28	32	36	44	60
0.30	0.05	0.025	11	16	17	20	22	25	29	36	51
0.30	0.10	0.050	8	12	13	15	17	20	24	30	44
0.30	0.20	0.100	5	8	9	11	13	15	19	24	36
0.35	0.01	0.005	15	19	21	23	26	29	33	39	52
0.35	0.02	0.010	12	16	18	20	22	25	29	35	47
0.35	0.05	0.025	9	12	14	15	17	20	23	29	40
0.35	0.10	0.050	6	9	11	12	14	16	19	24	35
0.35	0.20	0.100	4	7	8	9	10	12	15	19	29
0.40	0.01	0.005	12	16	18	19	21	24	27	32	43
0.40	0.02	0.010	10	14	15	16	18	21	24	28	39
0.40	0.05	0.025	7	10	11	13	14	16	19	23	33
0.40	0.10	0.050	5	8	9	10	11	13	16	20	28
0.40	0.20	0.100	3	5	6	7	8	10	12	16	24
0.45	0.01	0.005	10	14	15	16	18	20	23	27	36
0.45	0.02	0.010	9	11	13	14	15	17	20	24	33
0.45	0.05	0.025	6	9	10	11	12	14	16	20	28
0.45	0.10	0.050	5	7	8	9	10	11	13	17	24
0.45	0.20	0.100	3	5	5	6	7	9	10	13	20

Table 9.1

Table 9.1 Continued.

π_2	α 2-Sided	1-Sided	0.50	0.65	0.70	0.75	0.80	0.85	0.90	0.95	0.99
						Power $1-\beta$					

$\pi_1 = 0.05$

π_2	2-Sided	1-Sided	0.50	0.65	0.70	0.75	0.80	0.85	0.90	0.95	0.99
0.50	0.01	0.005	9	12	13	14	15	17	20	23	31
0.50	0.02	0.010	7	10	11	12	13	15	17	21	28
0.50	0.05	0.025	5	8	8	9	11	12	14	17	24
0.50	0.10	0.050	4	6	7	7	8	10	11	14	21
0.50	0.20	0.100	3	4	5	5	6	7	9	11	17
0.55	0.01	0.005	8	10	11	12	14	15	17	21	27
0.55	0.02	0.010	7	9	10	11	12	13	15	18	25
0.55	.05	0.025	5	7	7	8	9	11	12	15	21
0.55	.10	0.050	4	5	6	7	7	9	10	13	18
0.55	.20	0.100	2	4	4	5	6	7	8	10	15
0.60	.01	0.005	7	9	10	11	12	13	15	18	24
0.60	.02	0.010	6	8	9	9	10	12	13	16	22
0.60	.05	0.025	4	6	7	7	8	9	11	13	19
0.60	.10	0.050	3	5	5	6	7	8	9	11	16
0.60	.20	0.100	2	3	4	4	5	6	7	9	13
0.65	.01	0.005	6	8	9	10	11	12	14	16	22
0.65	.02	0.010	5	7	8	9	9	11	12	15	20
0.65	0.05	0.025	4	5	6	7	8	9	10	12	17
0.65	0.10	0.050	3	4	5	5	6	7	8	10	15
0.65	0.20	0.100	2	3	3	4	5	5	6	8	12
0.70	0.01	0.005	6	8	8	9	10	11	13	15	20
0.70	0.02	0.010	5	6	7	8	9	10	11	13	18
0.70	0.05	0.025	4	5	5	6	7	8	9	11	15
0.70	0.10	0.050	3	4	4	5	5	6	7	9	13
0.70	0.20	0.100	2	3	3	4	4	5	6	7	11
0.75	0.01	0.005	5	7	8	8	9	10	11	14	18
0.75	0.02	0.010	4	6	6	7	8	9	10	12	16
0.75	0.05	0.025	3	5	5	6	6	7	8	10	14
0.75	0.10	0.050	2	4	4	4	5	6	7	8	12
0.75	0.20	0.100	2	3	3	3	4	4	5	7	10
0.80	0.01	0.005	5	6	7	8	8	9	11	13	17
0.80	0.02	0.010	4	5	6	7	7	8	9	11	15
0.80	0.05	0.025	3	4	5	5	6	7	8	9	13
0.80	0.10	0.050	2	3	4	4	5	5	6	8	11
0.80	0.20	0.100	2	2	3	3	4	4	5	6	9
0.85	0.01	0.005	5	6	6	7	8	9	10	12	15
0.85	0.02	0.010	4	5	6	6	7	8	9	10	14
0.85	0.05	0.025	3	4	4	5	5	6	7	9	12
0.85	0.10	0.050	2	3	3	4	4	5	6	7	10
0.85	0.20	0.100	2	2	3	3	3	4	5	6	9
0.90	0.01	0.005	4	6	6	7	7	8	9	11	14
0.90	0.02	0.010	4	5	5	6	6	7	8	10	13
0.90	0.05	0.025	3	4	4	4	5	6	7	8	11
0.90	0.10	0.050	2	3	3	4	4	5	5	7	10
0.90	0.20	0.100	1	2	2	3	3	4	4	5	8
0.95	0.01	0.005	4	5	6	6	7	7	8	10	13
0.95	0.02	0.010	3	4	5	5	6	7	7	9	12
0.95	0.05	0.025	3	3	4	4	5	5	6	7	10
0.95	0.10	0.050	2	3	3	3	4	4	5	6	9
0.95	0.20	0.100	1	2	2	3	3	3	4	5	7

Table 9.1

Table 9.1 Continued

	α					Power $1-\beta$					
π_2	2-Sided	1-Sided	0.50	0.65	0.70	0.75	0.80	0.85	0.90	0.95	0.99

$\pi_1 = 0.10$

π_2	2-Sided	1-Sided	0.50	0.65	0.70	0.75	0.80	0.85	0.90	0.95	0.99
0.15	0.01	0.005	356	471	516	567	627	700	799	956	1290
0.15	0.02	0.010	291	395	436	484	539	607	699	846	1162
0.15	0.05	0.025	207	296	332	373	422	482	564	698	986
0.15	0.10	0.050	146	222	253	289	332	386	460	581	846
0.15	0.20	0.100	89	150	175	206	242	289	353	460	699
0.20	0.01	0.005	106	140	154	169	187	208	237	284	383
0.20	0.02	0.010	87	118	130	144	160	181	208	252	345
0.20	0.05	0.025	62	88	99	111	126	143	168	207	293
0.20	0.10	0.050	44	66	75	86	99	115	137	173	252
0.20	0.20	0.100	27	45	52	61	72	86	105	137	208
0.25	0.01	0.005	54	72	78	86	95	106	121	145	195
0.25	0.02	0.010	44	60	66	73	82	92	106	128	176
0.25	0.05	0.025	32	45	51	57	64	73	86	106	149
0.25	0.10	0.050	22	34	39	44	51	59	70	88	128
0.25	0.20	0.100	14	23	27	32	37	44	54	70	106
0.30	0.01	0.005	34	45	49	54	60	67	76	91	123
0.30	0.02	0.010	28	38	42	46	52	58	67	81	111
0.30	0.05	0.025	20	29	32	36	40	46	54	67	94
0.30	0.10	0.050	14	21	24	28	32	37	44	56	81
0.30	0.20	0.100	9	15	17	20	23	28	34	44	67
0.35	0.01	0.005	24	32	35	38	42	47	54	64	87
0.35	0.02	0.010	20	27	30	33	36	41	47	57	78
0.35	0.05	0.025	14	20	23	25	29	33	38	47	66
0.35	0.10	0.050	10	15	17	20	23	26	31	39	57
0.35	0.20	0.100	6	10	12	14	17	20	24	31	47
0.40	0.01	0.005	18	24	26	29	32	36	41	49	65
0.40	0.02	0.010	15	20	22	25	28	31	36	43	59
0.40	0.05	0.025	11	15	17	19	22	25	29	36	50
0.40	0.10	0.050	8	12	13	15	17	20	24	30	43
0.40	0.20	0.100	5	8	9	11	13	15	18	24	36
0.45	0.01	0.005	15	19	21	23	25	28	32	38	52
0.45	0.02	0.010	12	16	18	20	22	25	28	34	47
0.45	0.05	0.025	9	12	14	15	17	20	23	28	40
0.45	0.10	0.050	6	9	11	12	14	16	19	24	34
0.45	0.20	0.100	4	6	7	9	10	12	14	19	28
0.50	0.01	0.005	12	16	17	19	21	23	26	31	42
0.50	0.02	0.010	10	13	15	16	18	20	23	28	38
0.50	0.05	0.025	7	10	11	13	14	16	19	23	32
0.50	0.10	0.050	5	8	9	10	11	13	15	19	28
0.50	0.20	0.100	3	5	6	7	8	10	12	15	23
0.55	0.01	0.005	10	13	14	16	17	19	22	26	35
0.55	0.02	0.010	8	11	12	14	15	17	19	23	32
0.55	0.05	0.025	6	8	9	11	12	13	16	19	27
0.55	0.10	0.050	4	6	7	8	9	11	13	16	23
0.55	0.20	0.100	3	5	5	6	7	8	10	13	19
0.60	0.01	0.005	9	11	12	14	15	17	19	22	30
0.60	0.02	0.010	7	10	11	12	13	14	17	20	27
0.60	0.05	0.025	5	7	8	9	10	12	13	17	23
0.60	0.10	0.050	4	6	6	7	8	9	11	14	20
0.60	0.20	0.100	3	4	5	5	6	7	9	11	17
0.65	0.01	0.005	8	10	11	12	13	14	16	19	26
0.65	0.02	0.010	6	8	9	10	11	13	14	17	24
0.65	0.05	0.025	5	6	7	8	9	10	12	14	20
0.65	0.10	0.050	3	5	6	6	7	8	10	12	17
0.65	0.20	0.100	2	3	4	5	5	6	8	10	14

Table 9.1

100

Table 9.1 Continued.

π_2	α 2-Sided	1-Sided	0.50	0.65	0.70	0.75	0.80	0.85	0.90	0.95	0.99
						Power $1-\beta$					

$\pi_1 = 0.10$

π_2	2-Sided	1-Sided	0.50	0.65	0.70	0.75	0.80	0.85	0.90	0.95	0.99
0.70	0.01	0.005	7	9	9	10	11	13	14	17	23
0.70	0.02	0.010	6	7	8	9	10	11	13	15	21
0.70	0.05	0.025	4	6	6	7	8	9	10	13	18
0.70	0.10	0.050	3	4	5	6	6	7	8	11	15
0.70	0.20	0.100	2	3	4	4	5	6	7	8	13
0.75	0.01	0.005	6	8	8	9	10	11	13	15	20
0.75	0.02	0.010	5	7	7	8	9	10	11	14	18
0.75	0.05	0.025	4	5	6	6	7	8	9	11	16
0.75	0.10	0.050	3	4	4	5	6	6	8	9	14
0.75	0.20	0.100	2	3	3	4	4	5	6	8	11
0.80	0.01	0.005	5	7	8	8	9	10	11	14	18
0.80	0.02	0.010	4	6	6	7	8	9	10	12	16
0.80	0.05	0.025	3	5	5	6	6	7	8	10	14
0.80	0.10	0.050	2	4	4	4	5	6	7	8	12
0.80	0.20	0.100	2	3	3	3	4	4	5	7	10
0.85	0.01	0.005	5	6	7	8	8	9	10	12	16
0.85	0.02	0.010	4	5	6	6	7	8	9	11	15
0.85	0.05	0.025	3	4	5	5	6	6	7	9	13
0.85	0.10	0.050	2	3	4	4	5	5	6	8	11
0.85	0.20	0.100	2	2	3	3	3	4	5	6	9
0.90	0.01	0.005	4	6	6	7	8	8	9	11	15
0.90	0.02	0.010	4	5	5	6	7	7	8	10	13
0.90	0.05	0.025	3	4	4	5	5	6	7	8	12
0.90	0.10	0.050	2	3	3	4	4	5	6	7	10
0.90	0.20	0.100	1	2	2	3	3	4	4	6	8
0.95	0.01	0.005	4	5	6	6	7	8	9	10	14
0.95	0.02	0.010	3	5	5	5	6	7	8	9	12
0.95	0.05	0.025	3	4	4	4	5	5	6	8	11
0.95	0.10	0.050	2	3	3	3	4	4	5	6	9
0.95	0.20	0.100	1	2	2	3	3	3	4	5	8

$\pi_1 = 0.15$

π_2	2-Sided	1-Sided	0.50	0.65	0.70	0.75	0.80	0.85	0.90	0.95	0.99
0.20	0.01	0.005	493	652	714	785	868	970	1106	1324	1786
0.20	0.02	0.010	403	547	604	669	746	841	967	1172	1609
0.20	0.05	0.025	286	409	459	516	584	667	781	966	1365
0.20	0.10	0.050	201	307	350	400	460	535	637	804	1172
0.20	0.20	0.100	123	207	243	285	335	400	489	637	967
0.25	0.01	0.005	138	182	199	219	242	270	308	368	497
0.25	0.02	0.010	112	152	168	187	208	234	269	326	448
0.25	0.05	0.025	80	114	128	144	163	186	218	269	380
0.25	0.10	0.050	56	86	98	112	128	149	177	224	326
0.25	0.20	0.100	34	58	68	80	94	111	136	177	269
0.30	0.01	0.005	67	88	97	106	117	131	149	179	241
0.30	0.02	0.010	55	74	82	91	101	114	131	158	217
0.30	0.05	0.025	39	56	62	70	79	90	106	131	184
0.30	0.10	0.050	28	42	48	54	62	72	86	109	158
0.30	0.20	0.100	17	28	33	39	46	54	66	86	131
0.35	0.01	0.005	41	54	59	64	71	79	90	108	146
0.35	0.02	0.010	33	45	50	55	61	69	79	96	131
0.35	0.05	0.025	24	34	38	42	48	55	64	79	112
0.35	0.10	0.050	17	25	29	33	38	44	52	66	96
0.35	0.20	0.100	10	17	20	24	28	33	40	52	79

Table 9.1 Continued.

π_2	α 2-Sided	1-Sided	0.50	0.65	0.70	0.75	Power $1-\beta$ 0.80	0.85	0.90	0.95	0.99
						$\pi_1 = 0.15$					
0.40	0.01	0.005	28	37	40	44	49	54	62	74	99
0.40	0.02	0.010	23	31	34	38	42	47	54	65	90
0.40	0.05	0.025	16	23	26	29	33	37	44	54	76
0.40	0.10	0.050	12	17	20	23	26	30	36	45	65
0.40	0.20	0.100	7	12	14	16	19	23	28	36	54
0.45	0.01	0.005	20	27	29	32	36	40	45	54	73
0.45	0.02	0.010	17	23	25	28	31	35	40	48	66
0.45	0.05	0.025	12	17	19	21	24	28	32	40	56
0.45	0.10	0.050	9	13	15	17	19	22	26	33	48
0.45	0.20	0.100	5	9	10	12	14	17	20	26	40
0.50	0.01	0.005	16	21	23	25	28	31	35	42	56
0.50	0.02	0.010	13	18	19	21	24	27	31	37	51
0.50	0.05	0.025	9	13	15	17	19	21	25	31	43
0.50	0.10	0.050	7	10	11	13	15	17	20	26	37
0.50	0.20	0.100	4	7	8	9	11	13	16	20	31
0.55	0.01	0.005	13	17	18	20	22	25	28	33	45
0.55	0.02	0.010	10	14	15	17	19	21	24	30	40
0.55	0.05	0.025	8	11	12	13	15	17	20	24	34
0.55	0.10	0.050	5	8	9	10	12	14	16	20	30
0.55	0.20	0.100	4	6	7	8	9	10	13	16	24
0.60	0.01	0.005	11	14	15	16	18	20	23	27	37
0.60	0.02	0.010	9	12	13	14	16	18	20	24	33
0.60	0.05	0.025	6	9	10	11	12	14	16	20	28
0.60	0.10	0.050	5	7	8	9	10	11	13	17	24
0.60	0.20	0.100	3	5	5	6	7	9	10	13	20
0.65	0.01	0.005	9	12	13	14	15	17	19	23	31
0.65	0.02	0.010	7	10	11	12	13	15	17	20	28
0.65	0.05	0.025	5	7	8	9	10	12	14	17	24
0.65	0.10	0.050	4	6	6	7	8	10	11	14	20
0.65	0.20	0.100	3	4	5	5	6	7	9	11	17
0.70	0.01	0.005	8	10	11	12	13	14	16	20	26
0.70	0.02	0.010	6	8	9	10	11	13	14	17	24
0.70	0.05	0.025	5	6	7	8	9	10	12	14	20
0.70	0.10	0.050	3	5	6	6	7	8	10	12	17
0.70	0.20	0.100	2	3	4	5	5	6	8	10	14
0.75	0.01	0.005	7	9	9	10	11	13	14	17	23
0.75	0.02	0.010	5	7	8	9	10	11	12	15	20
0.75	0.05	0.025	4	6	6	7	8	9	10	12	17
0.75	0.10	0.050	3	4	5	5	6	7	8	10	15
0.75	0.20	0.100	2	3	4	4	5	5	7	8	12
0.80	0.01	0.005	6	8	8	9	10	11	12	15	20
0.80	0.02	0.010	5	6	7	8	9	10	11	13	18
0.80	0.05	0.025	4	5	5	6	7	8	9	11	15
0.80	0.10	0.050	3	4	4	5	5	6	7	9	13
0.80	0.20	0.100	2	3	3	4	4	5	6	7	11
0.85	0.01	0.005	5	7	7	8	9	10	11	13	17
0.85	0.02	0.010	4	6	6	7	8	8	10	12	16
0.85	0.05	0.025	3	4	5	5	6	7	8	10	13
0.85	0.10	0.050	2	3	4	4	5	6	7	8	12
0.85	0.20	0.100	2	2	3	3	4	4	5	7	10
0.90	0.01	0.005	5	6	7	7	8	9	10	12	16
0.90	0.02	0.010	4	5	6	6	7	8	9	10	14
0.90	0.05	0.025	3	4	4	5	5	6	7	9	12
0.90	0.10	0.050	2	3	3	4	4	5	6	7	10
0.90	0.20	0.100	2	2	3	3	3	4	5	6	9

Table 9.1

Table 9.1 Continued.

π_2	α 2-Sided	1-Sided	0.50	0.65	0.70	Power $1-\beta$ 0.75	0.80	0.85	0.90	0.95	0.99
0.95	0.01	0.005	4	5	6	6	7	8	9	10	14
0.95	0.02	0.010	4	5	5	6	6	7	8	9	13
0.95	0.05	0.025	3	4	4	4	5	6	6	8	11
0.95	0.10	0.050	2	3	3	3	4	5	5	7	9
0.95	0.20	0.100	1	2	2	3	3	3	4	5	8

$$\pi_1 = 0.20$$

π_2	α 2-Sided	1-Sided	0.50	0.65	0.70	0.75	0.80	0.85	0.90	0.95	0.99
0.25	0.01	0.005	598	791	867	953	1053	1176	1341	1606	2166
0.25	0.02	0.010	488	663	733	812	905	1020	1174	1422	1951
0.25	0.05	0.025	347	496	557	626	708	810	947	1172	1656
0.25	0.10	0.050	244	372	425	485	558	648	772	976	1422
0.25	0.20	0.100	149	251	294	345	407	485	593	772	1174
0.30	0.01	0.005	160	212	232	255	282	315	359	429	579
0.30	0.02	0.010	131	178	196	217	242	273	314	380	522
0.30	0.05	0.025	93	133	149	168	189	217	253	313	443
0.30	0.10	0.050	66	100	114	130	149	174	207	261	380
0.30	0.20	0.100	40	67	79	93	109	130	159	207	314
0.35	0.01	0.005	75	99	109	120	132	148	168	202	272
0.35	0.02	0.010	62	84	92	102	114	128	147	179	245
0.35	0.05	0.025	44	63	70	79	89	102	119	147	208
0.35	0.10	0.050	31	47	54	61	70	82	97	123	179
0.35	0.20	0.100	19	32	37	44	51	61	75	97	147
0.40	0.01	0.005	45	59	64	71	78	87	99	119	160
0.40	0.02	0.010	36	49	54	60	67	76	87	105	144
0.40	0.05	0.025	26	37	41	47	53	60	70	87	122
0.40	0.10	0.050	18	28	32	36	42	48	57	72	105
0.40	0.20	0.100	11	19	22	26	30	36	44	57	87
0.45	0.01	0.005	30	39	43	47	52	58	66	79	106
0.45	0.02	0.010	24	33	36	40	45	50	58	70	96
0.45	0.05	0.025	17	25	28	31	35	40	47	58	81
0.45	0.10	0.050	12	19	21	24	28	32	38	48	70
0.45	0.20	0.100	8	13	15	17	20	24	29	38	58
0.50	0.01	0.005	21	28	31	34	37	42	47	57	76
0.50	0.02	0.010	18	24	26	29	32	36	42	50	69
0.50	0.05	0.025	13	18	20	22	25	29	34	42	59
0.50	0.10	0.050	9	14	15	17	20	23	28	35	50
0.50	0.20	0.100	6	9	11	13	15	17	21	28	42
0.55	0.01	0.005	16	21	23	26	28	32	36	43	58
0.55	0.02	0.010	13	18	20	22	24	27	31	38	52
0.55	0.05	0.025	10	14	15	17	19	22	26	31	44
0.55	0.10	0.050	7	10	12	13	15	18	21	26	38
0.55	0.20	0.100	4	7	8	10	11	13	16	21	31
0.60	0.01	0.005	13	17	18	20	22	25	28	34	45
0.60	0.02	0.010	11	14	16	17	19	22	25	30	41
0.60	0.05	0.025	8	11	12	13	15	17	20	25	35
0.60	0.10	0.050	6	8	9	11	12	14	16	21	30
0.60	0.20	0.100	4	6	7	8	9	11	13	16	25
0.65	0.01	0.005	10	14	15	16	18	20	23	27	37
0.65	0.02	0.010	9	12	13	14	16	17	20	24	33
0.65	0.05	0.025	6	9	10	11	12	14	16	20	28
0.65	0.10	0.050	5	7	8	9	10	11	13	17	24
0.65	0.20	0.100	3	5	5	6	7	9	10	13	20
0.70	0.01	0.005	9	11	12	14	15	17	19	22	30
0.70	0.02	0.010	7	10	11	12	13	14	17	20	27
0.70	0.05	0.025	5	7	8	9	10	12	13	17	23
0.70	0.10	0.050	4	6	6	7	8	9	11	14	20
0.70	0.20	0.100	3	4	5	5	6	7	9	11	17

Table 9.1

Table 9.1 Continued.

π_2	α 2-Sided	1-Sided	0.50	0.65	0.70	Power $1-\beta$ 0.75	0.80	0.85	0.90	0.95	0.99
						$\pi_1 = 0.20$					
0.75	0.01	0.005	7	10	10	11	13	14	16	19	25
0.75	0.02	0.010	6	8	9	10	11	12	14	17	23
0.75	0.05	0.025	4	6	7	8	9	10	11	14	19
0.75	0.10	0.050	3	5	5	6	7	8	9	12	17
0.75	0.20	0.100	2	3	4	4	5	6	7	9	14
0.80	0.01	0.005	6	8	9	10	11	12	14	16	21
0.80	0.02	0.010	5	7	8	8	9	10	12	14	19
0.80	0.05	0.025	4	5	6	7	7	8	10	12	17
0.80	0.10	0.050	3	4	5	5	6	7	8	10	14
0.80	0.20	0.100	2	3	3	4	4	5	6	8	12
0.85	0.01	0.005	5	7	8	8	9	10	12	14	19
0.85	0.02	0.010	5	6	7	7	8	9	10	12	17
0.85	0.05	0.025	3	5	5	6	6	7	8	10	14
0.85	0.10	0.050	3	4	4	5	5	6	7	9	12
0.85	0.20	0.100	2	3	3	3	4	5	5	7	10
0.90	0.01	0.005	5	6	7	7	8	9	10	12	16
0.90	0.02	0.010	4	5	6	6	7	8	9	11	15
0.90	0.05	0.025	3	4	5	5	6	6	7	9	12
0.90	0.10	0.050	2	3	4	4	5	5	6	8	11
0.90	0.20	0.100	2	2	3	3	3	4	5	6	9
0.95	0.01	0.005	4	5	6	7	7	8	9	11	14
0.95	0.02	0.010	4	5	5	6	6	7	8	9	13
0.95	0.05	0.025	3	4	4	4	5	6	6	8	11
0.95	0.10	0.050	2	3	3	4	4	5	5	7	9
0.95	0.20	0.100	1	2	2	3	3	4	4	5	8
						$\pi_1 = 0.25$					
0.30	0.01	0.005	670	885	970	1067	1179	1317	1502	1798	2426
0.30	0.02	0.010	547	743	821	909	1013	1142	1314	1592	2185
0.30	0.05	0.025	388	556	623	701	793	907	1061	1312	1855
0.30	0.10	0.050	274	416	475	543	624	726	865	1093	1592
0.30	0.20	0.100	166	281	330	387	455	543	664	865	1314
0.35	0.01	0.005	174	230	252	277	307	343	390	467	630
0.35	0.02	0.010	142	193	214	237	264	297	342	414	568
0.35	0.05	0.025	101	145	162	182	206	236	276	341	482
0.35	0.10	0.050	71	109	124	141	163	189	225	284	414
0.35	0.20	0.100	44	73	86	101	119	141	173	225	342
0.40	0.01	0.005	80	106	116	127	141	157	179	214	289
0.40	0.02	0.010	65	89	98	109	121	136	157	190	260
0.40	0.05	0.025	47	67	75	84	95	108	127	156	221
0.40	0.10	0.050	33	50	57	65	75	87	103	130	190
0.40	0.20	0.100	20	34	40	46	55	65	79	103	157
0.45	0.01	0.005	46	61	67	73	81	91	103	124	167
0.45	0.02	0.010	38	51	57	63	70	79	90	109	150
0.45	0.05	0.025	27	38	43	48	55	63	73	90	127
0.45	0.10	0.050	19	29	33	38	43	50	60	75	109
0.45	0.20	0.100	12	20	23	27	32	38	46	60	90
0.50	0.01	0.005	30	40	44	48	53	59	67	81	109
0.50	0.02	0.010	25	34	37	41	46	51	59	71	98
0.50	0.05	0.025	18	25	28	32	36	41	48	59	83
0.50	0.10	0.050	13	19	22	25	28	33	39	49	71
0.50	0.20	0.100	8	13	15	18	21	25	30	39	59

Table 9.1

Table 9.1 Continued

π_2	α 2-Sided	1-Sided	0.50	0.65	0.70	Power $1-\beta$ 0.75	0.80	0.85	0.90	0.95	0.99
						$\pi_1 = 0.25$					
0.55	0.01	0.005	22	28	31	34	37	42	48	57	77
0.55	0.02	0.010	18	24	26	29	32	36	42	50	69
0.55	0.05	0.025	13	18	20	22	25	29	34	42	59
0.55	0.10	0.050	9	14	15	18	20	23	28	35	50
0.55	0.20	0.100	6	9	11	13	15	18	21	28	42
0.60	0.01	0.005	16	21	23	25	28	31	35	42	57
0.60	0.02	0.010	13	18	20	22	24	27	31	38	51
0.60	0.05	0.025	10	13	15	17	19	22	25	31	44
0.60	0.10	0.050	7	10	12	13	15	17	21	26	38
0.60	0.20	0.100	4	7	8	9	11	13	16	21	31
0.65	0.01	0.005	12	16	18	20	22	24	27	33	44
0.65	0.02	0.010	10	14	15	17	19	21	24	29	40
0.65	0.05	0.025	7	10	12	13	15	17	19	24	34
0.65	0.10	0.050	5	8	9	10	12	13	16	20	29
0.65	0.20	0.100	3	6	6	7	9	10	12	16	24
0.70	0.01	0.005	10	13	14	16	17	19	22	26	35
0.70	0.02	0.010	8	11	12	13	15	17	19	23	32
0.70	0.05	0.025	6	8	9	10	12	13	16	19	27
0.70	0.10	0.050	4	6	7	8	9	11	13	16	23
0.70	0.20	0.100	3	4	5	6	7	8	10	13	19
0.75	0.01	0.005	8	11	12	13	14	16	18	21	28
0.75	0.02	0.010	7	9	10	11	12	14	16	19	26
0.75	0.05	0.025	5	7	8	9	10	11	13	16	22
0.75	0.10	0.050	4	5	6	7	8	9	10	13	19
0.75	0.20	0.100	2	4	4	5	6	7	8	10	16
0.80	0.01	0.005	7	9	10	11	12	13	15	18	24
0.80	0.02	0.010	6	8	8	9	10	11	13	16	21
0.80	0.05	0.025	4	6	6	7	8	9	11	13	18
0.80	0.10	0.050	3	4	5	6	6	7	9	11	16
0.80	0.20	0.100	2	3	4	4	5	6	7	9	13
0.85	0.01	0.005	6	8	8	9	10	11	12	15	20
0.85	0.02	0.010	5	6	7	8	9	10	11	13	18
0.85	0.05	0.025	4	5	5	6	7	8	9	11	15
0.85	0.10	0.050	3	4	4	5	5	6	7	9	13
0.85	0.20	0.100	2	3	3	4	4	5	6	7	11
0.90	0.01	0.005	5	6	7	8	8	9	11	13	17
0.90	0.02	0.010	4	5	6	7	7	8	9	11	15
0.90	0.05	0.025	3	4	5	5	6	7	8	9	13
0.90	0.10	0.050	2	3	4	4	5	5	6	8	11
0.90	0.20	0.100	2	2	3	3	4	4	5	6	9
0.95	0.01	0.005	4	6	6	7	7	8	9	11	14
0.95	0.02	0.010	4	5	5	6	6	7	8	10	13
0.95	0.05	0.025	3	4	4	5	5	6	7	8	11
0.95	0.10	0.050	2	3	3	4	4	5	5	7	10
0.95	0.20	0.100	1	2	2	3	3	4	4	5	8

Table 9.1

Table 9.1 Continued.

π_2	α 2-Sided	1-Sided	0.50	0.65	0.70	Power $1 - \beta$ 0.75	0.80	0.85	0.90	0.95	0.99
						$\pi_1 = 0.30$					
0.35	0.01	0.005	710	938	1028	1130	1249	1395	1591	1905	2569
0.35	0.02	0.010	579	786	869	963	1073	1209	1392	1686	2314
0.35	0.05	0.025	411	588	660	742	839	960	1124	1389	1964
0.35	0.10	0.050	290	441	503	575	661	769	916	1157	1686
0.35	0.20	0.100	176	297	349	409	482	575	703	916	1392
0.40	0.01	0.005	181	239	262	287	318	355	405	484	653
0.40	0.02	0.010	147	200	221	245	273	308	354	429	588
0.40	0.05	0.025	105	150	168	189	214	244	286	353	499
0.40	0.10	0.050	74	112	128	147	168	196	233	294	429
0.40	0.20	0.100	45	76	89	104	123	146	179	233	354
0.45	0.01	0.005	81	107	118	129	143	160	182	218	294
0.45	0.02	0.010	67	90	100	110	123	138	159	193	265
0.45	0.05	0.025	47	68	76	85	96	110	129	159	225
0.45	0.10	0.050	33	51	58	66	76	88	105	132	193
0.45	0.20	0.100	21	34	40	47	55	66	81	105	159
0.50	0.01	0.005	46	61	67	73	81	90	103	123	166
0.50	0.02	0.010	38	51	57	63	70	78	90	109	150
0.50	0.05	0.025	27	38	43	48	55	62	73	90	127
0.50	0.10	0.050	19	29	33	38	43	50	60	75	109
0.50	0.20	0.100	12	20	23	27	32	38	46	60	90
0.55	0.01	0.005	30	39	43	47	52	58	66	79	107
0.55	0.02	0.010	24	33	36	40	45	50	58	70	96
0.55	0.05	0.025	17	25	28	31	35	40	47	58	82
0.55	0.10	0.050	12	19	21	24	28	32	38	48	70
0.55	0.20	0.100	8	13	15	17	20	24	30	38	58
0.60	0.01	0.005	21	27	30	33	36	40	46	55	74
0.60	0.02	0.010	17	23	25	28	31	35	40	49	67
0.60	0.05	0.025	12	17	19	22	25	28	33	40	57
0.60	0.10	0.050	9	13	15	17	19	23	27	34	49
0.60	0.20	0.100	6	9	10	12	14	17	21	27	40
0.65	0.01	0.005	15	20	22	24	27	30	34	40	54
0.65	0.02	0.010	13	17	19	21	23	26	30	36	49
0.65	0.05	0.025	9	13	14	16	18	21	24	30	42
0.65	0.10	0.050	7	10	11	13	14	17	20	25	36
0.65	0.20	0.100	4	7	8	9	11	13	15	20	30
0.70	0.01	0.005	12	15	17	18	20	23	26	31	41
0.70	0.02	0.010	10	13	14	16	18	20	23	27	37
0.70	0.05	0.025	7	10	11	12	14	16	18	23	32
0.70	0.10	0.050	5	7	8	10	11	13	15	19	27
0.70	0.20	0.100	3	5	6	7	8	10	12	15	23
0.75	0.01	0.005	9	12	13	14	16	18	20	24	32
0.75	0.02	0.010	8	10	11	12	14	15	18	21	29
0.75	0.05	0.025	6	8	9	10	11	12	14	18	25
0.75	0.10	0.050	4	6	7	8	9	10	12	15	21
0.75	0.20	0.100	3	4	5	6	6	8	9	12	18
0.80	0.01	0.005	8	10	11	12	13	14	16	19	26
0.80	0.02	0.010	6	8	9	10	11	12	14	17	23
0.80	0.05	0.025	5	6	7	8	9	10	12	14	20
0.80	0.10	0.050	3	5	5	6	7	8	10	12	17
0.80	0.20	0.100	2	3	4	5	5	6	7	10	14
0.85	0.01	0.005	6	8	9	10	11	12	13	16	21
0.85	0.02	0.010	5	7	7	8	9	10	12	14	19
0.85	0.05	0.025	4	5	6	6	7	8	10	12	16
0.85	0.10	0.050	3	4	5	5	6	7	8	10	14
0.85	0.20	0.100	2	3	3	4	4	5	6	8	12

Table 9.1

Table 9.1 Continued.

π_2	α 2-Sided	1-Sided	0.50	0.65	0.70	Power $1-\beta$ 0.75	0.80	0.85	0.90	0.95	0.99
						$\pi_1 = 0.30$					
0.90	0.01	0.005	5	7	7	8	9	10	11	13	18
0.90	0.02	0.010	4	6	6	7	8	9	10	12	16
0.90	0.05	0.025	3	4	5	5	6	7	8	10	14
0.90	0.10	0.050	2	3	4	4	5	6	7	8	12
0.90	0.20	0.100	2	2	3	3	4	4	5	7	10
0.95	0.01	0.005	4	6	6	7	7	8	9	11	15
0.95	0.02	0.010	4	5	5	6	6	7	8	10	13
0.95	0.05	0.025	3	4	4	5	5	6	7	8	11
0.95	0.10	0.050	2	3	3	4	4	5	6	7	10
0.95	0.20	0.100	1	2	2	3	3	4	4	6	8
						$\pi_1 = 0.35$					
0.40	0.01	0.005	720	951	1042	1146	1266	1415	1613	1932	2605
0.40	0.02	0.010	587	798	881	977	1088	1226	1412	1710	2347
0.40	0.05	0.025	417	597	670	753	851	974	1139	1409	1992
0.40	0.10	0.050	294	447	511	584	671	780	929	1174	1710
0.40	0.20	0.100	179	302	354	415	489	583	713	929	1412
0.45	0.01	0.005	180	238	260	286	316	353	403	482	650
0.45	0.02	0.010	147	199	220	244	272	306	353	427	586
0.45	0.05	0.025	104	149	167	188	213	243	285	352	497
0.45	0.10	0.050	74	112	128	146	168	195	232	293	427
0.45	0.20	0.100	45	76	89	104	122	146	178	232	353
0.50	0.01	0.005	80	105	115	127	140	156	178	213	287
0.50	0.02	0.010	65	88	98	108	120	136	156	189	259
0.50	0.05	0.025	46	66	74	83	94	108	126	156	220
0.50	0.10	0.050	33	50	57	65	74	86	103	130	189
0.50	0.20	0.100	20	34	39	46	54	65	79	103	156
0.55	0.01	0.005	45	59	64	71	78	87	99	119	160
0.55	0.02	0.010	36	49	54	60	67	76	87	105	144
0.55	0.05	0.025	26	37	42	47	53	60	70	87	123
0.55	0.10	0.050	18	28	32	36	42	48	57	72	105
0.55	0.20	0.100	11	19	22	26	30	36	44	57	87
0.60	0.01	0.005	28	37	41	45	49	55	63	75	101
0.60	0.02	0.010	23	31	35	38	43	48	55	67	91
0.60	0.05	0.025	17	24	26	30	33	38	45	55	78
0.60	0.10	0.050	12	18	20	23	26	31	36	46	67
0.60	0.20	0.100	7	12	14	17	19	23	28	36	55
0.65	0.01	0.005	19	26	28	31	34	38	43	51	69
0.65	0.02	0.010	16	22	24	26	29	33	38	46	62
0.65	0.05	0.025	11	16	18	20	23	26	31	38	53
0.65	0.10	0.050	8	12	14	16	18	21	25	31	46
0.65	0.20	0.100	5	8	10	11	13	16	19	25	38
0.70	0.01	0.005	14	19	20	22	25	27	31	37	50
0.70	0.02	0.010	12	16	17	19	21	24	27	33	45
0.70	0.05	0.025	8	12	13	15	17	19	22	27	38
0.70	0.10	0.050	6	9	10	12	13	15	18	23	33
0.70	0.20	0.100	4	6	7	8	10	12	14	18	27
0.75	0.01	0.005	11	14	15	17	18	21	23	28	38
0.75	0.02	0.010	9	12	13	14	16	18	21	25	34
0.75	0.05	0.025	6	9	10	11	13	14	17	21	29
0.75	0.10	0.050	5	7	8	9	10	12	14	17	25
0.75	0.20	0.100	3	5	6	6	7	9	11	14	21
0.80	0.01	0.005	8	11	12	13	14	16	18	22	29
0.80	0.02	0.010	7	9	10	11	12	14	16	19	26
0.80	0.05	0.025	5	7	8	9	10	11	13	16	22
0.80	0.10	0.050	4	5	6	7	8	9	11	13	19
0.80	0.20	0.100	2	4	4	5	6	7	8	11	16

Table 9.1

Table 9.1 Continued.

π_2	α 2-Sided	1-Sided	Power $1 - \beta$ 0.50	0.65	0.70	0.75	0.80	0.85	0.90	0.95	0.99
					$\pi_1 = 0.35$						
0.85	0.01	0.005	7	9	9	10	11	13	14	17	23
0.85	0.02	0.010	6	7	8	9	10	11	13	15	21
0.85	0.05	0.025	4	6	6	7	8	9	10	13	18
0.85	0.10	0.050	3	4	5	6	6	7	8	11	15
0.85	0.20	0.100	2	3	4	4	5	6	7	8	13
0.90	0.01	0.005	5	7	8	8	9	10	12	14	18
0.90	0.02	0.010	5	6	7	7	8	9	10	12	17
0.90	0.05	0.025	3	5	5	6	6	7	8	10	14
0.90	0.10	0.050	3	4	4	5	5	6	7	9	12
0.90	0.20	0.100	2	3	3	3	4	5	5	7	10
0.95	0.01	0.005	5	6	6	7	8	8	10	11	15
0.95	0.02	0.010	4	5	5	6	7	7	8	10	14
0.95	0.05	0.025	3	4	4	5	5	6	7	8	12
0.95	0.10	0.050	2	3	3	4	4	5	6	7	10
0.95	0.20	0.100	1	2	2	3	3	4	4	6	8
					$\pi_1 = 0.40$						
0.45	0.01	0.005	704	930	1019	1120	1238	1383	1577	1888	2547
0.45	0.02	0.010	574	780	862	955	1064	1199	1380	1672	2295
0.45	0.05	0.025	408	583	655	736	832	952	1114	1378	1948
0.45	0.10	0.050	287	437	499	571	656	762	908	1147	1672
0.45	0.20	0.100	175	295	346	406	478	570	697	908	1380
0.50	0.01	0.005	173	229	250	275	304	340	388	464	626
0.50	0.02	0.010	141	192	212	235	262	295	339	411	564
0.50	0.05	0.025	100	144	161	181	205	234	274	338	478
0.50	0.10	0.050	71	108	123	140	161	187	223	282	411
0.50	0.20	0.100	43	73	85	100	118	140	171	223	339
0.55	0.01	0.005	75	100	109	120	133	148	169	202	272
0.55	0.02	0.010	62	84	92	102	114	128	148	179	245
0.55	0.05	0.025	44	63	70	79	89	102	119	147	208
0.55	0.10	0.050	31	47	54	61	70	82	97	123	179
0.55	0.20	0.100	19	32	37	44	51	61	75	97	148
0.60	0.01	0.005	42	55	60	66	73	81	93	111	149
0.60	0.02	0.010	34	46	51	56	63	71	81	98	135
0.60	0.05	0.025	24	35	39	43	49	56	66	81	114
0.60	0.10	0.050	17	26	30	34	39	45	54	68	98
0.60	0.20	0.100	11	18	21	24	28	34	41	54	81
0.65	0.01	0.005	26	34	37	41	45	51	58	69	93
0.65	0.02	0.010	21	29	32	35	39	44	51	61	84
0.65	0.05	0.025	15	22	24	27	31	35	41	51	71
0.65	0.10	0.050	11	16	19	21	24	28	33	42	61
0.65	0.20	0.100	7	11	13	15	18	21	26	33	51
0.70	0.01	0.005	18	23	25	28	31	34	39	47	63
0.70	0.02	0.010	15	20	22	24	26	30	34	41	57
0.70	0.05	0.025	10	15	16	18	21	24	28	34	48
0.70	0.10	0.050	7	11	13	14	16	19	23	28	41
0.70	0.20	0.100	5	8	9	10	12	14	17	23	34
0.75	0.01	0.005	13	17	18	20	22	24	28	33	45
0.75	0.02	0.010	10	14	15	17	19	21	24	29	40
0.75	0.05	0.025	8	11	12	13	15	17	20	24	34
0.75	0.10	0.050	5	8	9	10	12	14	16	20	29
0.75	0.20	0.100	4	6	6	8	9	10	13	16	24
0.80	0.01	0.005	9	12	13	15	16	18	21	25	33
0.80	0.02	0.010	8	10	11	13	14	16	18	22	30
0.80	0.05	0.025	6	8	9	10	11	13	15	18	25
0.80	0.10	0.050	4	6	7	8	9	10	12	15	22
0.80	0.20	0.100	3	4	5	6	7	8	9	12	18

Table 9.1

Table 9.1 Continued.

π_2	α 2-Sided	1-Sided	Power $1 - \beta$ 0.50	0.65	0.70	0.75	0.80	0.85	0.90	0.95	0.99
			$\pi_1 = 0.40$								
0.85	0.01	0.005	7	9	10	11	12	14	16	19	25
0.85	0.02	0.010	6	8	9	10	11	12	14	17	23
0.85	0.05	0.025	4	6	7	8	9	10	11	14	19
0.85	0.10	0.050	3	5	5	6	7	8	9	12	17
0.85	0.20	0.100	2	3	4	4	5	6	7	9	14
0.90	0.01	0.005	6	7	8	9	10	11	12	15	20
0.90	0.02	0.010	5	6	7	8	8	9	11	13	18
0.90	0.05	0.025	4	5	5	6	7	8	9	11	15
0.90	0.10	0.050	3	4	4	5	5	6	7	9	13
0.90	0.20	0.100	2	3	3	4	4	5	6	7	11
0.95	0.01	0.005	5	6	7	7	8	9	10	12	16
0.95	0.02	0.010	4	5	6	6	7	8	9	10	14
0.95	0.05	0.025	3	4	4	5	5	6	7	9	12
0.95	0.10	0.050	2	3	3	4	4	5	6	7	10
0.95	0.20	0.100	2	2	3	3	3	4	5	6	9
			$\pi_1 = 0.45$								
0.50	0.01	0.005	665	879	964	1059	1171	1308	1492	1786	2409
0.50	0.02	0.010	543	737	815	903	1006	1134	1305	1581	2170
0.50	0.05	0.025	385	552	619	696	787	900	1054	1303	1842
0.50	0.10	0.050	272	414	472	540	620	721	859	1085	1581
0.50	0.20	0.100	165	279	327	384	452	539	659	859	1305
0.55	0.01	0.005	161	213	233	256	283	316	361	432	582
0.55	0.02	0.010	132	179	197	219	243	274	316	382	525
0.55	0.05	0.025	94	134	150	169	191	218	255	315	445
0.55	0.10	0.050	66	100	114	131	150	175	208	263	382
0.55	0.20	0.100	40	68	79	93	110	131	160	208	316
0.60	0.01	0.005	69	91	100	110	121	136	155	185	249
0.60	0.02	0.010	57	77	85	94	104	118	135	164	225
0.60	0.05	0.025	40	57	64	72	82	93	109	135	191
0.60	0.10	0.050	29	43	49	56	65	75	89	113	164
0.60	0.20	0.100	18	29	34	40	47	56	69	89	135
0.65	0.01	0.005	38	49	54	60	66	73	84	100	135
0.65	0.02	0.010	31	42	46	51	57	64	73	89	121
0.65	0.05	0.025	22	31	35	39	44	51	59	73	103
0.65	0.10	0.050	16	24	27	31	35	41	48	61	89
0.65	0.20	0.100	10	16	19	22	26	31	37	48	73
0.70	0.01	0.005	23	30	33	37	40	45	51	61	83
0.70	0.02	0.010	19	26	28	31	35	39	45	54	74
0.70	0.05	0.025	14	19	22	24	27	31	36	45	63
0.70	0.10	0.050	10	15	17	19	22	25	30	37	54
0.70	0.20	0.100	6	10	12	14	16	19	23	30	45
0.75	0.01	0.005	15	20	22	24	27	30	34	41	55
0.75	0.02	0.010	13	17	19	21	23	26	30	36	49
0.75	0.05	0.025	9	13	14	16	18	21	24	30	42
0.75	0.10	0.050	7	10	11	13	14	17	20	25	36
0.75	0.20	0.100	4	7	8	9	11	13	15	20	30
0.80	0.01	0.005	11	14	16	17	19	21	24	29	38
0.80	0.02	0.010	9	12	13	15	16	18	21	25	35
0.80	0.05	0.025	7	9	10	11	13	15	17	21	29
0.80	0.10	0.050	5	7	8	9	10	12	14	18	25
0.80	0.20	0.100	3	5	6	7	8	9	11	14	21
0.85	0.01	0.005	8	11	11	13	14	15	17	21	28
0.85	0.02	0.010	7	9	10	11	12	13	15	19	25
0.85	0.05	0.025	5	7	8	8	9	11	12	15	21
0.85	0.10	0.050	4	5	6	7	8	9	10	13	19
0.85	0.20	0.100	2	4	4	5	6	7	8	10	15

Table 9.1

Table 9.1 Continued.

π_2	α 2-Sided	1-Sided	Power $1-\beta$ 0.50	0.65	0.70	0.75	0.80	0.85	0.90	0.95	0.99
						$\pi_1 = 0.45$					
0.90	0.01	0.005	6	8	9	9	10	12	13	16	21
0.90	0.02	0.010	5	7	7	8	9	10	12	14	19
0.90	0.05	0.025	4	5	6	6	7	8	9	12	16
0.90	0.10	0.050	3	4	5	5	6	7	8	10	14
0.90	0.20	0.100	2	3	3	4	4	5	6	8	12
0.95	0.01	0.005	5	6	7	7	8	9	10	12	16
0.95	0.02	0.010	4	5	6	6	7	8	9	11	14
0.95	0.05	0.025	3	4	4	5	6	6	7	9	12
0.95	0.10	0.050	2	3	4	4	4	5	6	7	11
0.95	0.20	0.100	2	2	3	3	3	4	5	6	9
						$\pi_1 = 0.50$					
0.55	0.01	0.005	609	805	882	970	1072	1197	1365	1635	2205
0.55	0.02	0.010	497	675	746	827	921	1038	1195	1447	1986
0.55	0.05	0.025	353	505	567	637	721	824	964	1193	1686
0.55	0.10	0.050	249	379	432	494	568	660	786	993	1447
0.55	0.20	0.100	151	255	300	351	414	493	603	786	1195
0.60	0.01	0.005	145	192	210	231	255	285	325	389	524
0.60	0.02	0.010	118	161	178	197	219	247	284	344	472
0.60	0.05	0.025	84	120	135	152	172	196	230	284	401
0.60	0.10	0.050	59	90	103	118	135	157	187	236	344
0.60	0.20	0.100	36	61	72	84	99	118	144	187	284
0.65	0.01	0.005	61	81	89	97	108	120	137	164	221
0.65	0.02	0.010	50	68	75	83	93	104	120	145	199
0.65	0.05	0.025	36	51	57	64	73	83	97	120	169
0.65	0.10	0.050	25	38	44	50	57	66	79	100	145
0.65	0.20	0.100	16	26	30	36	42	50	61	79	120
0.70	0.01	0.005	33	43	47	52	57	64	73	87	117
0.70	0.02	0.010	27	36	40	44	49	56	64	77	106
0.70	0.05	0.025	19	27	31	34	39	44	52	64	90
0.70	0.10	0.050	14	21	23	27	31	35	42	53	77
0.70	0.20	0.100	8	14	16	19	22	27	32	42	64
0.75	0.01	0.005	20	26	29	31	35	39	44	53	71
0.75	0.02	0.010	16	22	24	27	30	34	39	47	64
0.75	0.05	0.025	12	17	19	21	23	27	31	39	54
0.75	0.10	0.050	8	13	14	16	19	22	26	32	47
0.75	0.20	0.100	5	9	10	12	14	16	20	26	39
0.80	0.01	0.005	13	17	19	21	23	25	29	34	46
0.80	0.02	0.010	11	14	16	18	20	22	25	30	42
0.80	0.05	0.025	8	11	12	14	15	18	20	25	35
0.80	0.10	0.050	6	8	9	11	12	14	17	21	30
0.80	0.20	0.100	4	6	7	8	9	11	13	17	25
0.85	0.01	0.005	9	12	13	14	16	17	20	24	32
0.85	0.02	0.010	8	10	11	12	14	15	17	21	29
0.85	0.05	0.025	5	8	9	10	11	12	14	17	24
0.85	0.10	0.050	4	6	7	7	9	10	12	15	21
0.85	0.20	0.100	3	4	5	5	6	7	9	12	17
0.90	0.01	0.005	7	9	9	10	11	13	14	17	23
0.90	0.02	0.010	5	7	8	9	10	11	13	15	20
0.90	0.05	0.025	4	6	6	7	8	9	10	12	17
0.90	0.10	0.050	3	4	5	5	6	7	8	10	15
0.90	0.20	0.100	2	3	4	4	5	5	7	8	13
0.95	0.01	0.005	5	6	7	8	8	9	11	12	17
0.95	0.02	0.010	4	5	6	7	7	8	9	11	15
0.95	0.05	0.025	3	4	5	5	6	7	8	9	13
0.95	0.10	0.050	2	3	4	4	5	5	6	8	11
0.95	0.20	0.100	2	2	3	3	4	4	5	6	9

Table 9.1

Table 9.1 Continued.

π_2	α 2-Sided	1-Sided	Power $1 - \beta$ 0.50	0.65	0.70	0.75	0.80	0.85	0.90	0.95	0.99
						$\pi_1 = 0.55$					
0.60	0.01	0.005	539	712	781	858	949	1060	1208	1447	1951
0.60	0.02	0.010	440	597	660	731	815	918	1057	1281	1758
0.60	0.05	0.025	312	447	502	564	638	729	853	1055	1492
0.60	0.10	0.050	220	335	382	437	502	584	696	879	1281
0.60	0.20	0.100	134	226	265	311	366	437	534	696	1057
0.65	0.01	0.005	126	167	183	201	222	248	283	338	456
0.65	0.02	0.010	103	140	155	171	191	215	247	299	411
0.65	0.05	0.025	73	105	118	132	149	171	200	247	349
0.65	0.10	0.050	52	79	90	102	118	137	163	206	299
0.65	0.20	0.100	32	53	62	73	86	102	125	163	247
0.70	0.01	0.005	52	69	76	83	92	103	117	140	189
0.70	0.02	0.010	43	58	64	71	79	89	102	124	170
0.70	0.05	0.025	31	44	49	55	62	71	83	102	144
0.70	0.10	0.050	22	33	37	43	49	57	68	85	124
0.70	0.20	0.100	13	22	26	30	36	43	52	68	102
0.75	0.01	0.005	28	36	40	44	48	54	61	73	98
0.75	0.02	0.010	23	30	34	37	41	47	54	65	89
0.75	0.05	0.025	16	23	26	29	32	37	43	53	75
0.75	0.10	0.050	12	17	20	22	26	30	35	45	65
0.75	0.20	0.100	7	12	14	16	19	22	27	35	54
0.80	0.01	0.005	16	22	24	26	29	32	36	43	58
0.80	0.02	0.010	13	18	20	22	25	28	32	38	52
0.80	0.05	0.025	10	14	15	17	19	22	26	32	45
0.80	0.10	0.050	7	10	12	13	15	18	21	26	38
0.80	0.20	0.100	4	7	8	10	11	13	16	21	32
0.85	0.01	0.005	11	14	15	17	18	20	23	28	37
0.85	0.02	0.010	9	12	13	14	16	18	20	25	34
0.85	0.05	0.025	6	9	10	11	12	14	17	20	29
0.85	0.10	0.050	5	7	8	9	10	11	14	17	25
0.85	0.20	0.100	3	5	5	6	7	9	11	14	20
0.90	0.01	0.005	7	9	10	11	12	14	16	19	25
0.90	0.02	0.010	6	8	9	10	11	12	14	17	23
0.90	0.05	0.025	4	6	7	8	9	10	11	14	19
0.90	0.10	0.050	3	5	5	6	7	8	9	12	17
0.90	0.20	0.100	2	3	4	4	5	6	7	9	14
0.95	0.01	0.005	5	7	7	8	9	10	11	13	17
0.95	0.02	0.010	4	6	6	7	8	8	10	12	16
0.95	0.05	0.025	3	4	5	5	6	7	8	10	13
0.95	0.10	0.050	2	3	4	4	5	6	7	8	12
0.95	0.20	0.100	2	2	3	3	4	4	5	7	10

Table 9.1

Table 9.1 Continued.

π_2	α 2-Sided	1-Sided	Power $1-\beta$ 0.50	0.65	0.70	0.75	0.80	0.85	0.90	0.95	0.99
						$\pi_1 = 0.60$					
0.65	0.01	0.005	460	607	666	732	809	903	1030	1233	1663
0.65	0.02	0.010	375	509	563	624	695	783	901	1092	1498
0.65	0.05	0.025	266	381	428	481	544	622	728	900	1272
0.65	0.10	0.050	188	286	326	373	428	498	593	749	1092
0.65	0.20	0.100	114	193	226	265	312	372	455	593	901
0.70	0.01	0.005	106	139	153	168	185	207	236	283	381
0.70	0.02	0.010	86	117	129	143	159	180	207	250	343
0.70	0.05	0.025	61	88	98	110	125	143	167	206	291
0.70	0.10	0.050	43	66	75	86	98	114	136	172	250
0.70	0.20	0.100	27	44	52	61	72	86	105	136	207
0.75	0.01	0.005	43	57	62	68	75	84	96	115	154
0.75	0.02	0.010	35	48	53	58	65	73	84	101	139
0.75	0.05	0.025	25	36	40	45	51	58	68	84	118
0.75	0.10	0.050	18	27	31	35	40	47	55	70	101
0.75	0.20	0.100	11	18	21	25	29	35	43	55	84
0.80	0.01	0.005	22	29	32	35	39	43	49	58	79
0.80	0.02	0.010	18	24	27	30	33	37	43	52	71
0.80	0.05	0.025	13	18	21	23	26	30	35	43	60
0.80	0.10	0.050	9	14	16	18	21	24	28	36	52
0.80	0.20	0.100	6	10	11	13	15	18	22	28	43
0.85	0.01	0.005	13	17	18	20	22	25	28	34	45
0.85	0.02	0.010	11	14	16	17	19	22	25	30	41
0.85	0.05	0.025	8	11	12	13	15	17	20	25	35
0.85	0.10	0.050	6	8	9	11	12	14	17	21	30
0.85	0.20	0.100	4	6	7	8	9	11	13	17	25
0.90	0.01	0.005	8	11	12	13	14	16	18	21	28
0.90	0.02	0.010	7	9	10	11	12	14	16	19	25
0.90	0.05	0.025	5	7	8	9	10	11	13	16	22
0.90	0.10	0.050	4	5	6	7	8	9	10	13	19
0.90	0.20	0.100	2	4	4	5	6	7	8	10	16
0.95	0.01	0.005	5	7	8	8	9	10	12	14	18
0.95	0.02	0.010	5	6	7	7	8	9	10	12	17
0.95	0.05	0.025	3	5	5	6	6	7	8	10	14
0.95	0.10	0.050	3	4	4	5	5	6	7	9	12
0.95	0.20	0.100	2	3	3	3	4	5	5	7	10
						$\pi_1 = 0.65$					
0.70	0.01	0.005	375	496	543	597	660	737	841	1006	1357
0.70	0.02	0.010	306	416	459	509	567	639	735	891	1223
0.70	0.05	0.025	217	311	349	392	444	507	594	734	1038
0.70	0.10	0.050	153	233	266	304	350	406	484	611	891
0.70	0.20	0.100	93	157	185	216	255	304	371	484	735
0.75	0.01	0.005	84	111	122	134	148	165	188	225	303
0.75	0.02	0.010	69	93	103	114	127	143	165	199	273
0.75	0.05	0.025	49	70	78	88	99	114	133	164	232
0.75	0.10	0.050	35	52	60	68	78	91	108	137	199
0.75	0.20	0.100	21	36	42	49	57	68	83	108	165
0.80	0.01	0.005	33	44	48	53	58	65	74	89	120
0.80	0.02	0.010	27	37	41	45	50	57	65	79	108
0.80	0.05	0.025	20	28	31	35	39	45	53	65	92
0.80	0.10	0.050	14	21	24	27	31	36	43	54	79
0.80	0.20	0.100	9	14	17	19	23	27	33	43	65
0.85	0.01	0.005	17	22	24	26	29	32	37	44	59
0.85	0.02	0.010	14	18	20	23	25	28	32	39	53
0.85	0.05	0.025	10	14	16	17	20	22	26	32	45
0.85	0.10	0.050	7	11	12	14	16	18	21	27	39
0.85	0.20	0.100	5	7	8	10	12	14	17	21	32

Table 9.1 112

Table 9.1 Continued.

π_2	α 2-Sided	1-Sided	0.50	0.65	0.70	Power 1−β 0.75	0.80	0.85	0.90	0.95	0.99
						$\pi_1 = 0.65$					
0.90	0.01	0.005	10	12	14	15	16	18	21	25	33
0.90	0.02	0.010	8	10	12	13	14	16	18	22	30
0.90	0.05	0.025	6	8	9	10	11	13	15	18	25
0.90	0.10	0.050	4	6	7	8	9	10	12	15	22
0.90	0.20	0.100	3	4	5	6	7	8	9	12	18
0.95	0.01	0.005	6	8	8	9	10	11	13	15	20
0.95	0.02	0.010	5	6	7	8	9	10	11	13	18
0.95	0.05	0.025	4	5	5	6	7	8	9	11	15
0.95	0.10	0.050	3	4	4	5	5	6	7	9	13
0.95	0.20	0.100	2	3	3	4	4	5	6	7	11
						$\pi_1 = 0.70$					
0.75	0.01	0.005	290	383	420	461	510	570	649	777	1049
0.75	0.02	0.010	237	321	355	393	438	494	568	688	945
0.75	0.05	0.025	168	240	270	303	343	392	459	567	802
0.75	0.10	0.050	118	180	206	235	270	314	374	472	688
0.75	0.20	0.100	72	122	143	167	197	235	287	374	568
0.80	0.01	0.005	63	83	91	100	111	124	141	168	227
0.80	0.02	0.010	52	70	77	85	95	107	123	149	205
0.80	0.05	0.025	37	52	59	66	74	85	100	123	174
0.80	0.10	0.050	26	39	45	51	59	68	81	103	149
0.80	0.20	0.100	16	27	31	37	43	51	62	81	123
0.85	0.01	0.005	24	32	35	38	42	47	54	64	86
0.85	0.02	0.010	20	27	30	33	36	41	47	57	78
0.85	0.05	0.025	14	20	23	25	29	33	38	47	66
0.85	0.10	0.050	10	15	17	20	23	26	31	39	57
0.85	0.20	0.100	6	10	12	14	17	20	24	31	47
0.90	0.01	0.005	12	15	17	18	20	23	26	31	41
0.90	0.02	0.010	10	13	14	16	17	20	22	27	37
0.90	0.05	0.025	7	10	11	12	14	16	18	22	32
0.90	0.10	0.050	5	7	8	10	11	13	15	19	27
0.90	0.20	0.100	3	5	6	7	8	10	12	15	22
0.95	0.01	0.005	6	8	9	10	11	12	14	16	22
0.95	0.02	0.010	5	7	8	9	9	11	12	15	20
0.95	0.05	0.025	4	5	6	7	8	9	10	12	17
0.95	0.10	0.050	3	4	5	5	6	7	8	10	15
0.95	0.20	0.100	2	3	3	4	5	5	6	8	12
						$\pi_1 = 0.75$					
0.80	0.01	0.005	208	275	302	331	366	409	467	559	753
0.80	0.02	0.010	170	231	255	283	315	355	408	494	679
0.80	0.05	0.025	121	173	194	218	246	282	330	408	576
0.80	0.10	0.050	85	130	148	169	194	226	269	339	494
0.80	0.20	0.100	52	88	103	120	142	169	206	269	408
0.85	0.01	0.005	43	57	63	69	76	85	97	116	156
0.85	0.02	0.010	36	48	53	59	65	74	85	103	141
0.85	0.05	0.025	25	36	40	45	51	59	68	85	119
0.85	0.10	0.050	18	27	31	35	40	47	56	71	103
0.85	0.20	0.100	11	18	22	25	30	35	43	56	85
0.90	0.01	0.005	16	21	23	25	28	31	35	42	56
0.90	0.02	0.010	13	18	19	21	24	27	31	37	51
0.90	0.05	0.025	9	13	15	17	19	21	25	31	43
0.90	0.10	0.050	7	10	11	13	15	17	20	26	37
0.90	0.20	0.100	4	7	8	9	11	13	16	20	31

Table 9.1

Table 9.1 Continued.

π_2	α 2-Sided	1-Sided	0.50	0.65	0.70	Power $1-\beta$ 0.75	0.80	0.85	0.90	0.95	0.99
					$\pi_1 = 0.75$						
0.95	0.01	0.005	7	10	10	11	13	14	16	19	25
0.95	0.02	0.010	6	8	9	10	11	12	14	17	23
0.95	0.05	0.025	4	6	7	8	9	10	11	14	19
0.95	0.10	0.050	3	5	5	6	7	8	9	12	17
0.95	0.20	0.100	2	3	4	4	5	6	7	9	14
					$\pi_1 = 0.80$						
0.85	0.01	0.005	135	178	195	214	237	265	302	361	487
0.85	0.02	0.010	110	149	165	183	204	229	264	320	439
0.85	0.05	0.025	78	112	125	141	159	182	213	263	372
0.85	0.10	0.050	55	84	96	109	126	146	174	219	320
0.85	0.20	0.100	34	57	66	78	92	109	133	174	264
0.90	0.01	0.005	26	35	38	42	46	51	58	70	94
0.90	0.02	0.010	22	29	32	36	40	44	51	62	85
0.90	0.05	0.025	15	22	25	27	31	35	41	51	72
0.90	0.10	0.050	11	17	19	21	25	28	34	43	62
0.90	0.20	0.100	7	11	13	15	18	21	26	34	51
0.95	0.01	0.005	9	12	13	14	15	17	19	23	31
0.95	0.02	0.010	7	10	11	12	13	15	17	21	28
0.95	0.05	0.025	5	8	8	9	11	12	14	17	24
0.95	0.10	0.050	4	6	7	7	8	10	11	14	21
0.95	0.20	0.100	3	4	5	5	6	7	9	11	17
					$\pi_1 = 0.85$						
0.90	0.01	0.005	73	97	106	117	129	144	164	196	264
0.90	0.02	0.010	60	81	90	99	111	125	143	174	238
0.90	0.05	0.025	43	61	68	77	87	99	116	143	202
0.90	0.10	0.050	30	46	52	60	68	79	95	119	174
0.90	0.20	0.100	19	31	36	43	50	60	73	95	143
0.95	0.01	0.005	13	17	18	20	22	25	28	33	45
0.95	0.02	0.010	10	14	16	17	19	21	25	30	40
0.95	0.05	0.025	8	11	12	13	15	17	20	25	34
0.95	0.10	0.050	5	8	9	10	12	14	16	20	30
0.95	0.20	0.100	4	6	7	8	9	10	13	16	25
					$\pi_1 = 0.90$						
0.95	0.01	0.005	28	37	41	45	50	55	63	75	101
0.95	0.02	0.010	23	31	35	38	43	48	55	67	91
0.95	0.05	0.025	17	24	26	30	33	38	45	55	78
0.95	0.10	0.050	12	18	20	23	26	31	36	46	67
0.95	0.20	0.100	7	12	14	17	19	23	28	36	55

Table 9.1

Table 9.2 Number of subjects for comparison of survival rates (logrank test).

π_2	α 2-sided	1-sided	0.50	0.65	0.70	0.75	Power $1-\beta$ 0.80	0.85	0.90	0.95	0.99
						$\pi_1 = 0.05$					
0.10	0.01	0.005	210	277	304	334	369	413	470	563	759
0.10	0.02	0.010	171	233	257	285	317	358	412	499	684
0.10	0.05	0.025	122	174	195	220	248	284	332	411	581
0.10	0.10	0.050	86	131	149	170	196	228	271	342	499
0.10	0.20	0.100	52	88	104	121	143	170	208	271	412
0.15	0.01	0.005	74	97	106	117	129	144	164	197	265
0.15	0.02	0.010	60	82	90	100	111	125	144	174	239
0.15	0.05	0.025	43	61	69	77	87	99	116	144	203
0.15	0.10	0.050	30	46	52	60	69	80	95	120	174
0.15	0.20	0.100	19	31	36	43	50	60	73	95	144
0.20	0.01	0.005	42	56	61	67	74	83	94	113	152
0.20	0.02	0.010	35	47	52	57	64	72	83	100	137
0.20	0.05	0.025	25	35	39	44	50	57	67	82	116
0.20	0.10	0.050	18	26	30	34	39	46	55	69	100
0.20	0.20	0.100	11	18	21	25	29	34	42	55	83
0.25	0.01	0.005	29	39	42	47	51	57	65	78	105
0.25	0.02	0.010	24	33	36	40	44	50	57	69	95
0.25	0.05	0.025	17	24	27	31	35	40	46	57	81
0.25	0.10	0.050	12	18	21	24	27	32	38	48	69
0.25	0.20	0.100	8	13	15	17	20	24	29	38	57
0.30	0.01	0.005	23	30	33	36	39	44	50	60	81
0.30	0.02	0.010	19	25	28	30	34	38	44	53	73
0.30	0.05	0.025	13	19	21	24	27	30	35	44	62
0.30	0.10	0.050	10	14	16	18	21	24	29	37	53
0.30	0.20	0.100	6	10	11	13	16	18	22	29	44
0.35	0.01	0.005	18	24	26	29	32	36	41	49	65
0.35	0.02	0.010	15	20	22	25	28	31	36	43	59
0.35	0.05	0.025	11	15	17	19	22	25	29	36	50
0.35	0.10	0.050	8	12	13	15	17	20	24	30	43
0.35	0.20	0.100	5	8	9	11	13	15	18	24	36
0.40	0.01	0.005	16	21	22	25	27	30	34	41	55
0.40	0.02	0.010	13	17	19	21	23	26	30	37	50
0.40	0.05	0.025	9	13	15	16	18	21	24	30	42
0.40	0.10	0.050	7	10	11	13	15	17	20	25	37
0.40	0.20	0.100	4	7	8	9	11	13	16	20	30
0.45	0.01	0.005	14	18	20	22	24	26	30	36	48
0.45	0.02	0.010	11	15	17	18	20	23	26	32	44
0.45	0.05	0.025	8	11	13	14	16	18	21	26	37
0.45	0.10	0.050	6	9	10	11	13	15	18	22	32
0.45	0.20	0.100	4	6	7	8	9	11	14	18	26
0.50	0.01	0.005	12	16	18	19	21	24	27	32	43
0.50	0.02	0.010	10	14	15	16	18	21	24	28	39
0.50	0.05	0.025	7	10	11	13	14	16	19	24	33
0.50	0.10	0.050	5	8	9	10	11	13	16	20	28
0.50	0.20	0.100	3	5	6	7	8	10	12	16	24
0.55	0.01	0.005	11	15	16	17	19	21	24	29	39
0.55	0.02	0.010	9	12	14	15	17	19	21	26	35
0.55	0.05	0.025	7	9	10	12	13	15	17	21	30
0.55	0.10	0.050	5	7	8	9	10	12	14	18	26
0.55	0.20	0.100	3	5	6	7	8	9	11	14	21
0.60	0.01	0.005	10	13	15	16	18	20	22	27	36
0.60	0.02	0.010	8	11	12	14	15	17	20	24	32
0.60	0.05	0.025	6	9	10	11	12	14	16	20	28
0.60	0.10	0.050	4	7	7	8	10	11	13	16	24
0.60	0.20	0.100	3	5	5	6	7	8	10	13	20

Table 9.2

Table 9.2 Continued.

π_2	α 2-sided	1-sided	Power $1-\beta$ 0.50	0.65	0.70	0.75	0.80	0.85	0.90	0.95	0.99
						$\pi_1 = 0.05$					
0.65	0.01	0.005	10	13	14	15	17	18	21	25	33
0.65	0.02	0.010	8	11	12	13	14	16	18	22	30
0.65	0.05	0.025	6	8	9	10	11	13	15	18	26
0.65	0.10	0.050	4	6	7	8	9	10	12	15	22
0.65	0.20	0.100	3	4	5	6	7	8	10	12	18
0.70	0.01	0.005	9	12	13	14	16	17	20	23	32
0.70	0.02	0.010	7	10	11	12	13	15	17	21	28
0.70	0.05	0.025	5	8	8	9	11	12	14	17	24
0.70	0.10	0.050	4	6	7	7	8	10	12	14	21
0.70	0.20	0.100	3	4	5	5	6	7	9	12	17
0.75	0.01	0.005	9	11	12	13	15	16	19	22	30
0.75	0.02	0.010	7	10	10	12	13	14	16	20	27
0.75	0.05	0.025	5	7	8	9	10	11	13	16	23
0.75	0.10	0.050	4	6	6	7	8	9	11	14	20
0.75	0.20	0.100	3	4	4	5	6	7	9	11	16
0.80	0.01	0.005	8	11	12	13	14	16	18	21	29
0.80	0.02	0.010	7	9	10	11	12	14	16	19	26
0.80	0.05	0.025	5	7	8	9	10	11	13	16	22
0.80	0.10	0.050	4	5	6	7	8	9	11	13	19
0.80	0.20	0.100	2	4	4	5	6	7	8	11	16
0.85	0.01	0.005	8	10	11	12	14	15	17	21	28
0.85	0.02	0.010	7	9	10	11	12	13	15	18	25
0.85	0.05	0.025	5	7	7	8	9	11	12	15	21
0.85	0.10	0.050	4	5	6	7	7	9	10	13	18
0.85	0.20	0.100	2	4	4	5	6	7	8	10	15
0.90	0.01	0.005	8	10	11	12	13	15	17	20	27
0.90	0.02	0.010	6	9	9	10	12	13	15	18	24
0.90	0.05	0.025	5	7	7	8	9	10	12	15	21
0.90	0.10	0.050	3	5	6	6	7	8	10	12	18
0.90	0.20	0.100	2	4	4	5	5	6	8	10	15
0.95	0.01	0.005	8	10	11	12	13	14	16	20	26
0.95	0.02	0.010	6	8	9	10	11	13	14	17	24
0.95	0.05	0.025	5	6	7	8	9	10	12	14	20
0.95	0.10	0.050	3	5	6	6	7	8	10	12	17
0.95	0.20	0.100	2	3	4	5	5	6	8	10	14
						$\pi_1 = 0.10$					
0.15	0.01	0.005	407	538	590	648	716	800	913	1093	1474
0.15	0.02	0.010	332	451	499	553	616	694	798	967	1328
0.15	0.05	0.025	236	338	379	426	482	551	645	797	1127
0.15	0.10	0.050	166	253	289	330	380	441	526	664	967
0.15	0.20	0.100	101	171	200	235	277	330	403	526	798
0.20	0.01	0.005	125	165	181	198	219	245	279	334	451
0.20	0.02	0.010	102	138	153	169	189	212	244	296	406
0.20	0.05	0.025	72	104	116	131	148	169	197	244	345
0.20	0.10	0.050	51	78	89	101	116	135	161	203	296
0.20	0.20	0.100	31	53	62	72	85	101	124	161	244
0.25	0.01	0.005	66	87	95	104	115	129	147	175	237
0.25	0.02	0.010	54	73	80	89	99	112	128	155	213
0.25	0.05	0.025	38	55	61	69	78	89	104	128	181
0.25	0.10	0.050	27	41	47	53	61	71	85	107	155
0.25	0.20	0.100	17	28	33	38	45	53	65	85	128
0.30	0.01	0.005	43	56	62	68	75	84	95	114	154
0.30	0.02	0.010	35	47	52	58	64	73	83	101	138
0.30	0.05	0.025	25	36	40	45	50	58	67	83	117
0.30	0.10	0.050	18	27	30	35	40	46	55	69	101
0.30	0.20	0.100	11	18	21	25	29	35	42	55	83

Table 9.2

116

Table 9.2 Continued.

π_2	α 2-sided	1-sided	Power $1 - \beta$ 0.50	0.65	0.70	0.75	0.80	0.85	0.90	0.95	0.99
					$\pi_1 = 0.10$						
0.35	0.01	0.005	31	41	45	49	54	61	69	83	112
0.35	0.02	0.010	26	34	38	42	47	53	61	73	101
0.35	0.05	0.025	18	26	29	33	37	42	49	61	85
0.35	0.10	0.050	13	20	22	25	29	34	40	50	73
0.35	0.20	0.100	8	13	16	18	21	25	31	40	61
0.40	0.01	0.005	24	32	35	38	42	47	54	65	87
0.40	0.02	0.010	20	27	30	33	37	41	47	57	78
0.40	0.05	0.025	14	20	23	25	29	33	38	47	67
0.40	0.10	0.050	10	15	17	20	23	26	31	39	57
0.40	0.20	0.100	6	10	12	14	17	20	24	31	47
0.45	0.01	0.005	20	26	29	31	35	39	44	53	71
0.45	0.02	0.010	16	22	24	27	30	34	39	47	64
0.45	0.05	0.025	12	17	19	21	24	27	31	39	54
0.45	0.10	0.050	8	13	14	16	19	22	26	32	47
0.45	0.20	0.100	5	9	10	12	14	16	20	26	39
0.50	0.01	0.005	17	22	24	27	29	33	37	45	60
0.50	0.02	0.010	14	19	21	23	25	28	33	40	54
0.50	0.05	0.025	10	14	16	18	20	23	27	33	46
0.50	0.10	0.050	7	11	12	14	16	18	22	27	40
0.50	0.20	0.100	5	7	9	10	12	14	17	22	33
0.55	0.01	0.005	15	19	21	23	26	28	32	39	52
0.55	0.02	0.010	12	16	18	20	22	25	28	34	47
0.55	0.05	0.025	9	12	14	15	17	20	23	28	40
0.55	0.10	0.050	6	9	11	12	14	16	19	24	34
0.55	0.20	0.100	4	6	7	9	10	12	15	19	28
0.60	0.01	0.005	13	17	19	21	23	25	29	34	46
0.60	0.02	0.010	11	14	16	18	20	22	25	30	42
0.60	0.05	0.025	8	11	12	14	15	18	20	25	35
0.60	0.10	0.050	6	8	9	11	12	14	17	21	30
0.60	0.20	0.100	4	6	7	8	9	11	13	17	25
0.65	0.01	0.005	12	15	17	19	20	23	26	31	41
0.65	0.02	0.010	10	13	14	16	18	20	23	27	37
0.65	0.05	0.025	7	10	11	12	14	16	18	23	32
0.65	0.10	0.050	5	8	9	10	11	13	15	19	27
0.65	0.20	0.100	3	5	6	7	8	10	12	15	23
0.70	0.01	0.005	11	14	15	17	19	21	24	28	38
0.70	0.02	0.010	9	12	13	15	16	18	21	25	34
0.70	0.05	0.025	6	9	10	11	13	14	17	21	29
0.70	0.10	0.050	5	7	8	9	10	12	14	17	25
0.70	0.20	0.100	3	5	6	6	8	9	11	14	21
0.75	0.01	0.005	10	13	14	16	17	19	22	26	35
0.75	0.02	0.010	8	11	12	13	15	17	19	23	32
0.75	0.05	0.025	6	8	9	10	12	13	16	19	27
0.75	0.10	0.050	4	6	7	8	9	11	13	16	23
0.75	0.20	0.100	3	4	5	6	7	8	10	13	19
0.80	0.01	0.005	9	12	13	15	16	18	20	24	33
0.80	0.02	0.010	8	10	11	13	14	16	18	22	30
0.80	0.05	0.025	6	8	9	10	11	13	15	18	25
0.80	0.10	0.050	4	6	7	8	9	10	12	15	22
0.80	0.20	0.100	3	4	5	6	7	8	9	12	18
0.85	0.01	0.005	9	12	13	14	15	17	19	23	31
0.85	0.02	0.010	7	10	11	12	13	15	17	20	28
0.85	0.05	0.025	5	7	8	9	10	12	14	17	24
0.85	0.10	0.050	4	6	6	7	8	10	11	14	20
0.85	0.20	0.100	3	4	5	5	6	7	9	11	17

Table 9.2

Table 9.2 Continued.

π_2	α 2-sided	1-sided	Power $1 - \beta$ 0.50	0.65	0.70	0.75	0.80	0.85	0.90	0.95	0.99
					$\pi_1 = 0.10$						
0.90	0.01	0.005	8	11	12	13	15	16	18	22	29
0.90	0.02	0.010	7	9	10	11	13	14	16	19	26
0.90	0.05	0.025	5	7	8	9	10	11	13	16	23
0.90	0.10	0.050	4	5	6	7	8	9	11	13	19
0.90	0.20	0.100	2	4	4	5	6	7	8	11	16
0.95	0.01	0.005	8	11	12	13	14	16	18	21	28
0.95	0.02	0.010	7	9	10	11	12	14	15	19	25
0.95	0.05	0.025	5	7	8	8	10	11	13	15	22
0.95	0.10	0.050	4	5	6	7	8	9	10	13	19
0.95	0.20	0.100	2	4	4	5	6	7	8	10	15
					$\pi_1 = 0.15$						
0.20	0.01	0.005	598	790	866	952	1052	1175	1340	1605	2164
0.20	0.02	0.010	488	663	732	811	904	1019	1173	1421	1950
0.20	0.05	0.025	346	496	556	625	707	809	947	1171	1655
0.20	0.10	0.050	244	372	424	485	557	648	772	975	1421
0.20	0.20	0.100	148	251	294	345	406	484	592	772	1173
0.25	0.01	0.005	172	227	249	273	302	337	385	460	621
0.25	0.02	0.010	140	190	210	233	260	293	337	408	559
0.25	0.05	0.025	100	143	160	180	203	232	272	336	475
0.25	0.10	0.050	70	107	122	139	160	186	222	280	408
0.25	0.20	0.100	43	72	85	99	117	139	170	222	337
0.30	0.01	0.005	86	114	125	137	151	169	193	231	311
0.30	0.02	0.010	70	95	105	117	130	147	169	204	280
0.30	0.05	0.025	50	72	80	90	102	116	136	168	238
0.30	0.10	0.050	35	54	61	70	80	93	111	140	204
0.30	0.20	0.100	22	36	43	50	59	70	85	111	169
0.35	0.01	0.005	54	71	78	86	95	106	120	144	194
0.35	0.02	0.010	44	60	66	73	81	92	105	128	175
0.35	0.05	0.025	31	45	50	56	64	73	85	105	149
0.35	0.10	0.050	22	34	38	44	50	58	70	88	128
0.35	0.20	0.100	14	23	27	31	37	44	53	70	105
0.40	0.01	0.005	38	50	55	60	67	75	85	102	137
0.40	0.02	0.010	31	42	47	52	57	65	74	90	123
0.40	0.05	0.025	22	32	36	40	45	51	60	74	105
0.40	0.10	0.050	16	24	27	31	36	41	49	62	90
0.40	0.20	0.100	10	16	19	22	26	31	38	49	74
0.45	0.01	0.005	29	38	42	46	51	57	64	77	104
0.45	0.02	0.010	24	32	35	39	44	49	56	68	94
0.45	0.05	0.025	17	24	27	30	34	39	46	56	80
0.45	0.10	0.050	12	18	21	24	27	31	37	47	68
0.45	0.20	0.100	8	12	15	17	20	24	29	37	56
0.50	0.01	0.005	23	31	33	37	41	45	52	62	83
0.50	0.02	0.010	19	26	28	31	35	39	45	55	75
0.50	0.05	0.025	14	19	22	24	27	31	37	45	63
0.50	0.10	0.050	10	15	17	19	22	25	30	38	55
0.50	0.20	0.100	6	10	12	14	16	19	23	30	45
0.55	0.01	0.005	19	25	28	30	34	38	43	51	69
0.55	0.02	0.010	16	21	24	26	29	33	37	45	62
0.55	0.05	0.025	11	16	18	20	23	26	30	37	53
0.55	0.10	0.050	8	12	14	16	18	21	25	31	45
0.55	0.20	0.100	5	8	10	11	13	16	19	25	37
0.60	0.01	0.005	17	22	24	26	29	32	36	43	59
0.60	0.02	0.010	14	18	20	22	25	28	32	39	53
0.60	0.05	0.025	10	14	15	17	19	22	26	32	45
0.60	0.10	0.050	7	10	12	13	15	18	21	27	39
0.60	0.20	0.100	4	7	8	10	11	13	16	21	32

Table 9.2

Table 9.2 Continued.

π_2	α 2-sided	1-sided	Power $1 - \beta$ 0.50	0.65	0.70	0.75	0.80	0.85	0.90	0.95	0.99
					$\pi_1 = 0.15$						
0.65	0.01	0.005	14	19	21	23	25	28	32	38	51
0.65	0.02	0.010	12	16	18	19	22	24	28	34	46
0.65	0.05	0.025	9	12	13	15	17	19	23	28	39
0.65	0.10	0.050	6	9	10	12	13	16	18	23	34
0.65	0.20	0.100	4	6	7	9	10	12	14	18	28
0.70	0.01	0.005	13	17	18	20	22	25	28	34	45
0.70	0.02	0.010	11	14	16	17	19	22	25	30	41
0.70	0.05	0.025	8	11	12	13	15	17	20	25	35
0.70	0.10	0.050	6	8	9	11	12	14	16	21	30
0.70	0.20	0.100	4	6	7	8	9	11	13	16	25
0.75	0.01	0.005	12	15	17	18	20	22	25	30	41
0.75	0.02	0.010	10	13	14	16	17	19	22	27	37
0.75	0.05	0.025	7	10	11	12	14	16	18	22	31
0.75	0.10	0.050	5	7	8	10	11	13	15	19	27
0.75	0.20	0.100	3	5	6	7	8	10	12	15	22
0.80	0.01	0.005	11	14	15	17	18	20	23	28	37
0.80	0.02	0.010	9	12	13	14	16	18	20	25	34
0.80	0.05	0.025	6	9	10	11	12	14	17	20	29
0.80	0.10	0.050	5	7	8	9	10	11	14	17	25
0.80	0.20	0.100	3	5	5	6	7	9	11	14	20
0.85	0.01	0.005	10	13	14	15	17	19	21	26	34
0.85	0.02	0.010	8	11	12	13	15	16	19	23	31
0.85	0.05	0.025	6	8	9	10	12	13	15	19	26
0.85	0.10	0.050	4	6	7	8	9	11	13	16	23
0.85	0.20	0.100	3	4	5	6	7	8	10	13	19
0.90	0.01	0.005	9	12	13	14	16	18	20	24	32
0.90	0.02	0.010	8	10	11	12	14	15	18	21	29
0.90	0.05	0.025	6	8	9	10	11	12	14	18	25
0.90	0.10	0.050	4	6	7	8	9	10	12	15	21
0.90	0.20	0.100	3	4	5	6	6	8	9	12	18
0.95	0.01	0.005	9	11	12	14	15	17	19	23	30
0.95	0.02	0.010	7	10	11	12	13	15	17	20	27
0.95	0.05	0.025	5	7	8	9	10	12	14	17	23
0.95	0.10	0.050	4	6	6	7	8	9	11	14	20
0.95	0.20	0.100	3	4	5	5	6	7	9	11	17
					$\pi_1 = 0.20$						
0.25	0.01	0.005	772	1020	1118	1229	1359	1518	1731	2072	2795
0.25	0.02	0.010	630	856	945	1048	1167	1315	1514	1834	2518
0.25	0.05	0.025	447	640	718	808	913	1045	1222	1512	2137
0.25	0.10	0.050	315	480	548	626	719	836	996	1259	1834
0.25	0.20	0.100	191	324	380	445	525	625	764	996	1514
0.30	0.01	0.005	213	282	309	340	375	419	478	572	772
0.30	0.02	0.010	174	237	261	290	323	363	418	507	695
0.30	0.05	0.025	124	177	199	223	252	289	338	418	590
0.30	0.10	0.050	87	133	152	173	199	231	275	348	507
0.30	0.20	0.100	53	90	105	123	145	173	211	275	418
0.35	0.01	0.005	104	137	150	165	182	204	232	278	375
0.35	0.02	0.010	85	115	127	141	157	177	203	246	338
0.35	0.05	0.025	60	86	97	109	123	140	164	203	287
0.35	0.10	0.050	43	65	74	84	97	112	134	169	246
0.35	0.20	0.100	26	44	51	60	71	84	103	134	203
0.40	0.01	0.005	63	84	92	101	111	124	142	169	228
0.40	0.02	0.010	52	70	78	86	96	108	124	150	206
0.40	0.05	0.025	37	53	59	66	75	86	100	124	175
0.40	0.10	0.050	26	40	45	52	59	69	82	103	150
0.40	0.20	0.100	16	27	31	37	43	51	63	82	124

Table 9.2

Table 9.2 Continued.

π_2	α 2-sided	1-sided	0.50	0.65	0.70	Power $1-\beta$ 0.75	0.80	0.85	0.90	0.95	0.99
						$\pi_1 = 0.20$					
0.45	0.01	0.005	44	58	63	69	77	86	98	117	157
0.45	0.02	0.010	36	49	54	59	66	74	86	103	142
0.45	0.05	0.025	26	36	41	46	52	59	69	85	120
0.45	0.10	0.050	18	27	31	36	41	47	56	71	103
0.45	0.20	0.100	11	19	22	25	30	36	43	56	86
0.50	0.01	0.005	33	43	47	52	57	64	73	87	117
0.50	0.02	0.010	27	36	40	44	49	55	64	77	106
0.50	0.05	0.025	19	27	30	34	39	44	52	64	90
0.50	0.10	0.050	14	21	23	27	31	35	42	53	77
0.50	0.20	0.100	8	14	16	19	22	27	32	42	64
0.55	0.01	0.005	26	34	37	41	45	50	57	68	92
0.55	0.02	0.010	21	29	31	35	39	44	50	61	83
0.55	0.05	0.025	15	21	24	27	30	35	41	50	70
0.55	0.10	0.050	11	16	18	21	24	28	33	42	61
0.55	0.20	0.100	7	11	13	15	18	21	26	33	50
0.60	0.01	0.005	21	28	30	33	37	41	47	56	75
0.60	0.02	0.010	17	23	26	28	32	36	41	49	68
0.60	0.05	0.025	12	18	20	22	25	28	33	41	58
0.60	0.10	0.050	9	13	15	17	20	23	27	34	49
0.60	0.20	0.100	6	9	11	12	14	17	21	27	41
0.65	0.01	0.005	18	23	26	28	31	34	39	47	63
0.65	0.02	0.010	15	20	22	24	27	30	34	42	57
0.65	0.05	0.025	11	15	17	19	21	24	28	34	48
0.65	0.10	0.050	8	11	13	15	17	19	23	29	42
0.65	0.20	0.100	5	8	9	10	12	14	18	23	34
0.70	0.01	0.005	15	20	22	24	27	30	34	40	54
0.70	0.02	0.010	13	17	19	21	23	26	30	36	49
0.70	0.05	0.025	9	13	14	16	18	21	24	30	42
0.70	0.10	0.050	7	10	11	13	14	17	20	25	36
0.70	0.20	0.100	4	7	8	9	11	13	15	20	30
0.75	0.01	0.005	14	18	19	21	23	26	30	35	48
0.75	0.02	0.010	11	15	16	18	20	23	26	31	43
0.75	0.05	0.025	8	11	13	14	16	18	21	26	37
0.75	0.10	0.050	6	9	10	11	13	15	17	22	31
0.75	0.20	0.100	4	6	7	8	9	11	13	17	26
0.80	0.01	0.005	12	16	17	19	21	23	27	32	42
0.80	0.02	0.010	10	13	15	16	18	20	23	28	38
0.80	0.05	0.025	7	10	11	13	14	16	19	23	33
0.80	0.10	0.050	5	8	9	10	11	13	15	19	28
0.80	0.20	0.100	3	5	6	7	8	10	12	15	23
0.85	0.01	0.005	11	14	16	17	19	21	24	29	38
0.85	0.02	0.010	9	12	13	15	16	18	21	25	35
0.85	0.05	0.025	7	9	10	11	13	15	17	21	30
0.85	0.10	0.050	5	7	8	9	10	12	14	18	25
0.85	0.20	0.100	3	5	6	7	8	9	11	14	21
0.90	0.01	0.005	10	13	14	16	17	19	22	26	35
0.90	0.02	0.010	8	11	12	14	15	17	19	23	32
0.90	0.05	0.025	6	8	9	11	12	13	16	19	27
0.90	0.10	0.050	4	6	7	8	9	11	13	16	23
0.90	0.20	0.100	3	5	5	6	7	8	10	13	19
0.95	0.01	0.005	9	12	13	15	16	18	20	24	33
0.95	0.02	0.010	8	10	11	13	14	16	18	22	29
0.95	0.05	0.025	6	8	9	10	11	12	15	18	25
0.95	0.10	0.050	4	6	7	8	9	10	12	15	22
0.95	0.20	0.100	3	4	5	6	7	8	9	12	18

Table 9.2

Table 9.2 Continued.

π_2	α 2-sided	1-sided	Power $1 - \beta$ 0.50	0.65	0.70	0.75	0.80	0.85	0.90	0.95	0.99
						$\pi_1 = 0.25$					
0.30	0.01	0.005	924	1221	1338	1471	1626	1817	2072	2480	3346
0.30	0.02	0.010	754	1024	1132	1254	1398	1575	1812	2196	3014
0.30	0.05	0.025	535	766	860	967	1093	1250	1463	1809	2558
0.30	0.10	0.050	377	574	656	749	861	1001	1193	1507	2196
0.30	0.20	0.100	229	387	455	533	628	748	915	1193	1812
0.35	0.01	0.005	249	329	360	396	438	489	558	668	900
0.35	0.02	0.010	203	276	305	338	376	424	488	591	811
0.35	0.05	0.025	144	206	232	260	294	337	394	487	688
0.35	0.10	0.050	102	155	177	202	232	270	321	406	591
0.35	0.20	0.100	62	105	123	144	169	202	246	321	488
0.40	0.01	0.005	118	156	171	188	208	232	265	317	428
0.40	0.02	0.010	97	131	145	161	179	202	232	281	385
0.40	0.05	0.025	69	98	110	124	140	160	187	232	327
0.40	0.10	0.050	49	74	84	96	110	128	153	193	281
0.40	0.20	0.100	30	50	58	69	81	96	117	153	232
0.45	0.01	0.005	71	94	103	113	125	139	159	190	256
0.45	0.02	0.010	58	79	87	96	107	121	139	168	231
0.45	0.05	0.025	41	59	66	74	84	96	112	139	196
0.45	0.10	0.050	29	44	51	58	66	77	92	116	168
0.45	0.20	0.100	18	30	35	41	48	58	70	92	139
0.50	0.01	0.005	48	64	70	77	85	94	108	129	174
0.50	0.02	0.010	39	53	59	65	73	82	94	114	156
0.50	0.05	0.025	28	40	45	50	57	65	76	94	133
0.50	0.10	0.050	20	30	34	39	45	52	62	78	114
0.50	0.20	0.100	12	21	24	28	33	39	48	62	94
0.55	0.01	0.005	36	47	51	56	62	69	79	95	127
0.55	0.02	0.010	29	39	43	48	53	60	69	84	115
0.55	0.05	0.025	21	30	33	37	42	48	56	69	97
0.55	0.10	0.050	15	22	25	29	33	38	46	58	84
0.55	0.20	0.100	9	15	18	21	24	29	35	46	69
0.60	0.01	0.005	28	36	40	44	48	54	61	73	99
0.60	0.02	0.010	23	31	34	37	41	47	54	65	89
0.60	0.05	0.025	16	23	26	29	33	37	43	54	76
0.60	0.10	0.050	12	17	20	22	26	30	35	45	65
0.60	0.20	0.100	7	12	14	16	19	22	27	35	54
0.65	0.01	0.005	22	29	32	35	39	43	49	59	80
0.65	0.02	0.010	18	25	27	30	33	38	43	52	72
0.65	0.05	0.025	13	19	21	23	26	30	35	43	61
0.65	0.10	0.050	9	14	16	18	21	24	29	36	52
0.65	0.20	0.100	6	10	11	13	15	18	22	29	43
0.70	0.01	0.005	19	24	27	29	32	36	41	49	66
0.70	0.02	0.010	15	21	23	25	28	31	36	44	60
0.70	0.05	0.025	11	16	17	19	22	25	29	36	51
0.70	0.10	0.050	8	12	13	15	17	20	24	30	44
0.70	0.20	0.100	5	8	9	11	13	15	18	24	36
0.75	0.01	0.005	16	21	23	25	28	31	35	42	56
0.75	0.02	0.010	13	18	19	21	24	27	31	37	51
0.75	0.05	0.025	9	13	15	17	19	21	25	31	43
0.75	0.10	0.050	7	10	11	13	15	17	20	26	37
0.75	0.20	0.100	4	7	8	9	11	13	16	20	31
0.80	0.01	0.005	14	18	20	22	24	27	30	36	49
0.80	0.02	0.010	11	15	17	19	21	23	27	32	44
0.80	0.05	0.025	8	12	13	14	16	19	22	27	38
0.80	0.10	0.050	6	9	10	11	13	15	18	22	32
0.80	0.20	0.100	4	6	7	8	10	11	14	18	27

Table 9.2

Table 9.2 Continued.

π_2	2-sided	1-sided	0.50	0.65	0.70	0.75	0.80	0.85	0.90	0.95	0.99
	α					Power $1 - \beta$					

$\pi_1 = 0.25$											
0.85	0.01	0.005	12	16	18	19	21	24	27	32	43
0.85	0.02	0.010	10	14	15	17	18	21	24	29	39
0.85	0.05	0.025	7	10	11	13	14	16	19	24	33
0.85	0.10	0.050	5	8	9	10	12	13	16	20	29
0.85	0.20	0.100	3	5	6	7	9	10	12	16	24
0.90	0.01	0.005	11	14	16	17	19	21	24	29	39
0.90	0.02	0.010	9	12	13	15	17	19	21	26	35
0.90	0.05	0.025	7	9	10	12	13	15	17	21	30
0.90	0.10	0.050	5	7	8	9	10	12	14	18	26
0.90	0.20	0.100	3	5	6	7	8	9	11	14	21
0.95	0.01	0.005	10	13	14	16	17	19	22	26	35
0.95	0.02	0.010	8	11	12	14	15	17	19	23	32
0.95	0.05	0.025	6	8	9	11	12	14	16	19	27
0.95	0.10	0.050	4	6	7	8	9	11	13	16	23
0.95	0.20	0.100	3	5	5	6	7	8	10	13	19

$\pi_1 = 0.30$											
0.35	0.01	0.005	1051	1389	1522	1673	1850	2067	2357	2821	3806
0.35	0.02	0.010	857	1165	1287	1426	1590	1791	2062	2498	3428
0.35	0.05	0.025	609	871	978	1099	1243	1422	1664	2058	2910
0.35	0.10	0.050	429	653	746	852	979	1139	1357	1714	2498
0.35	0.20	0.100	261	440	517	606	714	851	1041	1357	2062
0.40	0.01	0.005	278	367	402	442	488	546	622	745	1005
0.40	0.02	0.010	227	308	340	377	420	473	544	659	905
0.40	0.05	0.025	161	230	258	290	328	376	440	543	768
0.40	0.10	0.050	114	173	197	225	259	301	358	453	659
0.40	0.20	0.100	69	117	137	160	189	225	275	358	544
0.45	0.01	0.005	130	172	188	207	228	255	291	348	469
0.45	0.02	0.010	106	144	159	176	196	221	254	308	423
0.45	0.05	0.025	75	108	121	136	154	176	206	254	359
0.45	0.10	0.050	53	81	92	105	121	141	168	212	308
0.45	0.20	0.100	33	55	64	75	88	105	129	168	254
0.50	0.01	0.005	77	101	111	122	135	150	172	205	277
0.50	0.02	0.010	63	85	94	104	116	130	150	182	249
0.50	0.05	0.025	45	64	71	80	91	104	121	150	212
0.50	0.10	0.050	32	48	55	62	72	83	99	125	182
0.50	0.20	0.100	19	32	38	44	52	62	76	99	150
0.55	0.01	0.005	51	68	74	82	90	101	115	137	185
0.55	0.02	0.010	42	57	63	70	78	87	101	122	167
0.55	0.05	0.025	30	43	48	54	61	69	81	100	142
0.55	0.10	0.050	21	32	37	42	48	56	66	84	122
0.55	0.20	0.100	13	22	26	30	35	42	51	66	101
0.60	0.01	0.005	37	49	54	59	65	73	83	100	134
0.60	0.02	0.010	31	41	46	51	56	63	73	88	121
0.60	0.05	0.025	22	31	35	39	44	50	59	73	103
0.60	0.10	0.050	16	23	27	30	35	41	48	61	88
0.60	0.20	0.100	10	16	19	22	26	30	37	48	73
0.65	0.01	0.005	29	38	41	45	50	56	64	76	103
0.65	0.02	0.010	24	32	35	39	43	49	56	68	93
0.65	0.05	0.025	17	24	27	30	34	39	45	56	79
0.65	0.10	0.050	12	18	21	23	27	31	37	47	68
0.65	0.20	0.100	7	12	14	17	20	23	28	37	56
0.70	0.01	0.005	23	30	33	36	40	45	51	61	82
0.70	0.02	0.010	19	25	28	31	35	39	45	54	74
0.70	0.05	0.025	14	19	21	24	27	31	36	45	63
0.70	0.10	0.050	10	14	16	19	21	25	30	37	54
0.70	0.20	0.100	6	10	12	13	16	19	23	30	45

Table 9.2

Table 9.2 Continued.

π_2	α 2-sided	1-sided	0.50	0.65	0.70	0.75	0.80	0.85	0.90	0.95	0.99
						Power $1 - \beta$					

$\pi_1 = 0.30$

π_2	2-sided	1-sided	0.50	0.65	0.70	0.75	0.80	0.85	0.90	0.95	0.99
0.75	0.01	0.005	19	25	27	30	33	37	42	50	68
0.75	0.02	0.010	16	21	23	26	28	32	37	44	61
0.75	0.05	0.025	11	16	18	20	22	26	30	37	52
0.75	0.10	0.050	8	12	14	16	18	21	24	31	44
0.75	0.20	0.100	5	8	10	11	13	15	19	24	37
0.80	0.01	0.005	16	21	23	25	28	31	36	42	57
0.80	0.02	0.010	13	18	20	22	24	27	31	38	51
0.80	0.05	0.025	10	13	15	17	19	22	25	31	44
0.80	0.10	0.050	7	10	12	13	15	17	21	26	38
0.80	0.20	0.100	4	7	8	10	11	13	16	21	31
0.85	0.01	0.005	14	18	20	22	24	27	31	37	49
0.85	0.02	0.010	11	15	17	19	21	23	27	32	44
0.85	0.05	0.025	8	12	13	15	16	19	22	27	38
0.85	0.10	0.050	6	9	10	11	13	15	18	22	32
0.85	0.20	0.100	4	6	7	8	10	11	14	18	27
0.90	0.01	0.005	12	16	18	19	21	24	27	32	43
0.90	0.02	0.010	10	14	15	16	18	21	24	29	39
0.90	0.05	0.025	7	10	11	13	14	16	19	24	33
0.90	0.10	0.050	5	8	9	10	11	13	16	20	29
0.90	0.20	0.100	3	5	6	7	9	10	12	16	24
0.95	0.01	0.005	11	14	16	17	19	21	24	29	38
0.95	0.02	0.010	9	12	13	15	16	18	21	25	35
0.95	0.05	0.025	7	9	10	11	13	15	17	21	30
0.95	0.10	0.050	5	7	8	9	10	12	14	18	25
0.95	0.20	0.100	3	5	6	7	8	9	11	14	21

$\pi_1 = 0.35$

π_2	2-sided	1-sided	0.50	0.65	0.70	0.75	0.80	0.85	0.90	0.95	0.99
0.40	0.01	0.005	1151	1521	1667	1833	2026	2264	2581	3090	4168
0.40	0.02	0.010	939	1276	1410	1562	1741	1962	2258	2736	3755
0.40	0.05	0.025	667	954	1071	1204	1362	1558	1823	2254	3187
0.40	0.10	0.050	470	715	817	933	1073	1247	1486	1877	2736
0.40	0.20	0.100	285	482	566	664	782	932	1140	1486	2258
0.45	0.01	0.005	300	396	434	477	527	589	671	803	1084
0.45	0.02	0.010	244	332	367	406	453	510	587	711	976
0.45	0.05	0.025	174	248	279	313	354	405	474	586	829
0.45	0.10	0.050	122	186	213	243	279	325	387	488	711
0.45	0.20	0.100	75	126	148	173	204	243	297	387	587
0.50	0.01	0.005	138	183	200	220	243	271	309	370	500
0.50	0.02	0.010	113	153	169	187	209	235	271	328	450
0.50	0.05	0.025	80	115	129	145	163	187	219	270	382
0.50	0.10	0.050	57	86	98	112	129	150	178	225	328
0.50	0.20	0.100	35	58	68	80	94	112	137	178	271
0.55	0.01	0.005	81	106	117	128	142	158	180	216	291
0.55	0.02	0.010	66	89	99	109	122	137	158	191	262
0.55	0.05	0.025	47	67	75	84	95	109	127	157	222
0.55	0.10	0.050	33	50	57	65	75	87	104	131	191
0.55	0.20	0.100	20	34	40	47	55	65	80	104	158
0.60	0.01	0.005	53	71	77	85	94	105	119	143	192
0.60	0.02	0.010	44	59	65	72	81	91	104	126	173
0.60	0.05	0.025	31	44	50	56	63	72	84	104	147
0.60	0.10	0.050	22	33	38	43	50	58	69	87	126
0.60	0.20	0.100	14	23	27	31	36	43	53	69	104
0.65	0.01	0.005	38	51	55	61	67	75	86	102	138
0.65	0.02	0.010	31	43	47	52	58	65	75	91	124
0.65	0.05	0.025	22	32	36	40	45	52	61	75	106
0.65	0.10	0.050	16	24	27	31	36	42	49	62	91
0.65	0.20	0.100	10	16	19	22	26	31	38	49	75

Table 9.2

Table 9.2 Continued.

π_2	α 2-sided	1-sided	0.50	0.65	0.70	0.75	Power $1-\beta$ 0.80	0.85	0.90	0.95	0.99
						$\pi_1 = 0.35$					
0.70	0.01	0.005	29	39	42	46	51	57	65	78	105
0.70	0.02	0.010	24	32	36	40	44	50	57	69	94
0.70	0.05	0.025	17	24	27	31	35	39	46	57	80
0.70	0.10	0.050	12	18	21	24	27	32	38	47	69
0.70	0.20	0.100	8	13	15	17	20	24	29	38	57
0.75	0.01	0.005	23	31	33	37	40	45	51	61	83
0.75	0.02	0.010	19	26	28	31	35	39	45	54	75
0.75	0.05	0.025	14	19	22	24	27	31	36	45	63
0.75	0.10	0.050	10	15	17	19	22	25	30	38	54
0.75	0.20	0.100	6	10	12	14	16	19	23	30	45
0.80	0.01	0.005	19	25	27	30	33	37	42	50	68
0.80	0.02	0.010	16	21	23	26	28	32	37	44	61
0.80	0.05	0.025	11	16	18	20	22	26	30	37	52
0.80	0.10	0.050	8	12	14	16	18	21	24	31	44
0.80	0.20	0.100	5	8	10	11	13	15	19	24	37
0.85	0.01	0.005	16	21	23	25	28	31	35	42	57
0.85	0.02	0.010	13	18	19	22	24	27	31	37	51
0.85	0.05	0.025	9	13	15	17	19	21	25	31	43
0.85	0.10	0.050	7	10	11	13	15	17	20	26	37
0.85	0.20	0.100	4	7	8	9	11	13	16	20	31
0.90	0.01	0.005	14	18	20	22	24	27	30	36	48
0.90	0.02	0.010	11	15	17	18	21	23	26	32	44
0.90	0.05	0.025	8	11	13	14	16	18	21	26	37
0.90	0.10	0.050	6	9	10	11	13	15	18	22	32
0.90	0.20	0.100	4	6	7	8	9	11	14	18	26
0.95	0.01	0.005	12	16	17	19	21	23	26	31	42
0.95	0.02	0.010	10	13	15	16	18	20	23	28	38
0.95	0.05	0.025	7	10	11	13	14	16	19	23	32
0.95	0.10	0.050	5	8	9	10	11	13	15	19	28
0.95	0.20	0.100	3	5	6	7	8	10	12	15	23
						$\pi_1 = 0.40$					
0.45	0.01	0.005	1223	1617	1772	1948	2153	2406	2743	3284	4430
0.45	0.02	0.010	998	1356	1498	1660	1850	2085	2400	2907	3990
0.45	0.05	0.025	709	1014	1138	1280	1447	1655	1937	2396	3387
0.45	0.10	0.050	499	760	868	992	1140	1326	1579	1995	2907
0.45	0.20	0.100	303	513	602	706	831	991	1211	1579	2400
0.50	0.01	0.005	314	415	455	500	553	618	704	843	1137
0.50	0.02	0.010	256	348	385	426	475	535	616	746	1024
0.50	0.05	0.025	182	261	292	329	372	425	497	615	869
0.50	0.10	0.050	128	195	223	255	293	340	406	512	746
0.50	0.20	0.100	78	132	155	181	214	255	311	406	616
0.55	0.01	0.005	143	189	207	228	252	281	321	384	518
0.55	0.02	0.010	117	159	175	194	217	244	281	340	467
0.55	0.05	0.025	83	119	133	150	169	194	227	280	396
0.55	0.10	0.050	59	89	102	116	134	155	185	234	340
0.55	0.20	0.100	36	60	71	83	98	116	142	185	281
0.60	0.01	0.005	83	109	120	131	145	162	185	221	298
0.60	0.02	0.010	68	92	101	112	125	141	162	196	269
0.60	0.05	0.025	48	69	77	86	98	112	131	161	228
0.60	0.10	0.050	34	52	59	67	77	90	107	135	196
0.60	0.20	0.100	21	35	41	48	56	67	82	107	162
0.65	0.01	0.005	54	72	78	86	95	106	121	145	195
0.65	0.02	0.010	44	60	66	73	82	92	106	128	176
0.65	0.05	0.025	32	45	51	57	64	73	86	106	149
0.65	0.10	0.050	22	34	39	44	51	59	70	88	128
0.65	0.20	0.100	14	23	27	32	37	44	54	70	106

Table 9.2

Table 9.2 Continued.

π_2	α 2-sided	1-sided	Power $1 - \beta$ 0.50	0.65	0.70	0.75	0.80	0.85	0.90	0.95	0.99
						$\pi_1 = 0.40$					
0.70	0.01	0.005	39	51	56	61	68	76	86	103	139
0.70	0.02	0.010	32	43	47	52	58	66	75	91	125
0.70	0.05	0.025	23	32	36	40	46	52	61	75	106
0.70	0.10	0.050	16	24	28	31	36	42	50	63	91
0.70	0.20	0.100	10	16	19	22	26	31	38	50	75
0.75	0.01	0.005	29	38	42	46	51	57	65	77	104
0.75	0.02	0.010	24	32	36	39	44	49	57	69	94
0.75	0.05	0.025	17	24	27	30	34	39	46	57	80
0.75	0.10	0.050	12	18	21	24	27	32	37	47	69
0.75	0.20	0.100	8	12	15	17	20	24	29	37	57
0.80	0.01	0.005	23	30	33	36	40	45	51	61	82
0.80	0.02	0.010	19	25	28	31	34	39	44	54	74
0.80	0.05	0.025	13	19	21	24	27	31	36	44	63
0.80	0.10	0.050	10	14	16	19	21	25	29	37	54
0.80	0.20	0.100	6	10	12	13	16	19	23	29	44
0.85	0.01	0.005	19	24	27	29	32	36	41	49	66
0.85	0.02	0.010	15	21	23	25	28	31	36	44	60
0.85	0.05	0.025	11	16	17	19	22	25	29	36	51
0.85	0.10	0.050	8	12	13	15	17	20	24	30	44
0.85	0.20	0.100	5	8	9	11	13	15	18	24	36
0.90	.01	0.005	16	20	22	24	27	30	34	41	55
0.90	.02	0.010	13	17	19	21	23	26	30	36	50
0.90	.05	0.025	9	13	14	16	18	21	24	30	42
0.90	.10	0.050	7	10	11	13	15	17	20	25	36
0.90	.20	0.100	4	7	8	9	11	13	15	20	30
0.95	.01	0.005	13	17	19	21	23	26	29	35	47
0.95	.02	0.010	11	15	16	18	20	22	26	31	42
0.95	.05	0.025	8	11	12	14	16	18	21	26	36
0.95	.10	0.050	6	8	10	11	12	14	17	21	31
0.95	.20	0.100	4	6	7	8	9	11	13	17	26
						$\pi_1 = 0.45$					
0.50	0.01	0.005	1267	1674	1835	2017	2230	2491	2841	3401	4588
0.50	0.02	0.010	1034	1404	1552	1720	1916	2159	2485	3011	4133
0.50	0.05	0.025	734	1050	1179	1325	1499	1714	2006	2481	3508
0.50	0.10	0.050	517	787	899	1027	1181	1373	1635	2066	3011
0.50	0.20	0.100	314	531	623	731	861	1026	1255	1635	2485
0.55	0.01	0.005	322	425	466	512	566	632	721	863	1164
0.55	0.02	0.010	263	357	394	437	486	548	631	764	1049
0.55	0.05	0.025	187	267	299	337	381	435	509	630	890
0.55	0.10	0.050	132	200	228	261	300	349	415	525	764
0.55	0.20	0.100	80	135	158	186	219	261	319	415	631
0.60	0.01	0.005	145	192	210	231	255	285	325	389	524
0.60	0.02	0.010	119	161	178	197	219	247	284	344	473
0.60	0.05	0.025	84	120	135	152	172	196	230	284	401
0.60	0.10	0.050	59	90	103	118	135	157	187	236	344
0.60	0.20	0.100	36	61	72	84	99	118	144	187	284
0.65	0.01	0.005	83	109	120	132	146	163	185	222	299
0.65	0.02	0.010	68	92	101	112	125	141	162	196	269
0.65	0.05	0.025	48	69	77	87	98	112	131	162	229
0.65	0.10	0.050	34	52	59	67	77	90	107	135	196
0.65	0.20	0.100	21	35	41	48	56	67	82	107	162
0.70	0.01	0.005	54	71	78	85	94	105	120	144	194
0.70	0.02	0.010	44	60	66	73	81	91	105	127	175
0.70	0.05	0.025	31	45	50	56	64	73	85	105	148
0.70	0.10	0.050	22	34	38	44	50	58	69	88	127
0.70	0.20	0.100	14	23	27	31	37	44	53	69	105

Table 9.2

Table 9.2 Continued.

π_2	α 2-sided	1-sided	Power $1 - \beta$ 0.50	0.65	0.70	0.75	0.80	0.85	0.90	0.95	0.99
						$\pi_1 = 0.45$					
0.75	0.01	0.005	38	50	55	60	67	74	85	101	136
0.75	0.02	0.010	31	42	46	51	57	64	74	90	123
0.75	0.05	0.025	22	32	35	40	45	51	60	74	104
0.75	0.10	0.050	16	24	27	31	35	41	49	62	90
0.75	0.20	0.100	10	16	19	22	26	31	38	49	74
0.80	0.01	0.005	28	37	41	45	50	55	63	75	102
0.80	0.02	0.010	23	31	35	38	43	48	55	67	92
0.80	0.05	0.025	17	24	26	30	33	38	45	55	78
0.80	0.10	0.050	12	18	20	23	26	31	37	46	67
0.80	0.20	0.100	7	12	14	17	19	23	28	37	55
0.85	0.01	0.005	22	29	32	35	39	43	49	59	79
0.85	0.02	0.010	18	24	27	30	33	37	43	52	71
0.85	0.05	0.025	13	18	21	23	26	30	35	43	60
0.85	0.10	0.050	9	14	16	18	21	24	28	36	52
0.85	0.20	0.100	6	10	11	13	15	18	22	28	43
0.90	0.01	.005	18	23	26	28	31	35	39	47	63
0.90	0.02	.010	15	20	22	24	27	30	35	42	57
0.90	0.05	.025	11	15	17	19	21	24	28	34	49
0.90	0.10	.050	8	11	13	15	17	19	23	29	42
0.90	0.20	.100	5	8	9	11	12	15	18	23	35
0.95	0.01	.005	15	19	21	23	26	29	33	39	52
0.95	0.02	.010	12	16	18	20	22	25	29	34	47
0.95	0.05	0.025	9	12	14	15	17	20	23	29	40
0.95	0.10	0.050	6	9	11	12	14	16	19	24	34
0.95	0.20	0.100	4	6	8	9	10	12	15	19	29
						$\pi_1 = 0.50$					
0.55	0.01	0.005	1282	1694	1857	2041	2256	2520	2874	3441	4642
0.55	0.02	0.010	1046	1421	1570	1740	1939	2184	2514	3046	4181
0.55	0.05	0.025	742	1063	1192	1341	1516	1734	2030	2510	3549
0.55	0.10	0.050	523	796	909	1039	1195	1389	1654	2091	3046
0.55	0.20	0.100	318	537	630	739	871	1038	1269	1654	2514
0.60	0.01	0.005	322	425	466	512	566	633	721	864	1165
0.60	0.02	0.010	263	357	394	437	487	548	631	765	1049
0.60	0.05	0.025	187	267	300	337	381	436	510	630	891
0.60	0.10	0.050	132	200	229	261	300	349	415	525	765
0.60	0.20	0.100	80	135	159	186	219	261	319	415	631
0.65	0.01	0.005	144	190	208	229	253	282	322	385	519
0.65	0.02	0.010	117	159	176	195	217	245	282	341	468
0.65	0.05	0.025	83	119	134	150	170	194	227	281	397
0.65	0.10	0.050	59	89	102	117	134	156	185	234	341
0.65	0.20	0.100	36	60	71	83	98	117	142	185	282
0.70	0.01	0.005	81	107	117	129	143	159	182	217	293
0.70	0.02	0.010	66	90	99	110	123	138	159	192	264
0.70	0.05	0.025	47	67	76	85	96	110	128	159	224
0.70	0.10	0.050	33	51	58	66	76	88	105	132	192
0.70	0.20	0.100	20	34	40	47	55	66	80	105	159
0.75	0.01	0.005	52	69	75	83	92	102	117	139	188
0.75	0.02	0.010	43	58	64	71	79	89	102	124	169
0.75	0.05	0.025	30	43	49	55	62	71	82	102	144
0.75	0.10	0.050	22	33	37	42	49	57	67	85	124
0.75	0.20	0.100	13	22	26	30	36	42	52	67	102
0.80	0.01	0.005	37	48	53	58	64	71	81	97	131
0.80	0.02	0.010	30	40	45	49	55	62	71	86	118
0.80	0.05	0.025	21	30	34	38	43	49	58	71	100
0.80	0.10	0.050	15	23	26	30	34	40	47	59	86
0.80	0.20	0.100	9	16	18	21	25	30	36	47	71

Table 9.2

Table 9.2 Continued.

π_2	α 2-sided	1-sided	0.50	0.65	0.70	0.75	0.80	0.85	0.90	0.95	0.99
						Power $1 - \beta$					

$\pi_1 = 0.50$

π_2	2-sided	1-sided	0.50	0.65	0.70	0.75	0.80	0.85	0.90	0.95	0.99
0.85	0.01	0.005	27	36	39	43	47	53	60	72	97
0.85	0.02	0.010	22	30	33	37	41	46	53	64	87
0.85	0.05	0.025	16	23	25	28	32	36	43	52	74
0.85	0.10	0.050	11	17	19	22	25	29	35	44	64
0.85	0.20	0.100	7	12	14	16	19	22	27	35	53
0.90	0.01	0.005	21	27	30	33	36	41	46	55	74
0.90	0.02	0.010	17	23	25	28	31	35	41	49	67
0.90	0.05	0.025	12	17	19	22	25	28	33	40	57
0.90	0.10	0.050	9	13	15	17	20	23	27	34	49
0.90	0.20	0.100	6	9	11	12	14	17	21	27	41
0.95	0.01	0.005	17	22	24	26	29	32	37	44	59
0.95	0.02	0.010	14	18	20	23	25	28	32	39	53
0.95	0.05	0.025	10	14	16	17	20	22	26	32	45
0.95	0.10	0.050	7	11	12	14	16	18	21	27	39
0.95	0.20	0.100	5	7	8	10	12	14	17	21	32

$\pi_1 = 0.55$

π_2	2-sided	1-sided	0.50	0.65	0.70	0.75	0.80	0.85	0.90	0.95	0.99
0.60	0.01	0.005	1268	1675	1836	2018	2231	2493	2842	3403	4590
0.60	0.02	0.010	1034	1405	1553	1720	1917	2160	2487	3013	4135
0.60	0.05	0.025	734	1051	1179	1326	1500	1715	2007	2482	3510
0.60	0.10	0.050	517	788	899	1028	1181	1374	1636	2068	3013
0.60	0.20	0.100	314	531	623	731	861	1027	1255	1636	2487
0.65	0.01	0.005	315	416	456	501	554	619	706	845	1139
0.65	0.02	0.010	257	349	386	427	476	536	617	748	1026
0.65	0.05	0.025	183	261	293	329	372	426	498	616	871
0.65	0.10	0.050	129	196	224	255	294	341	406	513	748
0.65	0.20	0.100	78	132	155	182	214	255	312	406	617
0.70	0.01	0.005	139	184	201	221	244	273	311	373	502
0.70	0.02	0.010	114	154	170	189	210	237	272	330	453
0.70	0.05	0.025	81	115	129	145	164	188	220	272	384
0.70	0.10	0.050	57	87	99	113	130	151	179	227	330
0.70	0.20	0.100	35	59	69	80	95	113	138	179	272
0.75	0.01	0.005	78	103	112	124	137	152	174	208	280
0.75	0.02	0.010	64	86	95	105	117	132	152	184	253
0.75	0.05	0.025	45	65	72	81	92	105	123	152	214
0.75	0.10	0.050	32	48	55	63	72	84	100	127	184
0.75	0.20	0.100	20	33	38	45	53	63	77	100	152
0.80	0.01	0.005	50	65	71	79	87	97	110	132	178
0.80	0.02	0.010	40	55	61	67	75	84	97	117	160
0.80	0.05	0.025	29	41	46	52	58	67	78	96	136
0.80	0.10	0.050	20	31	35	40	46	54	64	80	117
0.80	0.20	0.100	13	21	25	29	34	40	49	64	97
0.85	0.01	0.005	34	45	49	54	60	67	76	91	123
0.85	0.02	0.010	28	38	42	46	52	58	67	81	111
0.85	0.05	0.025	20	28	32	36	40	46	54	67	94
0.85	0.10	0.050	14	21	24	28	32	37	44	56	81
0.85	0.20	0.100	9	15	17	20	23	28	34	44	67
0.90	0.01	0.005	25	33	36	40	44	49	56	67	90
0.90	0.02	0.010	21	28	31	34	38	42	49	59	81
0.90	0.05	0.025	15	21	23	26	30	34	39	49	69
0.90	0.10	0.050	11	16	18	20	23	27	32	41	59
0.90	0.20	0.100	7	11	13	15	17	20	25	32	49
0.95	0.01	0.005	19	25	28	30	33	37	42	51	68
0.95	0.02	0.010	16	21	23	26	29	32	37	45	62
0.95	0.05	0.025	11	16	18	20	23	26	30	37	52
0.95	0.10	0.050	8	12	14	16	18	21	25	31	45
0.95	0.20	0.100	5	8	10	11	13	16	19	25	37

Table 9.2

Table 9.2 Continued.

π_2	α 2-sided	1-sided	Power $1 - \beta$ 0.50	0.65	0.70	0.75	0.80	0.85	0.90	0.95	0.99
						$\pi_1 = 0.60$					
0.65	0.01	0.005	1225	1618	1774	1950	2155	2408	2746	3288	4435
0.65	0.02	0.010	999	1357	1500	1662	1852	2087	2402	2910	3995
0.65	0.05	0.025	709	1015	1139	1281	1449	1657	1939	2398	3391
0.65	0.10	0.050	500	761	869	993	1141	1327	1581	1997	2910
0.65	0.20	0.100	304	513	602	706	832	992	1213	1581	2402
0.70	0.01	0.005	301	397	435	478	529	591	674	806	1088
0.70	0.02	0.010	245	333	368	408	455	512	589	714	980
0.70	0.05	0.025	174	249	280	315	356	407	476	588	832
0.70	0.10	0.050	123	187	213	244	280	326	388	490	714
0.70	0.20	0.100	75	126	148	174	204	244	298	388	589
0.75	0.01	0.005	131	173	190	209	231	258	294	351	474
0.75	0.02	0.010	107	145	161	178	198	223	257	311	427
0.75	0.05	0.025	76	109	122	137	155	177	208	257	362
0.75	0.10	0.050	54	82	93	106	122	142	169	214	311
0.75	0.20	0.100	33	55	65	76	89	106	130	169	257
0.80	0.01	0.005	72	96	105	115	127	142	162	194	261
0.80	0.02	0.010	59	80	89	98	109	123	142	172	235
0.80	0.05	0.025	42	60	67	76	86	98	114	141	200
0.80	0.10	0.050	30	45	52	59	68	78	93	118	172
0.80	0.20	0.100	18	31	36	42	49	59	72	93	142
0.85	0.01	0.005	46	60	66	72	80	89	102	122	164
0.85	0.02	0.010	37	50	56	62	69	77	89	108	148
0.85	0.05	0.025	27	38	42	48	54	62	72	89	125
0.85	0.10	0.050	19	29	32	37	43	49	59	74	108
0.85	0.20	0.100	12	19	23	26	31	37	45	59	89
0.90	0.01	0.005	31	41	45	49	54	61	69	83	112
0.90	0.02	0.010	25	34	38	42	47	53	61	73	100
0.90	0.05	0.025	18	26	29	33	37	42	49	61	85
0.90	0.10	0.050	13	20	22	25	29	34	40	50	73
0.90	0.20	0.100	8	13	16	18	21	25	31	40	61
0.95	0.01	0.005	23	30	32	36	39	44	50	60	80
0.95	0.02	0.010	18	25	28	30	34	38	44	53	72
0.95	0.05	0.025	13	19	21	24	27	30	35	44	62
0.95	0.10	0.050	9	14	16	18	21	24	29	36	53
0.95	0.20	0.100	6	10	11	13	15	18	22	29	44
						$\pi_1 = 0.65$					
0.70	0.01	0.005	1153	1524	1670	1836	2029	2267	2585	3095	4175
0.70	0.02	0.010	941	1278	1412	1565	1744	1965	2262	2740	3761
0.70	0.05	0.025	668	956	1073	1206	1364	1560	1826	2258	3192
0.70	0.10	0.050	470	716	818	935	1074	1249	1488	1880	2740
0.70	0.20	0.100	286	483	567	665	784	934	1142	1488	2262
0.75	0.01	0.005	279	369	404	444	491	549	626	749	1010
0.75	0.02	0.010	228	309	342	379	422	476	547	663	910
0.75	0.05	0.025	162	232	260	292	330	378	442	546	772
0.75	0.10	0.050	114	174	198	226	260	303	360	455	663
0.75	0.20	0.100	69	117	138	161	190	226	276	360	547
0.80	0.01	0.005	120	159	174	191	211	236	269	322	434
0.80	0.02	0.010	98	133	147	163	181	204	235	285	391
0.80	0.05	0.025	70	100	112	126	142	162	190	235	332
0.80	0.10	0.050	49	75	85	98	112	130	155	196	285
0.80	0.20	0.100	30	51	59	69	82	97	119	155	235
0.85	0.01	0.005	65	86	95	104	115	128	146	175	236
0.85	0.02	0.010	53	72	80	89	99	111	128	155	212
0.85	0.05	0.025	38	54	61	68	77	88	103	128	180
0.85	0.10	0.050	27	41	47	53	61	71	84	106	155
0.85	0.20	0.100	17	28	32	38	45	53	65	84	128

Table 9.2

Table 9.2 Continued.

π_2	α 2-sided	1-sided	Power $1 - \beta$ 0.50	0.65	0.70	0.75	0.80	0.85	0.90	0.95	0.99
						$\pi_1 = 0.65$					
0.90	0.01	0.005	41	53	58	64	71	79	90	108	145
0.90	0.02	0.010	33	45	50	55	61	69	79	96	131
0.90	0.05	0.025	24	34	38	42	48	55	64	79	111
0.90	0.10	0.050	17	25	29	33	38	44	52	66	96
0.90	0.20	0.100	10	17	20	24	28	33	40	52	79
0.95	0.01	0.005	27	36	39	43	48	53	61	72	97
0.95	0.02	0.010	22	30	33	37	41	46	53	64	88
0.95	0.05	0.025	16	23	25	28	32	37	43	53	75
0.95	0.10	0.050	11	17	19	22	25	30	35	44	64
0.95	0.20	0.100	7	12	14	16	19	22	27	35	53
						$\pi_1 = 0.70$					
0.75	0.01	0.005	1053	1391	1525	1676	1853	2070	2360	2826	3812
0.75	0.02	0.010	859	1167	1289	1429	1592	1794	2065	2502	3434
0.75	0.05	0.025	610	873	979	1101	1245	1424	1667	2061	2914
0.75	0.10	0.050	430	654	747	854	981	1141	1359	1717	2502
0.75	0.20	0.100	261	441	518	607	715	853	1042	1359	2065
0.80	0.01	0.005	251	331	363	399	441	493	562	672	907
0.80	0.02	0.010	205	278	307	340	379	427	491	595	817
0.80	0.05	0.025	145	208	233	262	296	339	397	491	693
0.80	0.10	0.050	103	156	178	203	234	272	323	409	595
0.80	0.20	0.100	62	105	123	145	170	203	248	323	491
0.85	0.01	0.005	106	140	153	168	186	208	237	284	382
0.85	0.02	0.010	86	117	130	144	160	180	207	251	344
0.85	0.05	0.025	62	88	99	111	125	143	167	207	292
0.85	0.10	0.050	43	66	75	86	99	115	137	172	251
0.85	0.20	0.100	27	45	52	61	72	86	105	137	207
0.90	0.01	0.005	57	75	82	90	99	111	126	151	204
0.90	0.02	0.010	46	63	69	77	85	96	110	134	183
0.90	0.05	0.025	33	47	53	59	67	76	89	110	156
0.90	0.10	0.050	23	35	40	46	53	61	73	92	134
0.90	0.20	0.100	14	24	28	33	39	46	56	73	110
0.95	0.01	0.005	34	45	50	54	60	67	76	91	123
0.95	0.02	0.010	28	38	42	46	52	58	67	81	111
0.95	0.05	0.025	20	29	32	36	41	46	54	67	94
0.95	0.10	0.050	14	22	24	28	32	37	44	56	81
0.95	0.20	0.100	9	15	17	20	23	28	34	44	67
						$\pi_1 = 0.75$					
0.80	0.01	0.005	924	1221	1339	1471	1626	1817	2072	2481	3346
0.80	0.02	0.010	754	1024	1132	1254	1398	1575	1813	2196	3014
0.80	0.05	0.025	535	766	860	967	1093	1250	1463	1810	2558
0.80	0.10	0.050	377	574	656	749	861	1001	1193	1507	2196
0.80	0.20	0.100	229	387	455	533	628	749	915	1193	1813
0.85	0.01	0.005	215	284	311	342	378	423	482	577	778
0.85	0.02	0.010	176	238	263	292	325	366	422	511	701
0.85	0.05	0.025	125	178	200	225	254	291	340	421	595
0.85	0.10	0.050	88	134	153	174	200	233	277	351	511
0.85	0.20	0.100	54	90	106	124	146	174	213	277	422
0.90	0.01	0.005	89	117	128	141	156	174	198	237	320
0.90	0.02	0.010	72	98	108	120	134	151	173	210	288
0.90	0.05	0.025	52	74	82	93	105	120	140	173	244
0.90	0.10	0.050	36	55	63	72	83	96	114	144	210
0.90	0.20	0.100	22	37	44	51	60	72	88	114	173

Table 9.2

Table 9.2 Continued.

π_2	α 2-sided	1-sided	0.50	0.65	0.70	Power $1-\beta$ 0.75	0.80	0.85	0.90	0.95	0.99
						$\pi_1 = 0.75$					
0.95	0.01	0.005	46	61	66	73	81	90	102	123	165
0.95	0.02	0.010	38	51	56	62	69	78	90	109	149
0.95	0.05	0.025	27	38	43	48	54	62	73	90	126
0.95	0.10	0.050	19	29	33	37	43	50	59	75	109
0.95	0.20	0.100	12	20	23	27	31	37	46	59	90
						$\pi_1 = 0.80$					
0.85	0.01	0.005	768	1014	1112	1222	1351	1509	1721	2060	2779
0.85	0.02	0.010	626	851	940	1042	1161	1308	1506	1824	2503
0.85	0.05	0.025	445	636	714	803	908	1039	1215	1503	2125
0.85	0.10	0.050	313	477	545	622	715	832	991	1252	1824
0.85	0.20	0.100	190	322	378	443	522	622	760	991	1506
0.90	0.01	0.005	173	228	250	274	303	339	386	462	624
0.90	0.02	0.010	141	191	211	234	261	294	338	409	562
0.90	0.05	0.025	100	143	161	180	204	233	273	337	477
0.90	0.10	0.050	71	107	123	140	161	187	223	281	409
0.90	0.20	0.100	43	73	85	100	117	140	171	223	338
0.95	0.01	0.005	68	90	99	108	120	134	152	182	246
0.95	0.02	0.010	56	76	83	92	103	116	133	161	221
0.95	0.05	0.025	40	57	63	71	81	92	108	133	188
0.95	0.10	0.050	128	43	49	55	64	74	88	111	161
0.95	0.20	0.100	17	29	34	40	46	55	68	88	133
						$\pi_1 = 0.85$					
0.90	0.01	0.005	583	771	845	929	1027	1147	1308	1566	2112
0.90	0.02	0.010	476	647	714	792	882	994	1144	1386	1902
0.90	0.05	0.025	338	484	543	610	690	789	924	1142	1615
0.90	0.10	0.050	238	363	414	473	544	632	753	951	1386
0.90	0.20	0.100	145	245	287	337	397	473	578	753	1144
0.95	0.01	0.005	123	163	178	196	216	242	275	330	445
0.95	0.02	0.010	100	136	151	167	186	209	241	292	400
0.95	0.05	0.025	71	102	115	129	146	166	195	241	340
0.95	0.10	0.050	50	77	87	100	115	133	159	200	292
0.95	0.20	0.100	31	52	61	71	84	100	122	159	241
						$\pi_1 = 0.90$					
0.95	0.01	0.005	372	491	538	592	654	731	833	997	1345
0.95	0.02	0.010	303	412	455	504	562	633	729	883	1212
0.95	0.05	0.025	215	308	346	389	440	503	589	728	1029
0.95	0.10	0.050	152	231	264	302	347	403	480	606	883
0.95	0.20	0.100	92	156	183	215	253	301	368	480	729

Table 9.2

130

Table 9.3 Survival rate π_2 for given ratio of medians h and survival rate π_1.

h	π_1 0.05	0.10	0.15	0.20	0.25	0.30	0.35	0.40	0.45	0.50	0.55	0.60	0.65	0.70	0.75	0.80	0.85	0.90	0.95
1.1	0.066	0.123	0.178	0.232	0.284	0.335	0.385	0.435	0.484	0.533	0.581	0.629	0.676	0.723	0.770	0.816	0.863	0.909	0.954
1.2	0.082	0.147	0.206	0.262	0.315	0.367	0.417	0.466	0.514	0.561	0.608	0.653	0.698	0.743	0.787	0.830	0.873	0.916	0.958
1.3	0.100	0.170	0.232	0.290	0.344	0.396	0.446	0.494	0.541	0.587	0.631	0.675	0.718	0.760	0.801	0.842	0.882	0.922	0.961
1.4	0.118	0.193	0.258	0.317	0.371	0.423	0.472	0.520	0.565	0.610	0.652	0.694	0.735	0.775	0.814	0.853	0.890	0.928	0.964
1.5	0.136	0.215	0.282	0.342	0.397	0.448	0.497	0.543	0.587	0.630	0.671	0.711	0.750	0.788	0.825	0.862	0.897	0.932	0.966
1.6	0.154	0.237	0.306	0.366	0.420	0.471	0.519	0.564	0.607	0.648	0.688	0.727	0.764	0.800	0.835	0.870	0.903	0.936	0.968
1.7	0.172	0.258	0.328	0.388	0.442	0.493	0.539	0.583	0.625	0.665	0.704	0.740	0.776	0.811	0.844	0.877	0.909	0.940	0.970
1.8	0.189	0.278	0.349	0.409	0.463	0.512	0.558	0.601	0.642	0.680	0.717	0.753	0.787	0.820	0.852	0.883	0.914	0.943	0.972
1.9	0.207	0.298	0.368	0.429	0.482	0.531	0.575	0.617	0.657	0.694	0.730	0.764	0.797	0.829	0.859	0.889	0.918	0.946	0.973
2.0	0.224	0.316	0.387	0.447	0.500	0.548	0.592	0.632	0.671	0.707	0.742	0.775	0.806	0.837	0.866	0.894	0.922	0.949	0.975
2.1	0.240	0.334	0.405	0.465	0.517	0.564	0.607	0.646	0.684	0.719	0.752	0.784	0.815	0.844	0.872	0.899	0.926	0.951	0.976
2.2	0.256	0.351	0.422	0.481	0.533	0.579	0.621	0.659	0.696	0.730	0.762	0.793	0.822	0.850	0.877	0.904	0.929	0.953	0.977
2.3	0.272	0.367	0.438	0.497	0.547	0.592	0.634	0.671	0.707	0.740	0.771	0.801	0.829	0.856	0.882	0.908	0.932	0.955	0.978
2.4	0.287	0.383	0.454	0.511	0.561	0.606	0.646	0.683	0.717	0.749	0.780	0.808	0.836	0.862	0.887	0.911	0.935	0.957	0.979
2.5	0.302	0.398	0.468	0.525	0.574	0.618	0.657	0.693	0.727	0.758	0.787	0.815	0.842	0.867	0.891	0.915	0.937	0.959	0.980
2.6	0.316	0.412	0.482	0.538	0.587	0.629	0.668	0.703	0.736	0.766	0.795	0.822	0.847	0.872	0.895	0.918	0.939	0.960	0.980
2.7	0.330	0.426	0.495	0.551	0.598	0.640	0.678	0.712	0.744	0.774	0.801	0.828	0.853	0.876	0.899	0.921	0.942	0.962	0.981
2.8	0.343	0.439	0.508	0.563	0.610	0.651	0.687	0.721	0.752	0.781	0.808	0.833	0.857	0.880	0.902	0.923	0.944	0.963	0.982
2.9	0.356	0.452	0.520	0.574	0.620	0.660	0.696	0.729	0.759	0.787	0.814	0.838	0.862	0.884	0.906	0.926	0.946	0.964	0.982
3.0	0.368	0.464	0.531	0.585	0.630	0.669	0.705	0.737	0.766	0.794	0.819	0.843	0.866	0.888	0.909	0.928	0.947	0.965	0.983
3.5	0.425	0.518	0.582	0.631	0.673	0.709	0.741	0.770	0.796	0.820	0.843	0.864	0.884	0.903	0.921	0.938	0.955	0.970	0.985
4.0	0.473	0.562	0.622	0.669	0.707	0.740	0.769	0.795	0.819	0.841	0.861	0.880	0.898	0.915	0.931	0.946	0.960	0.974	0.987
4.5	0.514	0.599	0.656	0.699	0.735	0.765	0.792	0.816	0.837	0.857	0.876	0.893	0.909	0.924	0.938	0.952	0.965	0.977	0.989
5.0	0.549	0.631	0.684	0.725	0.758	0.786	0.811	0.833	0.852	0.871	0.887	0.903	0.917	0.931	0.944	0.956	0.968	0.979	0.990
10.0	0.741	0.794	0.827	0.851	0.871	0.887	0.900	0.912	0.923	0.933	0.942	0.950	0.958	0.965	0.972	0.978	0.984	0.990	0.995

Table 9.3

Chapter 10
Comparing Two Survival Curves (Exponential Survival)

10.1 INTRODUCTION

The logrank test used in the analysis of survival studies, and described in Chapter 9, is a non parametric test. In certain circumstances, however, it may be possible to postulate the underlying form of the survival distributions and make use of this knowledge. One form of the survival distribution which is often used for planning purposes is the exponential distribution. The exponential distribution has a hazard rate, the risk of failure per unit time, which is constant.

10.2 THEORY AND FORMULAE

Sample Size

The survivorship function for an exponential distribution is given by

$$S(t) = e^{-\lambda t}, t \geqslant 0 \tag{10.1}$$

where λ is the hazard function which does not change with time t.

It can be shown for an exponential distribution that the hazard λ and median survival time τ are related by

$$\tau = \log 2/\lambda. \tag{10.2}$$

Suppose that patients are recruited to a trial and are allocated at random to receive either one of two treatments. If we assume that survival time is exponentially distributed and that the median survival times are τ_1 and τ_2 ($\tau_1 > \tau_2$) and equal numbers of patients are to be entered on each treatment, then the number of events required in each patient group to give a test with significance α and power $1 - \beta$ is approximately

$$e = 2(z_{1-\alpha} + z_{1-\beta})^2/(\log \triangle)^2 \tag{10.3}$$

where $\triangle = \tau_2/\tau_1$ is the smallest ratio of medians that it is of interest to detect. Equation (10.3) can be easily evaluated by using Table 2.3 which gives $(z_{1-\alpha} + z_{1-\beta})^2$ for different values of α and β.

Power

In contrast if one were asked for the power of a trial for given observed number of events e, then equation (10.3) could be inverted to give

$$z_{1-\beta} = \sqrt{(\tfrac{1}{2} e)} \log \triangle - z_{1-\alpha}. \tag{10.4}$$

Duration of Study

One method of estimating the number of patients corresponding to the required number of events e, is to assume that the number of patients who enter the trial in a given time period can be regarded as a Poisson random variable with average entry rate per unit of time a. Patient entry is assumed to take place over a time period T at which time point patient recruitment is stopped, follow up is closed and the definitive analysis takes place. The required value for T is given as the solution to the following equation

$$\left(T - \frac{2e}{a}\right)\left(\frac{1}{\lambda_1} + \frac{1}{\lambda_2}\right) - \frac{\{1 - \exp(-\lambda_2 T)\}}{\lambda_1^2} - \frac{\{1 - \exp(-\lambda_2 T)\}}{\lambda_2^2} = 0 \tag{10.5}$$

where λ_1 and λ_2 are the corresponding hazard rates for the two exponential distributions.

To solve equation (10.5) it is necessary to specify both τ_1 and τ_2 and not just their ratio $\triangle$. However, if we assume that a, the entry rate per unit time, is equal to the number of patients recruited in a time interval of length τ_1, then equation (10.5) depends only on $\tau_2 = \triangle \tau_1$, the resulting T is then measured in time units which are multiples of τ_1. However, there is no explicit solution to equation (10.5) and the appropriate value of T is found using an iterative method. A lower limit to the solution for T is

$$T_L = 2e/a. \tag{10.6}$$

Thus T will be always at least as large as T_L. Having calculated T from (10.5) the number of subjects to be recruited to each treatment group is

$$n = aT/2. \tag{10.7}$$

It should be noted that patients arc recruited over the whole period T but there is no follow up beyond T for any patient.

Withdrawals

One aspect of a trial which can effect the number of patients recruited is the proportion of patients who are lost to follow-up during the trial. Such patients have censored observations as do those for whom the event of interest has not occurred at the end of the trial. If the anticipated withdrawal rate is $x\%$ then the required number of patients given in (10.7) should be modified to

$$n' = \frac{100n}{(100 - x)}. \tag{10.8}$$

Comparison With the Logrank Test

If we can assume cxponential survival distributions, then correspondingly fewer subjects are required than would be suggested by the non parametric logrank test.

10.3 BIBLIOGRAPHY

Theory, formulae and tables corresponding to equations (10.3) and (10.5) are given by George & Desu (1974). A useful discussion is given by Lee (1980) who describes the exponential distribution in detail. Nomograms for calculating the number of patients needed for a clinical trial with survival assumed exponential have been published by Schoenfeld & Richter (1982).

10.4 DESCRIPTION OF THE TABLES

Table 10.1

Table 10.1 gives the number of critical events, e, to be observed in one patient group, for two-sided test size α, ranging from 0.01 to 0.20, one-sided from 0.005 to 0.10 and power $1 - \beta$ from 0.50 to 0.99 and ratio of medians $\triangle$ from 1.05 to 4.25.

Table 10.2

Table 10.2 gives the duration of study, in pre-specified time units, in order to observe the required total number of events $2e$ on the assumption that the median survival of the control group is equal to the pre-specified time unit. The entry rate a over this period ranges from 20 to 200 and $\triangle$ from 1.1 to 4.0.

10.5 USE OF THE TABLES

Table 10.1

Example 10.1

An adjuvant study of the drug Levamisole is proposed for patients with resectable cancer of the colon (Duke's C) in which the primary objective of the study is to compare the efficacy of Levamisole against a placebo control with respect to relapse free survival. Assuming relapse times have an exponential distribution, how many relapses need to be observed in the trial if it is anticipated that the median relapse time is likely to be increased from 1 year to 1.35 years, and a power of 80% required?

In this example $\tau_1 = 1$, $\tau_2 = 1.35$ therefore $\triangle = 1.35$. Now with one-sided $\alpha = 0.05$, $1 - \beta = 0.8$ Table 10.1 gives $e = 137$, thus approximately $2e = 274$ relapses would need to be observed. It should be noted that this value is very close to that given in Example 9.1.

Example 10.2

Two drugs ampicillin and metronidazole are to be compared for their differing effects on the post-operative recovery of patients having an appendicectomy as measured by the duration of post-operative pyrexia. Previous studies (Chant, Turner & Machin, 1983) suggest the median duration of pyrexia with metronidazole to be approximately 80 hours. Duration of post-operative pyrexia is assumed to follow an exponential distribution. How many patients should be recruited to a trial to demonstrate a clinically worthwhile reduction in post-operative pyrexia of 10 hours at one-sided test size $\alpha = 0.01$ and power $1 - \beta = 0.90$?

It should be noted that such a trial does not require prolonged follow up of patients and the event of interest, i.e. return to normal temperature, will be observed in almost all patients. There will be very few patient losses as the follow up time is only a few days.

Here $\tau_1 = 70$, $\tau_2 = 80$ and $\triangle = 80/70 = 1.14$, with one-sided $\alpha = 0.01$, $1 - \beta = 0.90$. Table 10.1 gives for $\triangle = 1.15$, $e = 876$ and one would need to observe approximately 1750 events.

Table 10.2

Example 10.3

In the trial described in Example 10.1 it is anticipated that the recruitment rate will be approximately 80 patients per year. For what period should the trial be conducted and how many patients should be recruited?

With $\triangle = 1.35$, $a = 80$, one-sided $\alpha = 0.05$ and $1 - \beta = 0.80$. Table 10.2 gives the total duration of the trial as

10.5   134

$T = 5.1$ time units. Now the median relapse free survival is $\tau_1 = 1$ year and so recruitment time is 5.1 years. The total number of subjects required per treatment group is obtained from (10.7), as $n = 80 \times 5.1/2 = 204$, hence a total of approximately 400 patients would need to be recruited to the trial.

Example 10.4

The Multicenter Study Group (1980) describe a double blind controlled study of long-term oral acetylcysteine against placebo in chronic bronchitis. How many patients should be recruited to such a trial if it were possible to assume that the exacerbation-free times induced by treatment have exponential distributions. Approximately 200 patients could be recruited in each 13 week period which corresponds to the median exacerbation-free time, a doubling of this time with the active treatment is anticipated and the power is set at 90% and the one-sided test size at 5%.

Using Table 10.2 with $\triangle = 2$, $a = 200$, one-sided $\alpha = 0.05$ and $1 - \beta = 0.90$ gives $T = 1.4$ time units. Thus the trial duration is set at $13 \times 1.4 = 18.2$ weeks or approximately 4 months. The number of patients to be recruited to each treatment group in such a trial would be, from equation (10.7), 280 or a total of almost 600 patients in all.

Example 10.5

Suppose that the trial planned in Example 10.4 was prematurely closed after only observing a total of $2e = 50$ events. What is the power of this trial to detect a doubling of the median exacerbation-free time?

With $e = 25$, $\triangle = 2$, and from Table 2.2 $z_{1-\alpha} = 1.6445$ equation (10.4) gives $z_{1-\beta} = 0.8061$. From Table 2.1 the power $1-\beta = 0.79$

10.6 REFERENCES

Chant A.D.B., Turner D.T.L. & Machin D. (1983). Metronidazole v Ampicillin : differing effects on the post-operative recovery. *Ann. Roy. Coll. Surg. Engl.* **66,** 96–97.

George S. L. & Desu M. M. (1974). Planning the size and duration of a clinical trial studying the time to some critical event. *J. Chron. Dis,* **27,** 15-24.

Lee E. T. (1980) *Statistical Methods for Survival Data Analysis.* Lifetime Learning Publications, Belmont, California.

Multicenter Study Group (1980) Long-term oral acetycysteine in chronic bronchitis. A double-blind controlled study. *Europ. J. Resp. Dis,* **61,** Supp. 111, 93–108.

Schoenfeld D. A. & Richter J. R. (1982). Nomograms for calculating the number of patients needed for a clinical trial with survival as an endpoint. *Biometrics,* **38,** 163-170.

Table 10.1 Sample sizes for comparison of two exponential distributions. $\triangle$ is the ratio of the two median survivals.

$\triangle$	α 2-sided	1-sided	Power $1 - \beta$ 0.50	0.65	0.70	0.75	0.80	0.85	0.90	0.95	0.99
1.05	0.01	0.005	5574	7366	8075	8876	9812	10962	12501	14966	20190
1.05	0.02	0.010	4546	6177	6827	7565	8431	9500	10936	13249	18187
1.05	0.05	0.025	3227	4621	5185	5831	6594	7543	8827	10917	15435
1.05	0.10	0.050	2273	3462	3953	4519	5194	6040	7195	9092	13249
1.05	0.20	0.100	1379	2334	2740	3214	3787	4514	5519	7195	10936
1.10	0.01	0.005	1460	1930	2116	2325	2571	2872	3275	3922	5290
1.10	0.02	0.010	1191	1618	1789	1982	2209	2489	2865	3472	4766
1.10	0.05	0.025	845	1210	1358	1528	1728	1976	2313	2861	4045
1.10	0.10	0.050	595	907	1036	1184	1361	1582	1885	2382	3472
1.10	0.20	0.100	361	611	718	842	992	1182	1446	1885	2865
1.15	0.01	0.005	679	897	984	1081	1195	1336	1523	1823	2460
1.15	0.02	0.010	554	752	832	922	1027	1157	1332	1614	2216
1.15	0.05	0.025	393	563	631	710	803	919	1075	1330	1881
1.15	0.10	0.050	277	422	481	550	633	736	876	1108	1614
1.15	0.20	0.100	168	284	333	391	461	550	672	876	1332
1.20	0.01	0.005	399	527	578	635	702	785	895	1071	1445
1.20	0.02	0.010	325	442	488	541	603	680	783	948	1302
1.20	0.05	0.025	231	330	371	417	472	540	632	781	1105
1.20	0.10	0.050	162	247	283	323	371	432	515	651	948
1.20	0.20	0.100	98	167	196	230	271	323	395	515	783
1.25	0.01	0.005	266	352	386	424	469	524	597	715	965
1.25	0.02	0.010	217	295	326	361	403	454	522	633	869
1.25	0.05	0.025	154	220	247	278	315	360	422	521	737
1.25	0.10	0.050	108	165	189	216	248	288	343	434	633
1.25	0.20	0.100	65	111	131	153	181	215	263	343	522
1.30	0.01	0.005	192	254	279	306	339	379	432	517	698
1.30	0.02	0.010	157	213	236	261	291	328	378	458	628
1.30	0.05	0.025	111	159	179	201	228	260	305	377	533
1.30	0.10	0.050	78	119	136	156	179	208	248	314	458
1.30	0.20	0.100	47	80	94	111	130	156	190	248	378
1.35	0.01	0.005	147	194	213	234	259	289	330	395	533
1.35	0.02	0.010	120	163	180	199	222	251	289	350	480
1.35	0.05	0.025	85	122	137	154	174	199	233	288	407
1.35	0.10	0.050	60	91	104	119	137	159	190	240	350
1.35	0.20	0.100	36	61	72	84	100	119	145	190	289
1.40	0.01	0.005	117	154	169	186	206	230	262	314	424
1.40	0.02	0.010	95	129	143	159	177	199	229	278	382
1.40	0.05	0.025	67	97	109	122	138	158	185	229	324
1.40	0.10	0.050	47	72	83	95	109	127	151	191	278
1.40	0.20	0.100	29	49	57	67	79	94	116	151	229
1.45	0.01	0.005	96	127	139	153	169	189	215	258	348
1.45	0.02	0.010	78	106	117	130	145	163	188	228	313
1.45	0.05	0.025	55	79	89	100	113	130	152	188	266
1.45	0.10	0.050	39	59	68	77	89	104	124	156	228
1.45	0.20	0.100	23	40	47	55	65	77	95	124	188
1.50	0.01	0.005	80	106	116	128	142	158	181	216	292
1.50	0.02	0.010	65	89	98	109	122	137	158	191	263
1.50	0.05	0.025	46	66	75	84	95	109	127	158	223
1.50	0.10	0.050	32	50	57	65	75	87	104	131	191
1.50	0.20	0.100	19	33	39	46	54	65	79	104	158
1.55	0.01	0.005	69	91	100	110	121	135	154	185	250
1.55	0.02	0.010	56	76	84	93	104	117	135	164	225
1.55	0.05	0.025	40	57	64	72	81	93	109	135	191
1.55	0.10	0.050	28	42	49	56	64	74	89	112	164
1.55	0.20	0.100	17	28	33	39	46	55	68	89	135

Table 10.1

Table 10.1 Continued.

△	α 2-sided	1-sided	0.50	0.65	0.70	0.75	0.80	0.85	0.90	0.95	0.99
							Power $1 - \beta$				
1.60	0.01	0.005	60	79	87	95	105	118	134	161	217
1.60	0.02	0.010	48	66	73	81	90	102	117	142	195
1.60	0.05	0.025	34	49	55	62	71	81	95	117	166
1.60	0.10	0.050	24	37	42	48	55	65	77	97	142
1.60	0.20	0.100	14	25	29	34	40	48	59	77	117
1.65	0.01	0.005	52	69	76	84	93	104	118	142	191
1.65	0.02	0.010	43	58	64	71	80	90	103	125	172
1.65	0.05	0.025	30	43	49	55	62	71	83	103	146
1.65	0.10	0.050	21	32	37	42	49	57	68	86	125
1.65	0.20	0.100	13	22	26	30	35	42	52	68	103
1.70	0.01	0.005	47	62	68	75	82	92	105	126	170
1.70	0.02	0.010	38	52	57	63	71	80	92	112	153
1.70	0.05	0.025	27	39	43	49	55	63	74	92	130
1.70	0.10	0.050	19	29	33	38	43	51	60	76	112
1.70	0.20	0.100	11	19	23	27	32	38	46	60	92
1.75	0.01	0.005	42	55	61	67	74	83	95	113	153
1.75	0.02	0.010	34	46	51	57	64	72	83	100	138
1.75	0.05	0.025	24	35	39	44	50	57	67	82	117
1.75	0.10	0.050	17	26	30	34	39	45	54	69	100
1.75	0.20	0.100	10	17	20	24	28	34	41	54	83
1.80	0.01	0.005	38	50	55	61	67	75	86	103	139
1.80	0.02	0.010	31	42	47	52	58	65	75	91	125
1.80	0.05	0.025	22	31	35	40	45	51	60	75	106
1.80	0.10	0.050	15	23	27	31	35	41	49	62	91
1.80	0.20	0.100	9	16	18	22	26	31	38	49	75
1.85	0.01	0.005	35	46	50	55	61	68	78	94	126
1.85	0.02	0.010	28	38	42	47	53	59	68	83	114
1.85	0.05	0.025	20	29	32	36	41	47	55	68	97
1.85	0.10	0.050	14	21	24	28	32	37	45	57	83
1.85	0.20	0.100	8	14	17	20	23	28	34	45	68
1.90	0.01	0.005	32	42	46	51	56	63	72	86	116
1.90	0.02	0.010	26	35	39	43	48	54	63	76	105
1.90	0.05	0.025	18	26	29	33	38	43	51	63	89
1.90	0.10	0.050	13	20	22	26	30	34	41	52	76
1.90	0.20	0.100	7	13	15	18	21	26	31	41	63
1.95	0.01	0.005	29	39	43	47	52	58	66	79	107
1.95	0.02	0.010	24	32	36	40	45	50	58	70	97
1.95	0.05	0.025	17	24	27	31	35	40	47	58	82
1.95	0.10	0.050	12	18	21	24	27	32	38	48	70
1.95	0.20	0.100	7	12	14	17	20	24	29	38	58
2.00	0.01	0.005	27	36	40	43	48	54	61	74	100
2.00	0.02	0.010	22	30	33	37	41	47	54	65	90
2.00	0.05	0.025	15	22	25	28	32	37	43	54	76
2.00	0.10	0.050	11	17	19	22	25	29	35	45	65
2.00	0.20	0.100	6	11	13	15	18	22	27	35	54
2.25	0.01	0.005	20	26	29	32	35	39	45	54	73
2.25	0.02	0.010	16	22	24	27	30	34	39	47	65
2.25	0.05	0.025	11	16	18	21	23	27	31	39	55
2.25	0.10	0.050	8	12	14	16	18	21	26	32	47
2.25	0.20	0.100	4	8	9	11	13	16	19	26	39
2.50	0.01	0.005	15	20	22	25	27	31	35	42	57
2.50	0.02	0.010	12	17	19	21	23	26	31	37	51
2.50	0.05	0.025	9	13	14	16	18	21	25	30	43
2.50	0.10	0.050	6	9	11	12	14	17	20	25	37
2.50	0.20	0.100	3	6	7	9	10	12	15	20	31

Table 10.1

Table 10.1 Continued.

△	α 2-sided	1-sided	Power $1 - \beta$ 0.50	0.65	0.70	0.75	0.80	0.85	0.90	0.95	0.99
2.75	0.01	0.005	12	17	18	20	22	25	29	34	46
2.75	0.02	0.010	10	14	15	17	19	22	25	30	42
2.75	0.05	0.025	7	10	12	13	15	17	20	25	35
2.75	0.10	0.050	5	8	9	10	12	14	16	21	30
2.75	0.20	0.100	3	5	6	7	8	10	12	16	25
3.00	0.01	0.005	10	14	15	17	19	21	24	29	39
3.00	0.02	0.010	8	12	13	14	16	18	21	26	35
3.00	0.05	0.025	6	9	10	11	13	14	17	21	30
3.00	0.10	0.050	4	6	7	8	10	11	14	17	26
3.00	0.20	0.100	2	4	5	6	7	8	10	14	21
3.25	0.01	0.005	9	12	13	15	16	18	21	25	34
3.25	0.02	0.010	7	10	11	12	14	16	18	22	31
3.25	0.05	0.025	5	7	8	9	11	12	15	18	26
3.25	0.10	0.050	3	5	6	7	8	10	12	15	22
3.25	0.20	0.100	2	4	4	5	6	7	9	12	18
3.50	0.01	0.005	8	11	12	13	14	16	18	22	30
3.50	0.02	0.010	6	9	10	11	12	14	16	20	27
3.50	0.05	0.025	4	7	7	8	10	11	13	16	23
3.50	0.10	0.050	3	5	5	6	7	9	10	13	20
3.50	0.20	0.100	2	3	4	4	5	6	8	10	16
3.75	0.01	0.005	7	10	11	12	13	14	17	20	27
3.75	0.02	0.010	6	8	9	10	11	12	14	18	24
3.75	0.05	0.025	4	6	7	7	8	10	12	14	21
3.75	0.10	0.050	3	4	5	6	7	8	9	12	18
3.75	0.20	0.100	1	3	3	4	5	6	7	9	14
4.00	0.01	0.005	6	9	10	10	12	13	15	18	25
4.00	0.02	0.010	5	7	8	9	10	11	13	16	22
4.00	0.05	0.025	3	5	6	7	8	9	10	13	19
4.00	0.10	0.050	2	4	4	5	6	7	8	11	16
4.00	0.20	0.100	1	2	3	3	4	5	6	8	13
4.25	0.01	0.005	6	8	9	10	11	12	14	17	22
4.25	0.02	0.010	5	7	7	8	9	10	12	15	20
4.25	0.05	0.025	3	5	5	6	7	8	10	12	17
4.25	0.10	0.050	2	3	4	5	5	6	8	10	15
4.25	0.20	0.100	1	2	3	3	4	5	6	8	12

Table 10.1　　　　138

Table 10.2 Duration of study for comparison of two exponential distributions. $\triangle$ is the ratio of the two median survivals: the median of the control group is one time unit.

$\triangle$	α 2-sided	1-sided	Power $1-\beta$ 0.50	0.65	0.70	0.75	0.80	0.85	0.90	0.95	0.99
						**20 subjects entered per unit time**					
1.05	0.01	0.005	557.4	736.7	807.5	887.6	981.2	1096.3	1250.1	1496.7	2019.0
1.05	0.02	0.010	454.7	617.8	682.8	756.6	843.2	950.1	1093.6	1325.0	1818.8
1.05	0.05	0.025	322.7	462.1	518.6	583.1	659.4	754.3	882.8	1091.8	1543.6
1.05	0.10	0.050	228.8	346.3	395.4	452.0	519.4	604.0	719.5	909.2	1325.0
1.05	0.20	0.100	139.5	234.9	274.0	321.5	378.7	451.4	551.9	719.5	1093.6
1.10	0.01	0.005	147.6	194.6	213.1	234.1	258.7	287.3	327.6	392.2	529.1
1.10	0.02	0.010	120.7	163.4	180.4	199.8	222.5	250.5	286.6	347.2	476.6
1.10	0.05	0.025	86.1	122.6	137.4	154.3	174.3	199.2	232.9	286.1	404.5
1.10	0.10	0.050	61.1	92.3	105.1	120.0	137.6	159.8	190.1	239.8	347.2
1.10	0.20	0.100	37.7	62.7	73.3	85.8	100.8	119.8	146.2	190.1	286.6
1.15	0.01	0.005	69.5	91.3	100.0	109.7	121.1	135.2	153.9	184.0	247.6
1.15	0.02	0.010	57.0	76.8	84.8	93.8	104.3	117.3	134.8	163.0	223.2
1.15	0.05	0.025	40.9	57.9	64.8	72.6	81.9	93.5	109.1	134.6	189.7
1.15	0.10	0.050	29.3	43.8	49.7	56.6	64.9	75.2	89.2	112.4	163.0
1.15	0.20	0.100	18.4	30.0	35.0	40.7	47.7	56.6	68.8	89.2	134.8
1.20	0.01	0.005	41.5	54.4	59.4	65.2	71.9	80.1	91.1	108.8	146.2
1.20	0.02	0.010	34.2	45.8	50.5	55.8	62.0	69.6	79.9	96.5	131.8
1.20	0.05	0.025	24.7	34.7	38.7	43.4	48.8	55.6	64.8	79.8	112.1
1.20	0.10	0.050	17.9	26.4	29.9	34.0	38.8	44.9	53.1	66.7	96.5
1.20	0.20	0.100	11.5	18.3	21.2	24.6	28.7	33.9	41.1	53.1	79.9
1.25	0.01	0.005	28.3	36.9	40.2	44.1	48.6	54.1	61.4	73.2	98.2
1.25	0.02	0.010	23.4	31.2	34.3	37.8	42.0	47.1	53.9	65.0	88.6
1.25	0.05	0.025	17.1	23.7	26.4	29.5	33.2	37.7	43.8	53.8	75.4
1.25	0.10	0.050	12.5	18.2	20.5	23.2	26.5	30.5	36.0	45.1	65.0
1.25	0.20	0.100	8.2	12.8	14.7	17.0	19.7	23.2	28.0	36.0	53.9
1.30	0.01	0.005	21.0	27.2	29.6	32.4	35.6	39.6	44.9	53.4	71.5
1.30	0.02	0.010	17.4	23.1	25.3	27.9	30.8	34.5	39.5	47.5	64.6
1.30	0.05	0.025	12.8	17.7	19.6	21.9	24.5	27.8	32.2	39.4	55.1
1.30	0.10	0.050	9.5	13.7	15.4	17.3	19.7	22.6	26.6	33.1	47.5
1.30	0.20	0.100	6.4	9.8	11.2	12.8	14.8	17.3	20.8	26.6	39.5
1.35	0.01	0.005	16.5	21.2	23.1	25.2	27.7	30.7	34.8	41.3	55.1
1.35	0.02	0.010	13.8	18.1	19.8	21.7	24.0	26.8	30.6	36.8	49.8
1.35	0.05	0.025	10.3	13.9	15.4	17.1	19.2	21.7	25.1	30.6	42.5
1.35	0.10	0.050	7.7	10.9	12.2	13.7	15.5	17.7	20.8	25.8	36.8
1.35	0.20	0.100	5.3	7.9	9.0	10.2	11.7	13.7	16.3	20.8	30.6
1.40	0.01	0.005	13.5	17.3	18.8	20.4	22.4	24.8	28.1	33.2	44.2
1.40	0.02	0.010	11.3	14.8	16.1	17.7	19.5	21.8	24.8	29.6	40.0
1.40	0.05	0.025	8.5	11.5	12.7	14.0	15.6	17.6	20.3	24.7	34.2
1.40	0.10	0.050	6.5	9.0	10.1	11.3	12.7	14.5	16.9	20.9	29.6
1.40	0.20	0.100	4.5	6.6	7.5	8.5	9.7	11.3	13.4	16.9	24.8
1.45	0.01	0.005	11.4	14.5	15.8	17.1	18.7	20.7	23.4	27.6	36.6
1.45	0.02	0.010	9.7	12.5	13.6	14.9	16.4	18.2	20.7	24.7	33.2
1.45	0.05	0.025	7.4	9.8	10.8	11.9	13.2	14.8	17.0	20.7	28.4
1.45	0.10	0.050	5.7	7.8	8.6	9.6	10.8	12.2	14.2	17.5	24.7
1.45	0.20	0.100	4.0	5.8	6.5	7.3	8.3	9.6	11.3	14.2	20.7
1.50	0.01	0.005	9.9	12.5	13.6	14.7	16.1	17.7	20.0	23.5	31.1
1.50	0.02	0.010	8.4	10.8	11.8	12.8	14.1	15.6	17.7	21.1	28.2
1.50	0.05	0.025	6.5	8.5	9.4	10.3	11.4	12.8	14.7	17.7	24.2
1.50	0.10	0.050	5.0	6.8	7.6	8.4	9.4	10.6	12.3	15.0	21.1
1.50	0.20	0.100	3.6	5.1	5.7	6.5	7.3	8.4	9.9	12.3	17.7
1.55	0.01	0.005	8.8	11.1	11.9	12.9	14.1	15.5	17.4	20.5	26.9
1.55	0.02	0.010	7.5	9.6	10.4	11.3	12.4	13.7	15.5	18.3	24.5
1.55	0.05	0.025	5.8	7.6	8.3	9.1	10.1	11.3	12.9	15.5	21.1
1.55	0.10	0.050	4.5	6.1	6.8	7.5	8.3	9.4	10.8	13.2	18.3
1.55	0.20	0.100	3.3	4.6	5.2	5.8	6.5	7.5	8.7	10.8	15.5

Table 10.2

Table 10.2 Continued.

△	α 2-sided	α 1-sided	Power 1−β 0.50	0.65	0.70	0.75	0.80	0.85	0.90	0.95	0.99
					20 subjects entered per unit time						
1.60	0.01	0.005	7.9	9.9	10.7	11.5	12.5	13.8	15.4	18.1	23.7
1.60	0.02	0.010	6.8	8.6	9.3	10.1	11.1	12.2	13.8	16.3	21.6
1.60	0.05	0.025	5.3	6.9	7.5	8.2	9.1	10.1	11.5	13.7	18.6
1.60	0.10	0.050	4.2	5.6	6.1	6.8	7.5	8.4	9.7	11.8	16.3
1.60	0.20	0.100	3.0	4.2	4.7	5.3	5.9	6.8	7.9	9.7	13.8
1.65	0.01	0.005	7.2	9.0	9.7	10.4	11.3	12.4	13.9	16.2	21.2
1.65	0.02	0.010	6.2	7.8	8.5	9.2	10.0	11.0	12.4	14.6	19.3
1.65	0.05	0.025	4.9	6.3	6.9	7.5	8.2	9.2	10.4	12.4	16.7
1.65	0.10	0.050	3.9	5.1	5.6	6.2	6.9	7.7	8.8	10.6	14.6
1.65	0.20	0.100	2.8	3.9	4.4	4.9	5.5	6.2	7.2	8.8	12.4
1.70	0.01	0.005	6.7	8.3	8.9	9.6	10.4	11.3	12.6	14.7	19.1
1.70	0.02	0.010	5.8	7.2	7.8	8.4	9.2	10.1	11.3	13.3	17.5
1.70	0.05	0.025	4.5	5.8	6.3	6.9	7.6	8.4	9.5	11.3	15.1
1.70	0.10	0.050	3.6	4.8	5.2	5.7	6.3	7.1	8.1	9.7	13.3
1.70	0.20	0.100	2.6	3.7	4.1	4.5	5.1	5.7	6.6	8.1	11.3
1.75	0.01	0.005	6.2	7.7	8.2	8.8	9.6	10.4	11.6	13.5	17.5
1.75	0.02	0.010	5.4	6.7	7.2	7.8	8.5	9.3	10.4	12.2	16.0
1.75	0.05	0.025	4.3	5.4	5.9	6.4	7.0	7.8	8.8	10.4	13.9
1.75	0.10	0.050	3.4	4.5	4.9	5.4	5.9	6.6	7.5	9.0	12.2
1.75	0.20	0.100	2.5	3.5	3.8	4.3	4.7	5.4	6.2	7.5	10.4
1.80	0.01	0.005	5.8	7.2	7.7	8.2	8.9	9.7	10.8	12.5	16.1
1.80	0.02	0.010	5.1	6.3	6.8	7.3	7.9	8.7	9.7	11.3	14.7
1.80	0.05	0.025	4.0	5.1	5.5	6.0	6.6	7.3	8.2	9.7	12.8
1.80	0.10	0.050	3.2	4.2	4.6	5.0	5.6	6.2	7.0	8.4	11.3
1.80	0.20	0.100	2.4	3.3	3.6	4.0	4.5	5.0	5.8	7.0	9.7
1.85	0.01	0.005	5.5	6.7	7.2	7.7	8.3	9.1	10.1	11.6	14.9
1.85	0.02	0.010	4.8	5.9	6.4	6.9	7.4	8.1	9.1	10.5	13.7
1.85	0.05	0.025	3.8	4.8	5.2	5.7	6.2	6.8	7.7	9.0	11.9
1.85	0.10	0.050	3.1	4.0	4.4	4.8	5.3	5.8	6.6	7.9	10.5
1.85	0.20	0.100	2.3	3.1	3.4	3.8	4.2	4.8	5.5	6.6	9.1
1.90	0.01	0.005	5.2	6.4	6.8	7.3	7.9	8.5	9.5	10.9	14.0
1.90	0.02	0.010	4.6	5.6	6.0	6.5	7.0	7.7	8.5	9.9	12.8
1.90	0.05	0.025	3.6	4.6	5.0	5.4	5.9	6.5	7.3	8.5	11.2
1.90	0.10	0.050	2.9	3.8	4.2	4.5	5.0	5.5	6.3	7.4	9.9
1.90	0.20	0.100	2.2	3.0	3.3	3.6	4.0	4.5	5.2	6.3	8.5
1.95	0.01	0.005	5.0	6.1	6.5	6.9	7.5	8.1	8.9	10.3	13.1
1.95	0.02	0.010	4.4	5.4	5.7	6.2	6.7	7.3	8.1	9.4	12.0
1.95	0.05	0.025	3.5	4.4	4.8	5.2	5.6	6.2	6.9	8.1	10.5
1.95	0.10	0.050	2.8	3.7	4.0	4.3	4.8	5.3	6.0	7.0	9.4
1.95	0.20	0.100	2.1	2.9	3.1	3.5	3.9	4.3	5.0	6.0	8.1
2.00	0.01	0.005	4.8	5.8	6.2	6.6	7.1	7.7	8.5	9.8	12.4
2.00	0.02	0.010	4.2	5.1	5.5	5.9	6.4	6.9	7.7	8.9	11.4
2.00	0.05	0.025	3.4	4.2	4.6	4.9	5.4	5.9	6.6	7.7	10.0
2.00	0.10	0.050	2.7	3.5	3.8	4.2	4.6	5.0	5.7	6.7	8.9
2.00	0.20	0.100	2.0	2.7	3.0	3.3	3.7	4.2	4.8	5.7	7.7
2.25	0.01	0.005	4.0	4.8	5.1	5.5	5.9	6.3	6.9	7.9	9.9
2.25	0.02	0.010	3.5	4.3	4.6	4.9	5.3	5.7	6.3	7.2	9.1
2.25	0.05	0.025	2.9	3.6	3.9	4.2	4.5	4.9	5.5	6.3	8.1
2.25	0.10	0.050	2.3	3.0	3.2	3.5	3.9	4.3	4.8	5.6	7.2
2.25	0.20	0.100	1.7	2.4	2.6	2.9	3.2	3.5	4.0	4.8	6.3
2.50	0.01	0.005	3.6	4.3	4.5	4.8	5.1	5.5	6.0	6.8	8.5
2.50	0.02	0.010	3.2	3.8	4.1	4.3	4.7	5.0	5.5	6.3	7.9
2.50	0.05	0.025	2.6	3.2	3.4	3.7	4.0	4.3	4.8	5.5	7.0
2.50	0.10	0.050	2.1	2.7	2.9	3.1	3.4	3.8	4.2	4.9	6.3
2.50	0.20	0.100	1.6	2.1	2.3	2.6	2.8	3.1	3.6	4.2	5.5

Table 10.2

Table 10.2 Continued.

$\triangle$	α 2-sided	1-sided	Power $1-\beta$ 0.50	0.65	0.70	0.75	0.80	0.85	0.90	0.95	0.99

$\triangle$	2-sided	1-sided	0.50	0.65	0.70	0.75	0.80	0.85	0.90	0.95	0.99

20 subjects entered per unit time

$\triangle$	2-sided	1-sided	0.50	0.65	0.70	0.75	0.80	0.85	0.90	0.95	0.99
2.75	0.01	0.005	3.3	3.9	4.1	4.4	4.7	5.0	5.5	6.2	7.6
2.75	0.02	0.010	2.9	3.5	3.7	4.0	4.2	4.6	5.0	5.7	7.0
2.75	0.05	0.025	2.4	2.9	3.1	3.4	3.6	3.9	4.4	5.0	6.3
2.75	0.10	0.050	1.9	2.5	2.7	2.9	3.1	3.4	3.8	4.4	5.7
2.75	0.20	0.100	1.5	2.0	2.2	2.4	2.6	2.9	3.3	3.8	5.0
3.00	0.01	0.005	3.1	3.6	3.8	4.1	4.3	4.6	5.1	5.7	6.9
3.00	0.02	0.010	2.7	3.3	3.5	3.7	3.9	4.2	4.6	5.2	6.5
3.00	0.05	0.025	2.2	2.7	2.9	3.1	3.4	3.7	4.1	4.6	5.8
3.00	0.10	0.050	1.8	2.3	2.5	2.7	2.9	3.2	3.6	4.1	5.2
3.00	0.20	0.100	1.4	1.9	2.0	2.2	2.4	2.7	3.0	3.6	4.6
3.25	0.01	0.005	2.9	3.4	3.6	3.8	4.1	4.4	4.8	5.3	6.5
3.25	0.02	0.010	2.6	3.1	3.3	3.5	3.7	4.0	4.4	4.9	6.1
3.25	0.05	0.025	2.1	2.6	2.8	3.0	3.2	3.5	3.8	4.4	5.4
3.25	0.10	0.050	1.7	2.2	2.4	2.6	2.8	3.0	3.4	3.9	4.9
3.25	0.20	0.100	1.3	1.8	1.9	2.1	2.3	2.6	2.9	3.4	4.4
3.50	0.01	0.005	2.8	3.3	3.5	3.7	3.9	4.2	4.5	5.1	6.1
3.50	0.02	0.010	2.5	3.0	3.1	3.3	3.6	3.8	4.2	4.7	5.7
3.50	0.05	0.025	2.0	2.5	2.7	2.9	3.1	3.3	3.7	4.2	5.2
3.50	0.10	0.050	1.7	2.1	2.3	2.5	2.7	2.9	3.2	3.7	4.7
3.50	0.20	0.100	1.3	1.7	1.9	2.0	2.2	2.5	2.8	3.2	4.2
3.75	0.01	0.005	2.7	3.2	3.3	3.5	3.8	4.0	4.4	4.9	5.9
3.75	0.02	0.010	2.4	2.9	3.0	3.2	3.4	3.7	4.0	4.5	5.5
3.75	0.05	0.025	2.0	2.4	2.6	2.8	3.0	3.2	3.5	4.0	5.0
3.75	0.10	0.050	1.6	2.0	2.2	2.4	2.6	2.8	3.1	3.6	4.5
3.75	0.20	0.100	1.2	1.6	1.8	2.0	2.2	2.4	2.7	3.1	4.0
4.00	0.01	0.005	2.6	3.1	3.2	3.4	3.6	3.9	4.2	4.7	5.7
4.00	0.02	0.010	2.3	2.8	2.9	3.1	3.3	3.6	3.9	4.4	5.3
4.00	0.05	0.025	1.9	2.4	2.5	2.7	2.9	3.1	3.4	3.9	4.8
4.00	0.10	0.050	1.6	2.0	2.1	2.3	2.5	2.7	3.0	3.5	4.4
4.00	0.20	0.100	1.2	1.6	1.7	1.9	2.1	2.3	2.6	3.0	3.9
4.25	0.01	0.005	2.6	3.0	3.2	3.3	3.5	3.8	4.1	4.6	5.5
4.25	0.02	0.010	2.3	2.7	2.9	3.0	3.2	3.5	3.8	4.2	5.2
4.25	0.05	0.025	1.9	2.3	2.4	2.6	2.8	3.0	3.3	3.8	4.7
4.25	0.10	0.050	1.5	1.9	2.1	2.3	2.5	2.7	3.0	3.4	4.2
4.25	0.20	0.100	1.2	1.6	1.7	1.9	2.0	2.3	2.5	3.0	3.8

30 subjects entered per unit time

$\triangle$	2-sided	1-sided	0.50	0.65	0.70	0.75	0.80	0.85	0.90	0.95	0.99
1.05	0.01	0.005	371.6	491.1	538.3	591.7	654.2	730.9	833.4	997.8	1346.0
1.05	0.02	0.010	303.1	411.9	455.2	504.4	562.1	633.4	729.1	883.3	1212.5
1.05	0.05	0.025	216.6	308.1	345.7	388.7	439.6	502.9	588.5	727.8	1029.1
1.05	0.10	0.050	153.0	232.3	263.6	301.3	346.3	402.7	479.7	606.2	883.3
1.05	0.20	0.100	93.5	157.1	184.2	215.8	254.0	301.0	368.0	479.7	729.1
1.10	0.01	0.005	98.9	130.2	142.6	156.6	172.9	193.0	219.9	263.0	352.7
1.10	0.02	0.010	81.0	109.4	120.8	133.7	148.8	167.5	192.6	233.0	317.7
1.10	0.05	0.025	57.9	82.3	92.1	103.4	116.7	133.3	155.7	192.3	271.2
1.10	0.10	0.050	41.2	62.0	70.6	80.5	92.3	107.0	127.2	160.4	233.0
1.10	0.20	0.100	25.6	42.3	49.4	57.7	67.7	80.4	97.9	127.2	192.6
1.15	0.01	0.005	46.8	61.4	67.2	73.7	81.3	90.6	103.1	123.2	165.6
1.15	0.02	0.010	38.5	51.8	57.0	63.0	70.1	78.7	90.4	109.2	149.3
1.15	0.05	0.025	27.8	39.1	43.7	48.9	55.1	62.8	73.3	90.3	127.0
1.15	0.10	0.050	20.0	29.7	33.7	38.3	43.8	50.6	60.0	75.4	109.2
1.15	0.20	0.100	12.8	20.5	23.8	27.7	32.3	38.2	46.4	60.0	90.4
1.20	0.01	0.005	28.2	36.8	40.2	44.0	48.4	53.9	61.3	73.1	98.0
1.20	0.02	0.010	23.3	31.1	34.2	37.7	41.9	47.0	53.8	64.9	88.4
1.20	0.05	0.025	17.0	23.7	26.4	29.4	33.1	37.6	43.7	53.7	75.3
1.20	0.10	0.050	12.5	18.1	20.5	23.2	26.4	30.4	36.0	45.0	64.9
1.20	0.20	0.100	8.2	12.7	14.7	16.9	19.7	23.2	28.0	36.0	53.8

Table 10.2

Table 10.2 Continued.

$\triangle$	α 2-sided	1-sided	Power $1-\beta$ 0.50	0.65	0.70	0.75	0.80	0.85	0.90	0.95	0.99
						30 subjects entered per unit time					
1.25	0.01	0.005	19.4	25.1	27.4	29.9	32.9	36.6	41.5	49.3	66.0
1.25	0.02	0.010	16.1	21.3	23.4	25.8	28.5	31.9	36.5	43.9	59.6
1.25	0.05	0.025	11.9	16.4	18.2	20.2	22.7	25.7	29.8	36.4	50.8
1.25	0.10	0.050	8.9	12.7	14.2	16.0	18.2	20.9	24.6	30.6	43.9
1.25	0.20	0.100	6.0	9.1	10.4	11.9	13.7	16.0	19.2	24.6	36.5
1.30	0.01	0.005	14.5	18.7	20.3	22.2	24.3	27.0	30.5	36.2	48.2
1.30	0.02	0.010	12.2	15.9	17.4	19.1	21.1	23.6	26.9	32.2	43.6
1.30	0.05	0.025	9.1	12.3	13.6	15.1	16.9	19.1	22.0	26.9	37.3
1.30	0.10	0.050	6.9	9.7	10.8	12.1	13.7	15.6	18.3	22.6	32.2
1.30	0.20	0.100	4.8	7.0	8.0	9.1	10.4	12.1	14.4	18.3	26.9
1.35	0.01	0.005	11.6	14.7	16.0	17.4	19.0	21.1	23.8	28.1	37.3
1.35	0.02	0.010	9.7	12.6	13.8	15.1	16.6	18.5	21.0	25.1	33.8
1.35	0.05	0.025	7.4	9.9	10.9	12.0	13.4	15.0	17.3	21.0	28.9
1.35	0.10	0.050	5.7	7.8	8.7	9.7	10.9	12.4	14.4	17.8	25.1
1.35	0.20	0.100	4.0	5.8	6.5	7.4	8.4	9.7	11.5	14.4	21.0
1.40	0.01	0.005	9.6	12.1	13.1	14.2	15.5	17.1	19.3	22.8	30.1
1.40	0.02	0.010	8.1	10.4	11.3	12.4	13.6	15.1	17.1	20.4	27.3
1.40	0.05	0.025	6.2	8.2	9.0	10.0	11.0	12.4	14.2	17.1	23.4
1.40	0.10	0.050	4.8	6.6	7.3	8.1	9.0	10.2	11.9	14.5	20.4
1.40	0.20	0.100	3.4	4.9	5.5	6.2	7.0	8.1	9.5	11.9	17.1
1.45	0.01	0.005	8.2	10.3	11.1	12.0	13.1	14.4	16.2	19.0	25.0
1.45	0.02	0.010	7.0	8.9	9.7	10.5	11.5	12.7	14.4	17.1	22.7
1.45	0.05	0.025	5.4	7.1	7.8	8.5	9.4	10.5	12.0	14.4	19.6
1.45	0.10	0.050	4.2	5.7	6.3	7.0	7.8	8.7	10.1	12.3	17.1
1.45	0.20	0.100	3.1	4.3	4.8	5.4	6.1	7.0	8.1	10.1	14.4
1.50	0.01	0.005	7.2	9.0	9.7	10.4	11.3	12.5	13.9	16.3	21.4
1.50	0.02	0.010	6.2	7.8	8.4	9.2	10.0	11.0	12.4	14.7	19.4
1.50	0.05	0.025	4.8	6.3	6.8	7.5	8.2	9.1	10.4	12.4	16.8
1.50	0.10	0.050	3.8	5.1	5.6	6.2	6.8	7.7	8.8	10.6	14.7
1.50	0.20	0.100	2.8	3.9	4.3	4.8	5.4	6.1	7.2	8.8	12.4
1.55	0.01	0.005	6.4	8.0	8.6	9.2	10.0	11.0	12.3	14.3	18.6
1.55	0.02	0.010	5.6	7.0	7.5	8.1	8.9	9.8	11.0	12.9	17.0
1.55	0.05	0.025	4.4	5.6	6.1	6.7	7.3	8.1	9.2	10.9	14.7
1.55	0.10	0.050	3.5	4.6	5.0	5.5	6.1	6.8	7.8	9.4	12.9
1.55	0.20	0.100	2.5	3.5	3.9	4.4	4.9	5.5	6.4	7.8	11.0
1.60	0.01	0.005	5.9	7.2	7.7	8.3	9.0	9.8	11.0	12.7	16.5
1.60	0.02	0.010	5.1	6.3	6.8	7.3	8.0	8.8	9.8	11.5	15.0
1.60	0.05	0.025	4.0	5.1	5.6	6.1	6.6	7.3	8.3	9.8	13.1
1.60	0.10	0.050	3.2	4.2	4.6	5.0	5.6	6.2	7.1	8.5	11.5
1.60	0.20	0.100	2.3	3.3	3.6	4.0	4.5	5.0	5.8	7.1	9.8
1.65	0.01	0.005	5.4	6.6	7.1	7.6	8.2	8.9	9.9	11.5	14.8
1.65	0.02	0.010	4.7	5.8	6.2	6.7	7.3	8.0	8.9	10.4	13.5
1.65	0.05	0.025	3.7	4.7	5.1	5.6	6.1	6.7	7.5	8.9	11.8
1.65	0.10	0.050	3.0	3.9	4.3	4.7	5.1	5.7	6.5	7.7	10.4
1.65	0.20	0.100	2.2	3.0	3.3	3.7	4.1	4.7	5.3	6.5	8.9
1.70	0.01	0.005	5.0	6.1	6.5	7.0	7.5	8.2	9.1	10.5	13.5
1.70	0.02	0.010	4.4	5.4	5.8	6.2	6.7	7.4	8.2	9.5	12.3
1.70	0.05	0.025	3.5	4.4	4.8	5.2	5.6	6.2	7.0	8.2	10.8
1.70	0.10	0.050	2.8	3.6	4.0	4.3	4.8	5.3	6.0	7.1	9.5
1.70	0.20	0.100	2.1	2.8	3.1	3.5	3.9	4.3	5.0	6.0	8.2
1.75	0.01	0.005	4.7	5.7	6.1	6.5	7.0	7.6	8.4	9.7	12.4
1.75	0.02	0.010	4.1	5.0	5.4	5.8	6.3	6.8	7.6	8.8	11.3
1.75	0.05	0.025	3.3	4.1	4.5	4.8	5.3	5.8	6.5	7.6	9.9
1.75	0.10	0.050	2.6	3.4	3.7	4.1	4.5	4.9	5.6	6.6	8.8
1.75	0.20	0.100	2.0	2.7	3.0	3.3	3.6	4.1	4.7	5.6	7.6

Table 10.2

Table 10.2 Continued.

$\triangle$	α 2-sided	1-sided	0.50	0.65	0.70	0.75	Power $1-\beta$ 0.80	0.85	0.90	0.95	0.99
					30 subjects entered per unit time						
1.80	0.01	0.005	4.4	5.3	5.7	6.1	6.6	7.1	7.8	9.0	11.4
1.80	0.02	0.010	3.9	4.7	5.1	5.4	5.9	6.4	7.1	8.2	10.5
1.80	0.05	0.025	3.1	3.9	4.2	4.6	4.9	5.4	6.1	7.1	9.2
1.80	0.10	0.050	2.5	3.3	3.5	3.8	4.2	4.7	5.3	6.2	8.2
1.80	0.20	0.100	1.9	2.5	2.8	3.1	3.4	3.8	4.4	5.3	7.1
1.85	0.01	0.005	4.2	5.1	5.4	5.8	6.2	6.7	7.4	8.4	10.7
1.85	0.02	0.010	3.7	4.5	4.8	5.1	5.5	6.0	6.7	7.7	9.8
1.85	0.05	0.025	3.0	3.7	4.0	4.3	4.7	5.1	5.7	6.7	8.6
1.85	0.10	0.050	2.4	3.1	3.4	3.7	4.0	4.4	5.0	5.8	7.7
1.85	0.20	0.100	1.8	2.4	2.7	2.9	3.3	3.7	4.2	5.0	6.7
1.90	0.01	0.005	4.0	4.8	5.1	5.5	5.9	6.3	7.0	8.0	10.0
1.90	0.02	0.010	3.5	4.3	4.6	4.9	5.3	5.7	6.3	7.3	9.2
1.90	0.05	0.025	2.8	3.5	3.8	4.1	4.5	4.9	5.4	6.3	8.1
1.90	0.10	0.050	2.3	3.0	3.2	3.5	3.8	4.2	4.7	5.5	7.3
1.90	0.20	0.100	1.7	2.3	2.6	2.8	3.1	3.5	4.0	4.7	6.3
1.95	0.01	0.005	3.8	4.6	4.9	5.2	5.6	6.0	6.6	7.5	9.5
1.95	0.02	0.010	3.4	4.1	4.4	4.7	5.0	5.5	6.0	6.9	8.7
1.95	0.05	0.025	2.7	3.4	3.7	3.9	4.3	4.7	5.2	6.0	7.7
1.95	0.10	0.050	2.2	2.8	3.1	3.3	3.7	4.0	4.5	5.3	6.9
1.95	0.20	0.100	1.7	2.2	2.5	2.7	3.0	3.3	3.8	4.5	6.0
2.00	0.01	0.005	3.7	4.4	4.7	5.0	5.3	5.8	6.3	7.2	9.0
2.00	0.02	0.010	3.2	3.9	4.2	4.5	4.8	5.2	5.7	6.6	8.3
2.00	0.05	0.025	2.6	3.3	3.5	3.8	4.1	4.5	5.0	5.7	7.3
2.00	0.10	0.050	2.1	2.7	3.0	3.2	3.5	3.9	4.3	5.1	6.6
2.00	0.20	0.100	1.6	2.2	2.4	2.6	2.9	3.2	3.7	4.3	5.7
2.25	0.01	0.005	3.1	3.7	3.9	4.2	4.5	4.8	5.3	5.9	7.3
2.25	0.02	0.010	2.8	3.3	3.5	3.8	4.1	4.4	4.8	5.5	6.8
2.25	0.05	0.025	2.3	2.8	3.0	3.2	3.5	3.8	4.2	4.8	6.1
2.25	0.10	0.050	1.8	2.4	2.5	2.8	3.0	3.3	3.7	4.3	5.5
2.25	0.20	0.100	1.4	1.9	2.1	2.3	2.5	2.8	3.1	3.7	4.8
2.50	0.01	0.005	2.8	3.3	3.5	3.7	4.0	4.2	4.6	5.2	6.4
2.50	0.02	0.010	2.5	3.0	3.2	3.4	3.6	3.9	4.2	4.8	5.9
2.50	0.05	0.025	2.0	2.5	2.7	2.9	3.1	3.4	3.7	4.2	5.3
2.50	0.10	0.050	1.7	2.1	2.3	2.5	2.7	2.9	3.3	3.8	4.8
2.50	0.20	0.100	1.3	1.7	1.9	2.0	2.2	2.5	2.8	3.3	4.2
2.75	0.01	0.005	2.6	3.0	3.2	3.4	3.6	3.9	4.2	4.7	5.7
2.75	0.02	0.010	2.3	2.7	2.9	3.1	3.3	3.5	3.9	4.4	5.4
2.75	0.05	0.025	1.9	2.3	2.5	2.6	2.8	3.1	3.4	3.9	4.8
2.75	0.10	0.050	1.5	2.0	2.1	2.3	2.5	2.7	3.0	3.4	4.4
2.75	0.20	0.100	1.2	1.6	1.7	1.9	2.1	2.3	2.6	3.0	3.9
3.00	0.01	0.005	2.4	2.8	3.0	3.2	3.4	3.6	3.9	4.4	5.3
3.00	0.02	0.010	2.1	2.6	2.7	2.9	3.1	3.3	3.6	4.1	5.0
3.00	0.05	0.025	1.8	2.2	2.3	2.5	2.7	2.9	3.2	3.6	4.5
3.00	0.10	0.050	1.5	1.8	2.0	2.1	2.3	2.5	2.8	3.2	4.1
3.00	0.20	0.100	1.1	1.5	1.6	1.8	1.9	2.1	2.4	2.8	3.6
3.25	0.01	0.005	2.3	2.7	2.8	3.0	3.2	3.4	3.7	4.1	5.0
3.25	0.02	0.010	2.0	2.4	2.6	2.7	2.9	3.1	3.4	3.8	4.7
3.25	0.05	0.025	1.7	2.1	2.2	2.4	2.5	2.7	3.0	3.4	4.2
3.25	0.10	0.050	1.4	1.8	1.9	2.0	2.2	2.4	2.7	3.1	3.8
3.25	0.20	0.100	1.1	1.4	1.5	1.7	1.8	2.0	2.3	2.7	3.4
3.50	0.01	0.005	2.2	2.6	2.7	2.9	3.1	3.3	3.5	3.9	4.7
3.50	0.02	0.010	2.0	2.3	2.5	2.6	2.8	3.0	3.3	3.7	4.4
3.50	0.05	0.025	1.6	2.0	2.1	2.3	2.4	2.6	2.9	3.3	4.0
3.50	0.10	0.050	1.3	1.7	1.8	2.0	2.1	2.3	2.6	2.9	3.7
3.50	0.20	0.100	1.0	1.4	1.5	1.6	1.8	2.0	2.2	2.6	3.3

Table 10.2

Table 10.2 Continued.

$\triangle$	α 2-sided	α 1-sided	Power $1-\beta$ 0.50	0.65	0.70	0.75	0.80	0.85	0.90	0.95	0.99
			0.50	0.65	0.70	0.75	0.80	0.85	0.90	0.95	0.99

30 subjects entered per unit time

$\triangle$	2-sided	1-sided	0.50	0.65	0.70	0.75	0.80	0.85	0.90	0.95	0.99
3.75	0.01	0.005	2.1	2.5	2.6	2.8	3.0	3.2	3.4	3.8	4.6
3.75	0.02	0.010	1.9	2.3	2.4	2.5	2.7	2.9	3.1	3.5	4.3
3.75	0.05	0.025	1.6	1.9	2.0	2.2	2.4	2.5	2.8	3.1	3.9
3.75	0.10	0.050	1.3	1.6	1.8	1.9	2.0	2.2	2.5	2.8	3.5
3.75	0.20	0.100	1.0	1.3	1.4	1.6	1.7	1.9	2.1	2.5	3.1
4.00	0.01	0.005	2.1	2.4	2.6	2.7	2.9	3.1	3.3	3.7	4.4
4.00	0.02	0.010	1.9	2.2	2.3	2.5	2.6	2.8	3.1	3.4	4.1
4.00	0.05	0.025	1.5	1.9	2.0	2.1	2.3	2.5	2.7	3.1	3.7
4.00	0.10	0.050	1.3	1.6	1.7	1.8	2.0	2.2	2.4	2.7	3.4
4.00	0.20	0.100	1.0	1.3	1.4	1.5	1.7	1.8	2.1	2.4	3.1
4.25	0.01	0.005	2.0	2.4	2.5	2.6	2.8	3.0	3.2	3.6	4.3
4.25	0.02	0.010	1.8	2.1	2.3	2.4	2.6	2.8	3.0	3.3	4.0
4.25	0.05	0.025	1.5	1.8	1.9	2.1	2.2	2.4	2.6	3.0	3.7
4.25	0.10	0.050	1.2	1.6	1.7	1.8	1.9	2.1	2.3	2.7	3.3
4.25	0.20	0.100	1.0	1.3	1.4	1.5	1.6	1.8	2.0	2.3	3.0

40 subjects entered per unit time

$\triangle$	2-sided	1-sided	0.50	0.65	0.70	0.75	0.80	0.85	0.90	0.95	0.99
1.05	0.01	0.005	278.7	368.3	403.8	443.8	490.6	548.1	625.1	748.3	1009.5
1.05	0.02	0.010	228.8	308.9	341.4	378.3	421.6	475.0	546.8	662.5	909.4
1.05	0.05	0.025	162.9	232.5	259.3	291.6	329.7	377.2	441.4	545.9	771.8
1.05	0.10	0.050	115.1	174.6	199.2	227.5	259.7	302.0	359.8	454.6	662.5
1.05	0.20	0.100	70.5	118.2	138.5	162.2	190.8	227.2	276.0	359.8	546.8
1.10	0.01	0.005	74.6	98.0	107.3	117.8	130.1	145.2	165.3	197.6	266.1
1.10	0.02	0.010	61.1	82.5	91.0	100.6	112.0	126.0	144.8	175.1	239.8
1.10	0.05	0.025	43.8	62.1	69.5	77.9	87.9	100.4	117.2	144.6	203.8
1.10	0.10	0.050	31.3	46.9	53.3	60.7	69.6	80.7	95.8	120.7	175.1
1.10	0.20	0.100	19.6	32.1	37.4	43.6	51.1	60.7	73.8	95.8	144.8
1.15	0.01	0.005	35.5	46.4	50.8	55.6	61.3	68.4	77.7	92.8	124.6
1.15	0.02	0.010	29.3	39.2	43.2	47.7	52.9	59.5	68.2	82.3	112.4
1.15	0.05	0.025	21.2	29.7	33.2	37.1	41.7	47.5	55.4	68.1	95.6
1.15	0.10	0.050	15.4	22.7	25.6	29.1	33.2	38.4	45.4	57.0	82.3
1.15	0.20	0.100	10.0	15.8	18.3	21.1	24.6	29.1	35.2	45.4	68.2
1.20	0.01	0.005	21.6	28.0	30.5	33.4	36.7	40.9	46.4	55.2	73.9
1.20	0.02	0.010	17.9	23.7	26.0	28.7	31.8	35.6	40.8	49.0	66.7
1.20	0.05	0.025	13.2	18.1	20.2	22.5	25.2	28.6	33.2	40.7	56.9
1.20	0.10	0.050	9.7	14.0	15.8	17.8	20.2	23.2	27.4	34.2	49.0
1.20	0.20	0.100	6.5	10.0	11.4	13.1	15.2	17.8	21.4	27.4	40.8
1.25	0.01	0.005	15.0	19.3	20.9	22.9	25.1	27.8	31.5	37.4	49.9
1.25	0.02	0.010	12.5	16.4	18.0	19.7	21.8	24.4	27.8	33.3	45.1
1.25	0.05	0.025	9.4	12.7	14.0	15.6	17.4	19.7	22.7	27.7	38.5
1.25	0.10	0.050	7.1	9.9	11.1	12.4	14.1	16.1	18.8	23.4	33.3
1.25	0.20	0.100	4.9	7.2	8.2	9.3	10.7	12.4	14.8	18.8	27.8
1.30	0.01	0.005	11.3	14.4	15.7	17.0	18.7	20.6	23.3	27.6	36.6
1.30	0.02	0.010	9.5	12.4	13.5	14.8	16.3	18.1	20.6	24.6	33.1
1.30	0.05	0.025	7.2	9.7	10.7	11.8	13.1	14.7	17.0	20.6	28.4
1.30	0.10	0.050	5.5	7.7	8.5	9.5	10.7	12.1	14.1	17.4	24.6
1.30	0.20	0.100	3.9	5.7	6.4	7.2	8.2	9.5	11.2	14.1	20.6
1.35	0.01	0.005	9.1	11.5	12.4	13.5	14.7	16.2	18.3	21.5	28.4
1.35	0.02	0.010	7.7	9.9	10.8	11.7	12.9	14.3	16.2	19.2	25.8
1.35	0.05	0.025	5.9	7.8	8.6	9.4	10.4	11.7	13.4	16.2	22.1
1.35	0.10	0.050	4.6	6.3	6.9	7.7	8.6	9.7	11.2	13.7	19.2
1.35	0.20	0.100	3.3	4.7	5.3	5.9	6.7	7.7	9.0	11.2	16.2
1.40	0.01	0.005	7.6	9.5	10.3	11.1	12.1	13.3	14.9	17.5	23.0
1.40	0.02	0.010	6.5	8.3	8.9	9.7	10.6	11.8	13.3	15.7	20.9
1.40	0.05	0.025	5.1	6.6	7.2	7.9	8.7	9.7	11.1	13.3	18.0
1.40	0.10	0.050	4.0	5.3	5.9	6.5	7.2	8.1	9.3	11.3	15.7
1.40	0.20	0.100	2.9	4.0	4.5	5.0	5.7	6.5	7.6	9.3	13.3

Table 10.2

Table 10.2 Continued.

$\triangle$	α 2-sided	1-sided	Power $1-\beta$ 0.50	0.65	0.70	0.75	0.80	0.85	0.90	0.95	0.99
			40 subjects entered per unit time								
1.45	0.01	0.005	6.6	8.2	8.8	9.5	10.3	11.3	12.6	14.7	19.2
1.45	0.02	0.010	5.7	7.1	7.7	8.3	9.1	10.0	11.3	13.2	17.5
1.45	0.05	0.025	4.4	5.7	6.2	6.8	7.5	8.3	9.4	11.2	15.1
1.45	0.10	0.050	3.5	4.7	5.1	5.6	6.2	7.0	8.0	9.7	13.2
1.45	0.20	0.100	2.6	3.6	4.0	4.4	5.0	5.6	6.5	8.0	11.3
1.50	0.01	0.005	5.8	7.2	7.7	8.3	9.0	9.8	10.9	12.7	16.5
1.50	0.02	0.010	5.0	6.3	6.8	7.3	7.9	8.7	9.8	11.5	15.0
1.50	0.05	0.025	4.0	5.1	5.5	6.0	6.6	7.3	8.2	9.8	13.0
1.50	0.10	0.050	3.2	4.2	4.6	5.0	5.5	6.2	7.0	8.4	11.5
1.50	0.20	0.100	2.3	3.2	3.6	4.0	4.4	5.0	5.8	7.0	9.8
1.55	0.01	0.005	5.2	6.4	6.9	7.4	8.0	8.7	9.7	11.2	14.4
1.55	0.02	0.010	4.5	5.6	6.1	6.5	7.1	7.8	8.7	10.1	13.2
1.55	0.05	0.025	3.6	4.6	5.0	5.4	5.9	6.5	7.3	8.7	11.5
1.55	0.10	0.050	2.9	3.8	4.1	4.5	5.0	5.5	6.3	7.5	10.1
1.55	0.20	0.100	2.1	2.9	3.3	3.6	4.0	4.5	5.2	6.3	8.7
1.60	0.01	0.005	4.8	5.8	6.2	6.7	7.2	7.8	8.7	10.0	12.9
1.60	0.02	0.010	4.2	5.1	5.5	5.9	6.4	7.0	7.8	9.1	11.8
1.60	0.05	0.025	3.3	4.2	4.5	4.9	5.4	5.9	6.6	7.8	10.3
1.60	0.10	0.050	2.7	3.5	3.8	4.1	4.6	5.1	5.7	6.8	9.1
1.60	0.20	0.100	2.0	2.7	3.0	3.3	3.7	4.1	4.7	5.7	7.8
1.65	0.01	0.005	4.4	5.4	5.7	6.1	6.6	7.2	7.9	9.1	11.6
1.65	0.02	0.010	3.9	4.7	5.1	5.5	5.9	6.4	7.1	8.3	10.6
1.65	0.05	0.025	3.1	3.9	4.2	4.6	5.0	5.4	6.1	7.1	9.3
1.65	0.10	0.050	2.5	3.2	3.5	3.8	4.2	4.7	5.3	6.2	8.3
1.65	0.20	0.100	1.8	2.5	2.8	3.1	3.4	3.8	4.4	5.3	7.1
1.70	0.01	0.005	4.1	5.0	5.3	5.7	6.1	6.6	7.3	8.4	10.6
1.70	0.02	0.010	3.6	4.4	4.7	5.1	5.5	5.9	6.6	7.6	9.7
1.70	0.05	0.025	2.9	3.6	3.9	4.2	4.6	5.1	5.6	6.6	8.6
1.70	0.10	0.050	2.3	3.0	3.3	3.6	3.9	4.3	4.9	5.8	7.6
1.70	0.20	0.100	1.7	2.4	2.6	2.9	3.2	3.6	4.1	4.9	6.6
1.75	0.01	0.005	3.9	4.7	5.0	5.3	5.7	6.1	6.8	7.7	9.8
1.75	0.02	0.010	3.4	4.1	4.4	4.7	5.1	5.6	6.1	7.1	9.0
1.75	0.05	0.025	2.7	3.4	3.7	4.0	4.3	4.7	5.3	6.1	7.9
1.75	0.10	0.050	2.2	2.9	3.1	3.4	3.7	4.1	4.6	5.4	7.1
1.75	0.20	0.100	1.7	2.2	2.5	2.7	3.0	3.4	3.8	4.6	6.1
1.80	0.01	0.005	3.7	4.4	4.7	5.0	5.3	5.8	6.3	7.2	9.1
1.80	0.02	0.010	3.2	3.9	4.2	4.5	4.8	5.2	5.8	6.6	8.4
1.80	0.05	0.025	2.6	3.3	3.5	3.8	4.1	4.5	5.0	5.8	7.4
1.80	0.10	0.050	2.1	2.7	2.9	3.2	3.5	3.9	4.3	5.1	6.6
1.80	0.20	0.100	1.6	2.1	2.4	2.6	2.9	3.2	3.6	4.3	5.8
1.85	0.01	0.005	3.5	4.2	4.4	4.7	5.0	5.5	6.0	6.8	8.5
1.85	0.02	0.010	3.1	3.7	4.0	4.2	4.6	4.9	5.4	6.2	7.9
1.85	0.05	0.025	2.5	3.1	3.3	3.6	3.9	4.2	4.7	5.4	7.0
1.85	0.10	0.050	2.0	2.6	2.8	3.0	3.3	3.7	4.1	4.8	6.2
1.85	0.20	0.100	1.5	2.0	2.2	2.5	2.7	3.0	3.5	4.1	5.4
1.90	0.01	0.005	3.3	4.0	4.2	4.5	4.8	5.2	5.7	6.4	8.0
1.90	0.02	0.010	2.9	3.6	3.8	4.0	4.3	4.7	5.2	5.9	7.4
1.90	0.05	0.025	2.4	3.0	3.2	3.4	3.7	4.0	4.5	5.2	6.6
1.90	0.10	0.050	1.9	2.5	2.7	2.9	3.2	3.5	3.9	4.6	5.9
1.90	0.20	0.100	1.5	2.0	2.2	2.4	2.6	2.9	3.3	3.9	5.2
1.95	0.01	0.005	3.2	3.8	4.0	4.3	4.6	4.9	5.4	6.1	7.6
1.95	0.02	0.010	2.8	3.4	3.6	3.9	4.1	4.5	4.9	5.6	7.0
1.95	0.05	0.025	2.3	2.8	3.0	3.3	3.5	3.9	4.3	4.9	6.3
1.95	0.10	0.050	1.9	2.4	2.6	2.8	3.0	3.3	3.7	4.4	5.6
1.95	0.20	0.100	1.4	1.9	2.1	2.3	2.5	2.8	3.2	3.7	4.9

Table 10.2

Table 10.2 Continued.

$\triangle$	α 2-sided	1-sided	Power $1-\beta$ 0.50	0.65	0.70	0.75	0.80	0.85	0.90	0.95	0.99
						40 subjects entered per unit time					
2.00	0.01	0.005	3.1	3.7	3.9	4.1	4.4	4.7	5.2	5.9	7.2
2.00	0.02	0.010	2.7	3.3	3.5	3.7	4.0	4.3	4.7	5.4	6.7
2.00	0.05	0.025	2.2	2.7	2.9	3.1	3.4	3.7	4.1	4.7	6.0
2.00	0.10	0.050	1.8	2.3	2.5	2.7	2.9	3.2	3.6	4.2	5.4
2.00	0.20	0.100	1.4	1.8	2.0	2.2	2.4	2.7	3.0	3.6	4.7
2.25	0.01	0.005	2.6	3.1	3.3	3.5	3.7	4.0	4.3	4.9	6.0
2.25	0.02	0.010	2.3	2.8	3.0	3.2	3.4	3.6	4.0	4.5	5.6
2.25	0.05	0.025	1.9	2.4	2.5	2.7	2.9	3.2	3.5	4.0	5.0
2.25	0.10	0.050	1.6	2.0	2.1	2.3	2.5	2.8	3.1	3.5	4.5
2.25	0.20	0.100	1.2	1.6	1.7	1.9	2.1	2.3	2.6	3.1	4.0
2.50	0.01	0.005	2.4	2.8	2.9	3.1	3.3	3.5	3.8	4.3	5.2
2.50	0.02	0.010	2.1	2.5	2.7	2.8	3.0	3.2	3.5	4.0	4.9
2.50	0.05	0.025	1.7	2.1	2.3	2.4	2.6	2.8	3.1	3.5	4.4
2.50	0.10	0.050	1.4	1.8	1.9	2.1	2.3	2.5	2.7	3.2	4.0
2.50	0.20	0.100	1.1	1.4	1.6	1.7	1.9	2.1	2.3	2.7	3.5
2.75	0.01	0.005	2.2	2.6	2.7	2.9	3.0	3.2	3.5	3.9	4.7
2.75	0.02	0.010	1.9	2.3	2.4	2.6	2.8	3.0	3.2	3.6	4.4
2.75	0.05	0.025	1.6	2.0	2.1	2.2	2.4	2.6	2.8	3.2	4.0
2.75	0.10	0.050	1.3	1.7	1.8	1.9	2.1	2.3	2.5	2.9	3.6
2.75	0.20	0.100	1.0	1.3	1.5	1.6	1.8	1.9	2.2	2.5	3.2
3.00	0.01	0.005	2.1	2.4	2.5	2.7	2.8	3.0	3.3	3.7	4.4
3.00	0.02	0.010	1.8	2.2	2.3	2.4	2.6	2.8	3.0	3.4	4.1
3.00	0.05	0.025	1.5	1.8	2.0	2.1	2.3	2.4	2.7	3.0	3.7
3.00	0.10	0.050	1.2	1.6	1.7	1.8	2.0	2.1	2.4	2.7	3.4
3.00	0.20	0.100	1.0	1.3	1.4	1.5	1.6	1.8	2.0	2.4	3.0
3.25	0.01	0.005	2.0	2.3	2.4	2.5	2.7	2.9	3.1	3.5	4.2
3.25	0.02	0.010	1.7	2.1	2.2	2.3	2.5	2.6	2.9	3.2	3.9
3.25	0.05	0.025	1.4	1.8	1.9	2.0	2.1	2.3	2.5	2.9	3.5
3.25	0.10	0.050	1.2	1.5	1.6	1.7	1.9	2.0	2.3	2.6	3.2
3.25	0.20	0.100	0.9	1.2	1.3	1.4	1.6	1.7	1.9	2.3	2.9
3.50	0.01	0.005	1.9	2.2	2.3	2.4	2.6	2.8	3.0	3.3	4.0
3.50	0.02	0.010	1.7	2.0	2.1	2.2	2.4	2.5	2.8	3.1	3.7
3.50	0.05	0.025	1.4	1.7	1.8	1.9	2.1	2.2	2.4	2.8	3.4
3.50	0.10	0.050	1.1	1.4	1.5	1.7	1.8	2.0	2.2	2.5	3.1
3.50	0.20	0.100	0.9	1.2	1.3	1.4	1.5	1.7	1.9	2.2	2.8
3.75	0.01	0.005	1.8	2.1	2.2	2.4	2.5	2.7	2.9	3.2	3.8
3.75	0.02	0.010	1.6	1.9	2.0	2.2	2.3	2.5	2.7	3.0	3.6
3.75	0.05	0.025	1.3	1.6	1.7	1.9	2.0	2.1	2.4	2.7	3.3
3.75	0.10	0.050	1.1	1.4	1.5	1.6	1.7	1.9	2.1	2.4	3.0
3.75	0.20	0.100	0.9	1.1	1.2	1.3	1.5	1.6	1.8	2.1	2.7
4.00	0.01	0.005	1.8	2.1	2.2	2.3	2.4	2.6	2.8	3.1	3.7
4.00	0.02	0.010	1.6	1.9	2.0	2.1	2.2	2.4	2.6	2.9	3.5
4.00	0.05	0.025	1.3	1.6	1.7	1.8	1.9	2.1	2.3	2.6	3.2
4.00	0.10	0.050	1.1	1.4	1.5	1.6	1.7	1.8	2.0	2.3	2.9
4.00	0.20	0.100	0.8	1.1	1.2	1.3	1.4	1.6	1.8	2.0	2.6
4.25	0.01	0.005	1.7	2.0	2.1	2.2	2.4	2.5	2.7	3.0	3.6
4.25	0.02	0.010	1.5	1.8	1.9	2.0	2.2	2.3	2.5	2.8	3.4
4.25	0.05	0.025	1.3	1.6	1.7	1.8	1.9	2.0	2.2	2.5	3.1
4.25	0.10	0.050	1.1	1.3	1.4	1.5	1.7	1.8	2.0	2.3	2.8
4.25	0.20	0.100	0.8	1.1	1.2	1.3	1.4	1.5	1.7	2.0	2.5

Table 10.2   146

Table 10.2 Continued.

$\triangle$	α 2-sided	1-sided	0.50	0.65	0.70	Power $1-\beta$ 0.75	0.80	0.85	0.90	0.95	0.99
					50 subjects entered per unit time						
1.05	0.01	0.005	224.5	294.7	323.0	355.0	392.5	438.5	500.0	598.7	807.6
1.05	0.02	0.010	183.4	248.6	273.1	302.6	337.3	380.0	437.5	530.0	727.5
1.05	0.05	0.025	130.6	186.3	208.9	234.7	263.8	301.7	353.1	436.7	617.4
1.05	0.10	0.050	92.4	140.0	159.6	182.3	209.3	243.1	287.8	363.7	530.0
1.05	0.20	0.100	56.7	94.9	111.1	130.1	153.0	182.1	222.3	287.8	437.5
1.10	0.01	0.005	59.9	78.7	86.2	94.6	104.4	116.4	132.6	158.4	213.2
1.10	0.02	0.010	49.2	66.3	73.1	80.8	89.9	101.1	116.2	140.4	192.2
1.10	0.05	0.025	35.3	70.0	55.9	62.6	70.6	80.6	94.1	116.0	163.3
1.10	0.10	0.050	25.3	37.8	43.0	48.9	56.0	64.8	76.9	96.8	140.4
1.10	0.20	0.100	16.0	26.0	30.2	35.2	41.2	48.8	59.4	76.9	116.2
1.15	0.01	0.005	28.7	37.5	40.9	44.8	49.4	55.0	62.5	74.5	100.0
1.15	0.02	0.010	23.7	31.7	34.8	38.4	42.7	47.9	54.9	66.1	90.2
1.15	0.05	0.025	17.3	24.1	26.8	30.0	33.7	38.3	44.6	54.8	76.8
1.15	0.10	0.050	12.6	18.4	20.8	23.6	26.9	31.0	36.6	45.9	66.1
1.15	0.20	0.100	8.3	12.9	14.9	17.2	20.0	23.6	28.5	36.6	54.9
1.20	0.01	0.005	17.6	22.7	24.7	27.0	29.7	33.0	37.4	44.5	59.4
1.20	0.02	0.010	14.6	19.3	21.2	23.3	25.8	28.8	32.9	39.6	53.7
1.20	0.05	0.025	10.8	14.8	16.5	18.3	20.5	23.2	26.9	32.9	45.8
1.20	0.10	0.050	8.1	11.5	12.9	14.5	16.5	18.9	22.2	27.6	39.6
1.20	0.20	0.100	5.5	8.3	9.4	10.8	12.4	14.5	17.4	22.2	32.9
1.25	0.01	0.005	12.3	15.7	17.1	18.6	20.4	22.6	25.5	30.3	40.3
1.25	0.02	0.010	10.3	13.5	14.7	16.1	17.8	19.8	22.6	27.0	36.4
1.25	0.05	0.025	7.8	10.5	11.6	12.8	14.3	16.1	18.5	22.5	31.2
1.25	0.10	0.050	5.9	8.3	9.2	10.3	11.6	13.2	15.4	19.0	27.0
1.25	0.20	0.100	4.1	6.1	6.9	7.8	8.9	10.3	12.2	15.4	22.6
1.30	0.01	0.005	9.4	11.9	12.9	14.0	15.3	16.9	19.0	22.4	29.6
1.30	0.02	0.010	8.0	10.2	11.1	12.2	13.4	14.8	16.8	20.0	26.8
1.30	0.05	0.025	6.1	8.1	8.9	9.8	10.8	12.1	13.9	16.8	23.0
1.30	0.10	0.050	4.7	6.4	7.1	7.9	8.9	10.0	11.6	14.3	20.0
1.30	0.20	0.100	3.4	4.8	5.4	6.1	6.9	7.9	9.3	11.6	16.8
1.35	0.01	0.005	7.6	9.5	10.3	11.1	12.1	13.3	14.9	17.6	23.1
1.35	0.02	0.010	6.5	8.2	8.9	9.7	10.6	11.8	13.3	15.7	21.0
1.35	0.05	0.025	5.0	6.6	7.2	7.9	8.7	9.7	11.1	13.3	18.1
1.35	0.10	0.050	3.9	5.3	5.8	6.5	7.2	8.1	9.3	11.3	15.7
1.35	0.20	0.100	2.9	4.0	4.5	5.0	5.7	6.5	7.5	9.3	13.3
1.40	0.01	0.005	6.4	7.9	8.6	9.2	10.0	11.0	12.3	14.4	18.8
1.40	0.02	0.010	5.5	6.9	7.5	8.1	8.9	9.8	11.0	12.9	17.1
1.40	0.05	0.025	4.3	5.6	6.1	6.6	7.3	8.1	9.2	11.0	14.8
1.40	0.10	0.050	3.4	4.5	5.0	5.5	6.1	6.8	7.8	9.4	12.9
1.40	0.20	0.100	2.5	3.5	3.9	4.3	4.8	5.5	6.4	7.8	11.0
1.45	0.01	0.005	5.6	6.9	7.4	7.9	8.6	9.4	10.4	12.1	15.8
1.45	0.02	0.010	4.8	6.0	6.5	7.0	7.6	8.4	9.4	11.0	14.4
1.45	0.05	0.025	3.8	4.9	5.3	5.8	6.3	7.0	7.9	9.3	12.5
1.45	0.10	0.050	3.0	4.0	4.4	4.8	5.3	5.9	6.7	8.1	11.0
1.45	0.20	0.100	2.2	3.1	3.4	3.8	4.2	4.8	5.5	6.7	9.4
1.50	0.01	0.005	5.0	6.1	6.5	7.0	7.5	8.2	9.1	10.5	13.6
1.50	0.02	0.010	4.3	5.3	5.7	6.2	6.7	7.3	8.2	9.5	12.4
1.50	0.05	0.025	3.4	4.4	4.7	5.1	5.6	6.2	6.9	8.2	10.8
1.50	0.10	0.050	2.7	3.6	3.9	4.3	4.7	5.2	6.0	7.1	9.5
1.50	0.20	0.100	2.0	2.8	3.1	3.4	3.8	4.3	4.9	6.0	8.2
1.55	0.01	0.005	4.5	5.4	5.8	6.2	6.7	7.3	8.1	9.3	11.9
1.55	0.02	0.010	3.9	4.8	5.2	5.6	6.0	6.6	7.3	8.5	10.9
1.55	0.05	0.025	3.1	3.9	4.3	4.6	5.0	5.5	6.2	7.3	9.6
1.55	0.10	0.050	2.5	3.3	3.6	3.9	4.3	4.7	5.4	6.3	8.5
1.55	0.20	0.100	1.9	2.6	2.8	3.1	3.5	3.9	4.5	5.4	7.3

Table 10.2

Table 10.2 Continued.

$\triangle$	α 2-sided	1-sided	Power $1 - \beta$ 0.50	0.65	0.70	0.75	0.80	0.85	0.90	0.95	0.99
					50 subjects entered per unit time						
1.60	0.01	0.005	4.1	5.0	5.3	5.7	6.1	6.6	7.3	8.4	10.7
1.60	0.02	0.010	3.6	4.4	4.7	5.1	5.5	6.0	6.6	7.6	9.8
1.60	0.05	0.025	2.9	3.6	3.9	4.2	4.6	5.1	5.6	6.6	8.6
1.60	0.10	0.050	2.3	3.0	3.3	3.6	3.9	4.3	4.9	5.8	7.6
1.60	0.20	0.100	1.7	2.4	2.6	2.9	3.2	3.6	4.1	4.9	6.6
1.65	0.01	0.005	3.8	4.6	4.9	5.2	5.6	6.1	6.7	7.6	9.7
1.65	0.02	0.010	3.3	4.1	4.4	4.7	5.0	5.5	6.1	7.0	8.9
1.65	0.05	0.025	2.7	3.4	3.6	3.9	4.3	4.7	5.2	6.0	7.8
1.65	0.10	0.050	2.2	2.8	3.1	3.3	3.6	4.0	4.5	5.3	7.0
1.65	0.20	0.100	1.6	2.2	2.4	2.7	3.0	3.3	3.8	4.5	6.1
1.70	0.01	0.005	3.6	4.3	4.5	4.8	5.2	5.6	6.2	7.0	8.9
1.70	0.02	0.010	3.1	3.8	4.1	4.4	4.7	5.1	5.6	6.4	8.2
1.70	0.05	0.025	2.5	3.2	3.4	3.7	4.0	4.3	4.8	5.6	7.2
1.70	0.10	0.050	2.0	2.6	2.9	3.1	3.4	3.7	4.2	4.9	6.4
1.70	0.20	0.100	1.5	2.1	2.3	2.5	2.8	3.1	3.5	4.2	5.6
1.75	0.01	0.005	3.4	4.0	4.3	4.5	4.9	5.3	5.8	6.6	8.2
1.75	0.02	0.010	2.9	3.6	3.8	4.1	4.4	4.8	5.2	6.0	7.6
1.75	0.05	0.025	2.4	3.0	3.2	3.5	3.7	4.1	4.5	5.2	6.7
1.75	0.10	0.050	1.9	2.5	2.7	2.9	3.2	3.5	4.0	4.6	6.0
1.75	0.20	0.100	1.5	2.0	2.2	2.4	2.6	2.9	3.3	4.0	5.2
1.80	0.01	0.005	3.2	3.8	4.0	4.3	4.6	4.9	5.4	6.1	7.7
1.80	0.02	0.010	2.8	3.4	3.6	3.9	4.1	4.5	4.9	5.6	7.1
1.80	0.05	0.025	2.3	2.8	3.0	3.3	3.5	3.9	4.3	4.9	6.3
1.80	0.10	0.050	1.8	2.4	2.6	2.8	3.0	3.3	3.7	4.4	5.6
1.80	0.20	0.100	1.4	1.9	2.1	2.3	2.5	2.8	3.2	3.7	4.9
1.85	0.01	0.005	3.0	3.6	3.8	4.1	4.3	4.7	5.1	5.8	7.2
1.85	0.02	0.010	2.7	3.2	3.4	3.7	3.9	4.3	4.7	5.3	6.7
1.85	0.05	0.025	2.2	2.7	2.9	3.1	3.4	3.7	4.1	4.7	5.9
1.85	0.10	0.050	1.8	2.3	2.4	2.7	2.9	3.2	3.6	4.1	5.3
1.85	0.20	0.100	1.3	1.8	2.0	2.2	2.4	2.7	3.0	3.6	4.7
1.90	0.01	0.005	2.9	3.4	3.7	3.9	4.1	4.5	4.9	5.5	6.8
1.90	0.02	0.010	2.6	3.1	3.3	3.5	3.8	4.1	4.4	5.1	6.3
1.90	0.05	0.025	2.1	2.6	2.8	3.0	3.2	3.5	3.9	4.4	5.6
1.90	0.10	0.050	1.7	2.2	2.3	2.5	2.8	3.0	3.4	3.9	5.1
1.90	0.20	0.100	1.3	1.7	1.9	2.1	2.3	2.5	2.9	3.4	4.4
1.95	0.01	0.005	2.8	3.3	3.5	3.7	4.0	4.3	4.6	5.2	6.5
1.95	0.02	0.010	2.5	3.0	3.1	3.4	3.6	3.9	4.3	4.8	6.0
1.95	0.05	0.025	2.0	2.5	2.7	2.9	3.1	3.3	3.7	4.2	5.4
1.95	0.10	0.050	1.6	2.1	2.3	2.4	2.7	2.9	3.2	3.8	4.8
1.95	0.20	0.100	1.2	1.7	1.8	2.0	2.2	2.4	2.8	3.2	4.3
2.00	0.01	0.005	2.7	3.2	3.4	3.6	3.8	4.1	4.5	5.0	6.2
2.00	0.02	0.010	2.4	2.8	3.0	3.2	3.4	3.7	4.1	4.6	5.7
2.00	0.05	0.025	1.9	2.4	2.6	2.7	3.0	3.2	3.6	4.1	5.1
2.00	0.10	0.050	1.6	2.0	2.2	2.4	2.6	2.8	3.1	3.6	4.6
2.00	0.20	0.100	1.2	1.6	1.8	1.9	2.1	2.4	2.7	3.1	4.1
2.25	0.01	0.005	2.3	2.7	2.9	3.0	3.2	3.5	3.8	4.2	5.1
2.25	0.02	0.010	2.0	2.4	2.6	2.8	2.9	3.2	3.5	3.9	4.8
2.25	0.05	0.025	1.7	2.1	2.2	2.4	2.5	2.8	3.0	3.5	4.3
2.25	0.10	0.050	1.4	1.8	1.9	2.0	2.2	2.4	2.7	3.1	3.9
2.25	0.20	0.100	1.1	1.4	1.5	1.7	1.8	2.0	2.3	2.7	3.5
2.50	0.01	0.005	2.1	2.4	2.6	2.7	2.9	3.1	3.3	3.7	4.5
2.50	0.02	0.010	1.8	2.2	2.3	2.5	2.6	2.8	3.1	3.5	4.2
2.50	0.05	0.025	1.5	1.9	2.0	2.1	2.3	2.5	2.7	3.1	3.8
2.50	0.10	0.050	1.3	1.6	1.7	1.8	2.0	2.2	2.4	2.8	3.5
2.50	0.20	0.100	1.0	1.3	1.4	1.5	1.7	1.8	2.1	2.4	3.1

Table 10.2   148

Table 10.2 Continued.

△	α 2-sided	1-sided	Power 1−β 0.50	0.65	0.70	0.75	0.80	0.85	0.90	0.95	0.99
50 subjects entered per unit time											
2.75	0.01	0.005	1.9	2.2	2.4	2.5	2.7	2.8	3.1	3.4	4.1
2.75	0.02	0.010	1.7	2.0	2.2	2.3	2.4	2.6	2.8	3.2	3.9
2.75	0.05	0.025	1.4	1.7	1.8	2.0	2.1	2.3	2.5	2.8	3.5
2.75	0.10	0.050	1.2	1.5	1.6	1.7	1.8	2.0	2.2	2.5	3.2
2.75	0.20	0.100	0.9	1.2	1.3	1.4	1.5	1.7	1.9	2.2	2.8
3.00	0.01	0.005	1.8	2.1	2.2	2.4	2.5	2.7	2.9	3.2	3.8
3.00	0.02	0.010	1.6	1.9	2.0	2.1	2.3	2.4	2.7	3.0	3.6
3.00	0.05	0.025	1.3	1.6	1.7	1.9	2.0	2.1	2.3	2.7	3.3
3.00	0.10	0.050	1.1	1.4	1.5	1.6	1.7	1.9	2.1	2.4	3.0
3.00	0.20	0.100	0.8	1.1	1.2	1.3	1.5	1.6	1.8	2.1	2.7
3.25	0.01	0.005	1.7	2.0	2.1	2.2	2.4	2.5	2.7	3.0	3.6
3.25	0.02	0.010	1.5	1.8	1.9	2.0	2.2	2.3	2.5	2.8	3.4
3.25	0.05	0.025	1.3	1.5	1.7	1.8	1.9	2.0	2.2	2.5	3.1
3.25	0.10	0.050	1.1	1.3	1.4	1.5	1.7	1.8	2.0	2.3	2.8
3.25	0.20	0.100	0.8	1.1	1.2	1.3	1.4	1.5	1.7	2.0	2.5
3.50	0.01	0.005	1.7	1.9	2.0	2.2	2.3	2.4	2.6	2.9	3.5
3.50	0.02	0.010	1.5	1.8	1.9	2.0	2.1	2.2	2.4	2.7	3.3
3.50	0.05	0.025	1.2	1.5	1.6	1.7	1.8	2.0	2.1	2.4	3.0
3.50	0.10	0.050	1.0	1.3	1.4	1.5	1.6	1.7	1.9	2.2	2.7
3.50	0.20	0.100	0.8	1.0	1.1	1.2	1.3	1.5	1.6	1.9	2.4
3.75	0.01	0.005	1.6	1.9	2.0	2.1	2.2	2.3	2.5	2.8	3.3
3.75	0.02	0.010	1.4	1.7	1.8	1.9	2.0	2.2	2.3	2.6	3.1
3.75	0.05	0.025	1.2	1.4	1.5	1.6	1.8	1.9	2.1	2.3	2.9
3.75	0.10	0.050	1.0	1.2	1.3	1.4	1.5	1.7	1.8	2.1	2.6
3.75	0.20	0.100	0.8	1.0	1.1	1.2	1.3	1.4	1.6	1.8	2.3
4.00	0.01	0.005	1.6	1.8	1.9	2.0	2.1	2.3	2.5	2.7	3.2
4.00	0.02	0.010	1.4	1.7	1.7	1.8	2.0	2.1	2.3	2.5	3.1
4.00	0.05	0.025	1.2	1.4	1.5	1.6	1.7	1.8	2.0	2.3	2.8
4.00	0.10	0.050	1.0	1.2	1.3	1.4	1.5	1.6	1.8	2.0	2.5
4.00	0.20	0.100	0.7	1.0	1.1	1.2	1.3	1.4	1.6	1.8	2.3
4.25	0.01	0.005	1.5	1.8	1.9	2.0	2.1	2.2	2.4	2.7	3.2
4.25	0.02	0.010	1.4	1.6	1.7	1.8	1.9	2.1	2.2	2.5	3.0
4.25	0.05	0.025	1.1	1.4	1.5	1.6	1.7	1.8	2.0	2.2	2.7
4.25	0.10	0.050	1.0	1.2	1.3	1.4	1.5	1.6	1.8	2.0	2.5
4.25	0.20	0.100	0.7	1.0	1.0	1.1	1.2	1.4	1.5	1.8	2.2
60 subjects entered per unit time											
1.05	0.01	0.005	187.3	247.0	269.2	295.9	327.1	365.4	416.7	498.9	673.0
1.05	0.02	0.010	153.0	207.4	229.1	253.7	281.1	316.7	364.5	441.7	606.3
1.05	0.05	0.025	109.1	155.5	174.3	195.8	221.3	252.9	294.3	363.9	514.5
1.05	0.10	0.050	77.2	116.9	133.3	152.1	174.6	202.8	241.3	303.1	441.7
1.05	0.20	0.100	47.5	79.3	92.8	108.6	127.7	152.0	185.5	241.3	364.5
1.10	0.01	0.005	50.2	65.9	72.1	79.1	87.2	97.3	110.7	132.3	177.9
1.10	0.02	0.010	41.2	55.5	61.2	67.6	75.2	84.5	97.0	117.3	160.4
1.10	0.05	0.025	29.7	41.9	46.8	52.5	59.1	67.4	78.6	96.9	136.4
1.10	0.10	0.050	21.4	31.8	36.1	41.0	46.9	54.3	64.4	80.9	117.3
1.10	0.20	0.100	13.6	21.9	25.5	29.6	34.6	41.0	49.7	64.4	97.0
1.15	0.01	0.005	24.2	31.5	34.4	37.6	41.4	46.1	52.3	62.4	83.6
1.15	0.02	0.010	20.0	26.7	29.3	32.3	35.8	40.2	46.0	55.4	75.4
1.15	0.05	0.025	14.7	20.3	22.6	25.2	28.3	32.2	37.4	45.9	64.3
1.15	0.10	0.050	10.8	15.6	17.6	19.9	22.7	26.1	30.8	38.5	55.4
1.15	0.20	0.100	7.1	11.0	12.7	14.6	16.9	19.9	24.0	30.8	46.0
1.20	0.01	0.005	14.9	19.2	20.9	22.8	25.0	27.8	31.4	37.3	49.8
1.20	0.02	0.010	12.5	16.3	17.9	19.7	21.7	24.3	27.7	33.2	45.0
1.20	0.05	0.025	9.3	12.6	14.0	15.5	17.3	19.6	22.7	27.7	38.4
1.20	0.10	0.050	7.0	9.9	11.0	12.4	14.0	16.0	18.8	23.3	33.2
1.20	0.20	0.100	4.8	7.2	8.1	9.3	10.6	12.4	14.8	18.8	27.7

Table 10.2

Table 10.2 Continued.

△	α 2-sided	α 1-sided	Power 1 − β 0.50	0.65	0.70	0.75	0.80	0.85	0.90	0.95	0.99
					60 subjects entered per unit time						
1.25	0.01	0.005	10.5	13.4	14.5	15.8	17.3	19.1	21.6	25.5	33.8
1.25	0.02	0.010	8.9	11.5	12.5	13.7	15.1	16.8	19.1	22.8	30.6
1.25	0.05	0.025	6.8	9.0	9.9	10.9	12.2	13.7	15.7	19.0	26.2
1.25	0.10	0.050	5.2	7.1	7.9	8.8	9.9	11.3	13.1	16.1	22.8
1.25	0.20	0.100	3.7	5.3	6.0	6.7	7.7	8.8	10.4	13.1	19.1
1.30	0.01	0.005	8.1	10.2	11.0	11.9	13.0	14.3	16.1	18.9	25.0
1.30	0.02	0.010	6.9	8.8	9.6	10.4	11.4	12.6	14.3	17.0	22.7
1.30	0.05	0.025	5.3	7.0	7.6	8.4	9.3	10.4	11.9	14.3	19.5
1.30	0.10	0.050	4.2	5.6	6.2	6.9	7.7	8.6	10.0	12.2	17.0
1.30	0.20	0.100	3.0	4.2	4.7	5.3	6.0	6.9	8.0	10.0	14.3
1.35	0.01	0.005	6.6	8.2	8.8	9.6	10.4	11.4	12.7	14.9	19.5
1.35	0.02	0.010	5.7	7.1	7.7	8.4	9.2	10.1	11.4	13.4	17.8
1.35	0.05	0.025	4.4	5.7	6.2	6.8	7.5	8.4	9.5	11.4	15.3
1.35	0.10	0.050	3.5	4.7	5.1	5.6	6.3	7.0	8.1	9.7	13.4
1.35	0.20	0.100	2.5	3.6	4.0	4.4	5.0	5.6	6.6	8.1	11.4
1.40	0.01	0.005	5.6	6.9	7.4	8.0	8.6	9.5	10.5	12.3	15.9
1.40	0.02	0.010	4.8	6.0	6.5	7.0	7.7	8.4	9.4	11.1	14.5
1.40	0.05	0.025	3.8	4.9	5.3	5.8	6.3	7.0	7.9	9.4	12.6
1.40	0.10	0.050	3.0	4.0	4.4	4.8	5.3	5.9	6.8	8.1	11.1
1.40	0.20	0.100	2.2	3.1	3.4	3.8	4.3	4.8	5.6	6.8	9.4
1.45	0.01	0.005	4.9	6.0	6.4	6.9	7.4	8.1	9.0	10.4	13.4
1.45	0.02	0.010	4.2	5.3	5.7	6.1	6.6	7.2	8.1	9.4	12.3
1.45	0.05	0.025	3.4	4.3	4.7	5.0	5.5	6.1	6.8	8.1	10.7
1.45	0.10	0.050	2.7	3.5	3.9	4.2	4.7	5.2	5.9	7.0	9.4
1.45	0.20	0.100	2.0	2.8	3.0	3.4	3.8	4.2	4.9	5.9	8.1
1.50	0.01	0.005	4.4	5.3	5.7	6.1	6.5	7.1	7.9	9.1	11.6
1.50	0.02	0.010	3.8	4.7	5.0	5.4	5.8	6.4	7.1	8.2	10.7
1.50	0.05	0.025	3.1	3.8	4.2	4.5	4.9	5.4	6.0	7.1	9.3
1.50	0.10	0.050	2.4	3.2	3.5	3.8	4.2	4.6	5.2	6.2	8.2
1.50	0.20	0.100	1.8	2.5	2.7	3.0	3.4	3.8	4.3	5.2	7.1
1.55	0.01	0.005	4.0	4.8	5.1	5.5	5.9	6.4	7.0	8.1	10.3
1.55	0.02	0.010	3.5	4.2	4.5	4.9	5.3	5.7	6.4	7.3	9.4
1.55	0.05	0.025	2.8	3.5	3.8	4.1	4.4	4.9	5.4	6.3	8.3
1.55	0.10	0.050	2.2	2.9	3.2	3.5	3.8	4.2	4.7	5.6	7.3
1.55	0.20	0.100	1.7	2.3	2.5	2.8	3.1	3.5	3.9	4.7	6.4
1.60	0.01	0.005	3.6	4.4	4.7	5.0	5.3	5.8	6.4	7.3	9.2
1.60	0.02	0.010	3.2	3.9	4.2	4.5	4.8	5.2	5.8	6.6	8.5
1.60	0.05	0.025	2.6	3.2	3.5	3.8	4.1	4.5	5.0	5.8	7.5
1.60	0.10	0.050	2.1	2.7	2.9	3.2	3.5	3.8	4.3	5.1	6.6
1.60	0.20	0.100	1.6	2.1	2.3	2.6	2.8	3.2	3.6	4.3	5.8
1.65	0.01	0.005	3.4	4.1	4.3	4.6	4.9	5.3	5.8	6.7	8.4
1.65	0.02	0.010	3.0	3.6	3.9	4.1	4.4	4.8	5.3	6.1	7.7
1.65	0.05	0.025	2.4	3.0	3.2	3.5	3.8	4.1	4.6	5.3	6.8
1.65	0.10	0.050	2.0	2.5	2.7	3.0	3.2	3.6	4.0	4.7	6.1
1.65	0.20	0.100	1.5	2.0	2.2	2.4	2.7	3.0	3.4	4.0	5.3
1.70	0.01	0.005	3.2	3.8	4.0	4.3	4.6	4.9	5.4	6.2	7.7
1.70	0.02	0.010	2.8	3.4	3.6	3.9	4.1	4.5	4.9	5.6	7.1
1.70	0.05	0.025	2.3	2.8	3.0	3.3	3.5	3.8	4.3	4.9	6.3
1.70	0.10	0.050	1.8	2.4	2.6	2.8	3.0	3.3	3.7	4.4	5.6
1.70	0.20	0.100	1.4	1.9	2.1	2.3	2.5	2.8	3.2	3.7	4.9
1.75	0.01	0.005	3.0	3.6	3.8	4.0	4.3	4.6	5.1	5.7	7.1
1.75	0.02	0.010	2.6	3.2	3.4	3.6	3.9	4.2	4.6	5.3	6.6
1.75	0.05	0.025	2.1	2.7	2.9	3.1	3.3	3.6	4.0	4.6	5.9
1.75	0.10	0.050	1.7	2.2	2.4	2.6	2.9	3.1	3.5	4.1	5.3
1.75	0.20	0.100	1.3	1.8	1.9	2.1	2.4	2.6	3.0	3.5	4.6

Table 10.2

Table 10.2 Continued.

$\triangle$	α 2-sided	1-sided	0.50	0.65	0.70	Power $1 - \beta$ 0.75	0.80	0.85	0.90	0.95	0.99
					60 subjects entered per unit time						
1.80	0.01	0.005	2.8	3.4	3.6	3.8	4.1	4.4	4.8	5.4	6.7
1.80	0.02	0.010	2.5	3.0	3.2	3.4	3.7	4.0	4.4	5.0	6.2
1.80	0.05	0.025	2.0	2.5	2.7	2.9	3.1	3.4	3.8	4.4	5.5
1.80	0.10	0.050	1.7	2.1	2.3	2.5	2.7	3.0	3.3	3.9	5.0
1.80	0.20	0.100	1.3	1.7	1.9	2.0	2.2	2.5	2.8	3.3	4.4
1.85	0.01	0.005	2.7	3.2	3.4	3.6	3.9	4.1	4.5	5.1	6.3
1.85	0.02	0.010	2.4	2.9	3.1	3.3	3.5	3.8	4.1	4.7	5.9
1.85	0.05	0.025	2.0	2.4	2.6	2.8	3.0	3.3	3.6	4.1	5.2
1.85	0.10	0.050	1.6	2.0	2.2	2.4	2.6	2.8	3.2	3.7	4.7
1.85	0.20	0.100	1.2	1.6	1.8	1.9	2.1	2.4	2.7	3.2	4.1
1.90	0.01	0.005	2.6	3.1	3.2	3.5	3.7	4.0	4.3	4.9	6.0
1.90	0.02	0.010	2.3	2.8	2.9	3.1	3.3	3.6	3.9	4.5	5.5
1.90	0.05	0.025	1.9	2.3	2.5	2.7	2.9	3.1	3.4	3.9	5.0
1.90	0.10	0.050	1.5	2.0	2.1	2.3	2.5	2.7	3.0	3.5	4.5
1.90	0.20	0.100	1.2	1.6	1.7	1.9	2.1	2.3	2.6	3.0	3.9
1.95	0.01	0.005	2.5	2.9	3.1	3.3	3.5	3.8	4.1	4.6	5.7
1.95	0.02	0.010	2.2	2.6	2.8	3.0	3.2	3.4	3.8	4.3	5.3
1.95	0.05	0.025	1.8	2.2	2.4	2.6	2.7	3.0	3.3	3.8	4.7
1.95	0.10	0.050	1.5	1.9	2.0	2.2	2.4	2.6	2.9	3.4	4.3
1.95	0.20	0.100	1.1	1.5	1.6	1.8	2.0	2.2	2.5	2.9	3.8
2.00	0.01	0.005	2.4	2.8	3.0	3.2	3.4	3.6	4.0	4.4	5.4
2.00	0.02	0.010	2.1	2.5	2.7	2.9	3.1	3.3	3.6	4.1	5.1
2.00	0.05	0.025	1.7	2.1	2.3	2.5	2.6	2.9	3.2	3.6	4.5
2.00	0.10	0.050	1.4	1.8	2.0	2.1	2.3	2.5	2.8	3.2	4.1
2.00	0.20	0.100	1.1	1.5	1.6	1.7	1.9	2.1	2.4	2.8	3.6
2.25	0.01	0.005	2.1	2.4	2.6	2.7	2.9	3.1	3.4	3.8	4.6
2.25	0.02	0.010	1.8	2.2	2.3	2.5	2.6	2.8	3.1	3.5	4.3
2.25	0.05	0.025	1.5	1.9	2.0	2.1	2.3	2.5	2.7	3.1	3.8
2.25	0.10	0.050	1.2	1.6	1.7	1.8	2.0	2.2	2.4	2.8	3.5
2.25	0.20	0.100	1.0	1.3	1.4	1.5	1.7	1.8	2.1	2.4	3.1
2.50	0.01	0.005	1.9	2.2	2.3	2.4	2.6	2.8	3.0	3.3	4.0
2.50	0.02	0.010	1.7	2.0	2.1	2.2	2.4	2.5	2.8	3.1	3.8
2.50	0.05	0.025	1.4	1.7	1.8	1.9	2.1	2.2	2.4	2.8	3.4
2.50	0.10	0.050	1.1	1.4	1.5	1.7	1.8	2.0	2.2	2.5	3.1
2.50	0.20	0.100	0.9	1.2	1.3	1.4	1.5	1.7	1.9	2.2	2.8
2.75	0.01	0.005	1.7	2.0	2.1	2.3	2.4	2.5	2.8	3.1	3.7
2.75	0.02	0.010	1.5	1.8	1.9	2.1	2.2	2.3	2.5	2.8	3.4
2.75	0.05	0.025	1.3	1.6	1.7	1.8	1.9	2.1	2.2	2.5	3.1
2.75	0.10	0.050	1.1	1.3	1.4	1.5	1.7	1.8	2.0	2.3	2.8
2.75	0.20	0.100	0.8	1.1	1.2	1.3	1.4	1.5	1.7	2.0	2.5
3.00	0.01	0.005	1.6	1.9	2.0	2.1	2.2	2.4	2.6	2.9	3.4
3.00	0.02	0.010	1.5	1.7	1.8	1.9	2.1	2.2	2.4	2.7	3.2
3.00	0.05	0.025	1.2	1.5	1.6	1.7	1.8	1.9	2.1	2.4	2.9
3.00	0.10	0.050	1.0	1.3	1.3	1.4	1.6	1.7	1.9	2.1	2.7
3.00	0.20	0.100	0.8	1.0	1.1	1.2	1.3	1.4	1.6	1.9	2.4
3.25	0.01	0.005	1.6	1.8	1.9	2.0	2.1	2.3	2.5	2.7	3.3
3.25	0.02	0.010	1.4	1.6	1.7	1.8	2.0	2.1	2.3	2.5	3.1
3.25	0.05	0.025	1.2	1.4	1.5	1.6	1.7	1.8	2.0	2.3	2.8
3.25	0.10	0.050	1.0	1.2	1.3	1.4	1.5	1.6	1.8	2.0	2.5
3.25	0.20	0.100	0.7	1.0	1.1	1.2	1.3	1.4	1.5	1.8	2.3
3.50	0.01	0.005	1.5	1.7	1.8	1.9	2.1	2.2	2.4	2.6	3.1
3.50	0.02	0.010	1.3	1.6	1.7	1.8	1.9	2.0	2.2	2.4	2.9
3.50	0.05	0.025	1.1	1.3	1.4	1.5	1.6	1.8	1.9	2.2	2.7
3.50	0.10	0.050	0.9	1.2	1.2	1.3	1.4	1.6	1.7	2.0	2.4
3.50	0.20	0.100	0.7	0.9	1.0	1.1	1.2	1.3	1.5	1.7	2.2

Table 10.2

Table 10.2 Continued.

$\triangle$	α 2-sided	1-sided	Power $1 - \beta$ 0.50	0.65	0.70	0.75	0.80	0.85	0.90	0.95	0.99
					60 subjects entered per unit time						
3.75	0.01	0.005	1.5	1.7	1.8	1.9	2.0	2.1	2.3	2.5	3.0
3.75	0.02	0.010	1.3	1.5	1.6	1.7	1.8	1.9	2.1	2.4	2.8
3.75	0.05	0.025	1.1	1.3	1.4	1.5	1.6	1.7	1.9	2.1	2.6
3.75	0.10	0.050	0.9	1.1	1.2	1.3	1.4	1.5	1.7	1.9	2.4
3.75	0.20	0.100	0.7	0.9	1.0	1.1	1.2	1.3	1.4	1.7	2.1
4.00	0.01	0.005	1.4	1.6	1.7	1.8	1.9	2.1	2.2	2.5	2.9
4.00	0.02	0.010	1.3	1.5	1.6	1.7	1.8	1.9	2.1	2.3	2.7
4.00	0.05	0.025	1.1	1.3	1.4	1.4	1.5	1.7	1.8	2.0	2.5
4.00	0.10	0.050	0.9	1.1	1.2	1.3	1.4	1.5	1.6	1.9	2.3
4.00	0.20	0.100	0.7	0.9	1.0	1.1	1.1	1.3	1.4	1.6	2.1
4.25	0.01	0.005	1.4	1.6	1.7	1.8	1.9	2.0	2.2	2.4	2.8
4.25	0.02	0.010	1.2	1.5	1.5	1.6	1.7	1.9	2.0	2.2	2.7
4.25	0.05	0.025	1.0	1.3	1.3	1.4	1.5	1.6	1.8	2.0	2.4
4.25	0.10	0.050	0.9	1.1	1.2	1.2	1.3	1.4	1.6	1.8	2.2
4.25	0.20	0.100	0.7	0.9	1.0	1.0	1.1	1.2	1.4	1.6	2.0
					70 subjects entered per unit time						
1.05	0.01	0.005	160.7	212.0	232.2	255.1	280.4	313.2	357.2	427.6	576.9
1.05	0.02	0.010	131.4	178.0	196.6	217.6	242.4	271.5	312.5	378.6	519.6
1.05	0.05	0.025	93.7	133.5	149.6	168.1	189.9	217.0	253.7	311.9	441.0
1.05	0.10	0.050	66.4	100.4	114.4	130.6	149.9	174.1	207.1	259.8	378.6
1.05	0.20	0.100	40.9	68.2	79.8	93.3	109.7	130.5	159.2	207.1	312.5
1.10	0.01	0.005	43.3	56.7	62.0	68.0	75.0	83.6	95.1	113.6	152.7
1.10	0.02	0.010	35.6	47.8	52.6	58.2	64.6	72.7	83.4	100.7	137.7
1.10	0.05	0.025	25.7	36.1	40.3	45.2	50.9	58.0	67.6	83.3	117.1
1.10	0.10	0.050	18.5	27.4	31.1	35.4	40.4	46.7	55.4	69.6	100.7
1.10	0.20	0.100	11.8	19.0	22.0	25.6	29.9	35.3	42.8	55.4	83.4
1.15	0.01	0.005	21.0	27.2	29.7	32.5	35.7	39.7	45.1	53.7	71.9
1.15	0.02	0.010	17.4	23.1	25.3	27.9	30.9	34.6	39.6	47.7	64.9
1.15	0.05	0.025	12.8	17.6	19.6	21.9	24.5	27.8	32.3	39.6	55.3
1.15	0.10	0.050	9.5	13.6	15.3	17.3	19.6	22.6	26.6	33.2	47.7
1.15	0.20	0.100	6.3	9.7	11.1	12.8	14.7	17.3	20.8	26.6	39.6
1.20	0.01	0.005	13.0	16.7	18.1	19.8	21.7	24.0	27.2	32.2	42.9
1.20	0.02	0.010	10.9	14.2	15.6	17.1	18.9	21.0	24.0	28.7	38.8
1.20	0.05	0.025	8.2	11.1	12.2	13.5	15.1	17.0	19.7	23.9	33.2
1.20	0.10	0.050	6.2	8.7	9.7	10.8	12.2	14.0	16.3	20.2	28.7
1.20	0.20	0.100	4.3	6.3	7.2	8.2	9.3	10.8	12.9	16.3	24.0
1.25	0.01	0.005	9.3	11.7	12.7	13.8	15.0	16.6	18.7	22.1	29.2
1.25	0.02	0.010	7.8	10.1	11.0	12.0	13.2	14.6	16.6	19.7	26.5
1.25	0.05	0.025	6.0	7.9	8.7	9.6	10.7	11.9	13.7	16.6	22.7
1.25	0.10	0.050	4.6	6.3	7.0	7.8	8.7	9.9	11.5	14.1	19.7
1.25	0.20	0.100	3.3	4.7	5.3	6.0	6.8	7.8	9.2	11.5	16.6
1.30	0.01	0.005	7.2	9.0	9.7	10.5	11.4	12.5	14.0	16.5	21.6
1.30	0.02	0.010	6.1	7.8	8.4	9.2	10.0	11.1	12.5	14.8	19.7
1.30	0.05	0.025	4.8	6.2	6.8	7.4	8.2	9.1	10.4	12.5	16.9
1.30	0.10	0.050	3.7	5.0	5.5	6.1	6.8	7.6	8.8	10.7	14.8
1.30	0.20	0.100	2.7	3.8	4.3	4.8	5.4	6.1	7.1	8.8	12.5
1.35	0.01	0.005	5.9	7.3	7.8	8.4	9.1	10.0	11.2	13.0	17.0
1.35	0.02	0.010	5.1	6.3	6.9	7.4	8.1	8.9	10.0	11.7	15.5
1.35	0.05	0.025	4.0	5.1	5.6	6.1	6.7	7.4	8.4	10.0	13.4
1.35	0.10	0.050	3.2	4.2	4.6	5.0	5.6	6.2	7.1	8.6	11.7
1.35	0.20	0.100	2.3	3.2	3.6	4.0	4.5	5.0	5.8	7.1	10.0

Table 10.2

Table 10.2 Continued.

△	α 2-sided	1-sided	0.50	0.65	0.70	Power 1 − β 0.75	0.80	0.85	0.90	0.95	0.99
					70 subjects entered per unit time						
1.40	0.01	0.005	5.0	6.1	6.6	7.1	7.6	8.3	9.3	10.8	13.9
1.40	0.02	0.010	4.3	5.4	5.8	6.3	6.8	7.5	8.3	9.7	12.7
1.40	0.05	0.025	3.5	4.4	4.8	5.2	5.7	6.2	7.0	8.3	11.1
1.40	0.10	0.050	2.8	3.6	3.9	4.3	4.8	5.3	6.0	7.2	9.7
1.40	0.20	0.100	2.0	2.8	3.1	3.4	3.8	4.3	5.0	6.0	8.3
1.45	0.01	0.005	4.4	5.3	5.7	6.1	6.6	7.2	8.0	9.2	11.8
1.45	0.02	0.010	3.8	4.7	5.1	5.4	5.9	6.4	7.2	8.3	10.8
1.45	0.05	0.025	3.1	3.9	4.2	4.5	4.9	5.4	6.1	7.2	9.4
1.45	0.10	0.050	2.5	3.2	3.5	3.8	4.2	4.6	5.3	6.2	8.3
1.45	0.20	0.100	1.8	2.5	2.8	3.1	3.4	3.8	4.4	5.3	7.2
1.50	0.01	0.005	3.9	4.8	5.1	5.4	5.8	6.3	7.0	8.0	10.2
1.50	0.02	0.010	3.4	4.2	4.5	4.8	5.2	5.7	6.3	7.3	9.4
1.50	0.05	0.025	2.8	3.5	3.8	4.1	4.4	4.8	5.4	6.3	8.2
1.50	0.10	0.050	2.2	2.9	3.1	3.4	3.8	4.2	4.7	5.5	7.3
1.50	0.20	0.100	1.7	2.3	2.5	2.8	3.1	3.4	3.9	4.7	6.3
1.55	0.01	0.005	3.6	4.3	4.6	4.9	5.3	5.7	6.3	7.2	9.0
1.55	0.02	0.010	3.1	3.8	4.1	4.4	4.7	5.1	5.7	6.5	8.3
1.55	0.05	0.025	2.5	3.2	3.4	3.7	4.0	4.4	4.9	5.7	7.3
1.55	0.10	0.050	2.0	2.6	2.9	3.1	3.4	3.8	4.2	5.0	6.5
1.55	0.20	0.100	1.5	2.1	2.3	2.5	2.8	3.1	3.6	4.2	5.7
1.60	0.01	0.005	3.3	4.0	4.2	4.5	4.8	5.2	5.7	6.5	8.1
1.60	0.02	0.010	2.9	3.5	3.8	4.0	4.3	4.7	5.2	5.9	7.5
1.60	0.05	0.025	2.3	2.9	3.2	3.4	3.7	4.0	4.5	5.2	6.6
1.60	0.10	0.050	1.9	2.4	2.7	2.9	3.2	3.5	3.9	4.6	5.9
1.60	0.20	0.100	1.4	1.9	2.1	2.3	2.6	2.9	3.3	3.9	5.2
1.65	0.01	0.005	3.1	3.7	3.9	4.1	4.4	4.8	5.2	6.0	7.4
1.65	0.02	0.010	2.7	3.3	3.5	3.7	4.0	4.3	4.8	5.5	6.9
1.65	0.05	0.025	2.2	2.7	2.9	3.2	3.4	3.7	4.1	4.8	6.1
1.65	0.10	0.050	1.8	2.3	2.5	2.7	2.9	3.2	3.6	4.2	5.5
1.65	0.20	0.100	1.3	1.8	2.0	2.2	2.4	2.7	3.1	3.6	4.8
1.70	0.01	0.005	2.9	3.4	3.6	3.9	4.1	4.5	4.9	5.5	6.9
1.70	0.02	0.010	2.5	3.1	3.3	3.5	3.7	4.0	4.4	5.1	6.3
1.70	0.05	0.025	2.1	2.6	2.7	3.0	3.2	3.5	3.9	4.4	5.6
1.70	0.10	0.050	1.7	2.2	2.3	2.5	2.8	3.0	3.4	3.9	5.1
1.70	0.20	0.100	1.3	1.7	1.9	2.1	2.3	2.5	2.9	3.4	4.4
1.75	0.01	0.005	2.7	3.2	3.4	3.6	3.9	4.2	4.6	5.2	6.4
1.75	0.02	0.010	2.4	2.9	3.1	3.3	3.5	3.8	4.2	4.7	5.9
1.75	0.05	0.025	2.0	2.4	2.6	2.8	3.0	3.3	3.6	4.2	5.3
1.75	0.10	0.050	1.6	2.0	2.2	2.4	2.6	2.9	3.2	3.7	4.7
1.75	0.20	0.100	1.2	1.6	1.8	2.0	2.2	2.4	2.7	3.2	4.2
1.80	0.01	0.005	2.6	3.1	3.3	3.4	3.7	3.9	4.3	4.9	6.0
1.80	0.02	0.010	2.3	2.7	2.9	3.1	3.3	3.6	3.9	4.5	5.6
1.80	0.05	0.025	1.9	2.3	2.5	2.6	2.9	3.1	3.4	3.9	5.0
1.80	0.10	0.050	1.5	1.9	2.1	2.3	2.5	2.7	3.0	3.5	4.5
1.80	0.20	0.100	1.2	1.6	1.7	1.9	2.0	2.3	2.6	3.0	3.9
1.85	0.01	0.005	2.5	2.9	3.1	3.3	3.5	3.8	4.1	4.6	5.6
1.85	0.02	0.010	2.2	2.6	2.8	3.0	3.2	3.4	3.7	4.2	5.3
1.85	0.05	0.025	1.8	2.2	2.4	2.5	2.7	3.0	3.3	3.7	4.7
1.85	0.10	0.050	1.5	1.9	2.0	2.2	2.4	2.6	2.9	3.3	4.2
1.85	0.20	0.100	1.1	1.5	1.6	1.8	2.0	2.2	2.4	2.9	3.7
1.90	0.01	0.005	2.4	2.8	2.9	3.1	3.3	3.6	3.9	4.4	5.4
1.90	0.02	0.010	2.1	2.5	2.7	2.8	3.0	3.3	3.6	4.0	5.0
1.90	0.05	0.025	1.7	2.1	2.3	2.4	2.6	2.8	3.1	3.6	4.5
1.90	0.10	0.050	1.4	1.8	1.9	2.1	2.3	2.5	2.7	3.2	4.0
1.90	0.20	0.100	1.1	1.4	1.6	1.7	1.9	2.1	2.3	2.7	3.6

Table 10.2

Table 10.2 Continued.

△	α 2-sided	1-sided	0.50	0.65	0.70	0.75	0.80	0.85	0.90	0.95	0.99
						Power 1 − β					
					70 subjects entered per unit time						
1.95	0.01	0.005	2.3	2.7	2.8	3.0	3.2	3.4	3.7	4.2	5.1
1.95	0.02	0.010	2.0	2.4	2.6	2.7	2.9	3.1	3.4	3.9	4.8
1.95	0.05	0.025	1.7	2.0	2.2	2.3	2.5	2.7	3.0	3.4	4.3
1.95	0.10	0.050	1.4	1.7	1.9	2.0	2.2	2.4	2.6	3.0	3.9
1.95	0.20	0.100	1.0	1.4	1.5	1.6	1.8	2.0	2.3	2.6	3.4
2.00	0.01	0.005	2.2	2.6	2.7	2.9	3.1	3.3	3.6	4.0	4.9
2.00	0.02	0.010	1.9	2.3	2.5	2.6	2.8	3.0	3.3	3.7	4.6
2.00	0.05	0.025	1.6	2.0	2.1	2.2	2.4	2.6	2.9	3.3	4.1
2.00	0.10	0.050	1.3	1.7	1.8	1.9	2.1	2.3	2.5	2.9	3.7
2.00	0.20	0.100	1.0	1.3	1.5	1.6	1.7	1.9	2.2	2.5	3.3
2.25	0.01	0.005	1.9	2.2	2.4	2.5	2.6	2.8	3.1	3.4	4.1
2.25	0.02	0.010	1.7	2.0	2.1	2.3	2.4	2.6	2.8	3.2	3.9
2.25	0.05	0.025	1.4	1.7	1.8	1.9	2.1	2.3	2.5	2.8	3.5
2.25	0.10	0.050	1.1	1.5	1.6	1.7	1.8	2.0	2.2	2.5	3.2
2.25	0.20	0.100	0.9	1.2	1.3	1.4	1.5	1.7	1.9	2.2	2.8
2.50	0.01	0.005	1.7	2.0	2.1	2.2	2.4	2.5	2.7	3.0	3.7
2.50	0.02	0.010	1.5	1.8	1.9	2.0	2.2	2.3	2.5	2.8	3.4
2.50	0.05	0.025	1.3	1.5	1.6	1.8	1.9	2.0	2.2	2.5	3.1
2.50	0.10	0.050	1.0	1.3	1.4	1.5	1.7	1.8	2.0	2.3	2.8
2.50	0.20	0.100	0.8	1.1	1.2	1.3	1.4	1.5	1.7	2.0	2.5
2.75	0.01	0.005	1.6	1.9	2.0	2.1	2.2	2.3	2.5	2.8	3.3
2.75	0.02	0.010	1.4	1.7	1.8	1.9	2.0	2.1	2.3	2.6	3.1
2.75	0.05	0.025	1.2	1.4	1.5	1.6	1.7	1.9	2.1	2.3	2.8
2.75	0.10	0.050	1.0	1.2	1.3	1.4	1.5	1.7	1.8	2.1	2.6
2.75	0.20	0.100	0.7	1.0	1.1	1.2	1.3	1.4	1.6	1.8	2.3
3.00	0.01	0.005	1.5	1.8	1.8	1.9	2.1	2.2	2.4	2.6	3.1
3.00	0.02	0.010	1.3	1.6	1.7	1.8	1.9	2.0	2.2	2.4	2.9
3.00	0.05	0.025	1.1	1.3	1.4	1.5	1.6	1.8	1.9	2.2	2.7
3.00	0.10	0.050	0.9	1.2	1.2	1.3	1.4	1.6	1.7	2.0	2.4
3.00	0.20	0.100	0.7	0.9	1.0	1.1	1.2	1.3	1.5	1.7	2.2
3.25	0.01	0.005	1.4	1.7	1.8	1.8	2.0	2.1	2.2	2.5	3.0
3.25	0.02	0.010	1.3	1.5	1.6	1.7	1.8	1.9	2.1	2.3	2.8
3.25	0.05	0.025	1.1	1.3	1.4	1.5	1.6	1.7	1.8	2.1	2.5
3.25	0.10	0.050	0.9	1.1	1.2	1.3	1.4	1.5	1.6	1.9	2.3
3.25	0.20	0.100	0.7	0.9	1.0	1.1	1.2	1.3	1.4	1.6	2.1
3.50	0.01	0.005	1.4	1.6	1.7	1.8	1.9	2.0	2.2	2.4	2.8
3.50	0.02	0.010	1.2	1.5	1.5	1.6	1.7	1.8	2.0	2.2	2.7
3.50	0.05	0.025	1.0	1.2	1.3	1.4	1.5	1.6	1.8	2.0	2.4
3.50	0.10	0.050	0.9	1.1	1.1	1.2	1.3	1.4	1.6	1.8	2.2
3.50	0.20	0.100	0.7	0.9	0.9	1.0	1.1	1.2	1.4	1.6	2.0
3.75	0.01	0.005	1.3	1.6	1.6	1.7	1.8	1.9	2.1	2.3	2.7
3.75	0.02	0.010	1.2	1.4	1.5	1.6	1.7	1.8	1.9	2.2	2.6
3.75	0.05	0.025	1.0	1.2	1.3	1.4	1.5	1.6	1.7	1.9	2.4
3.75	0.10	0.050	0.8	1.0	1.1	1.2	1.3	1.4	1.5	1.7	2.2
3.75	0.20	0.100	0.6	0.8	0.9	1.0	1.1	1.2	1.3	1.5	1.9
4.00	0.01	0.005	1.3	1.5	1.6	1.7	1.8	1.9	2.0	2.2	2.7
4.00	0.02	0.010	1.2	1.4	1.5	1.5	1.6	1.7	1.9	2.1	2.5
4.00	0.05	0.025	1.0	1.2	1.3	1.3	1.4	1.5	1.7	1.9	2.3
4.00	0.10	0.050	0.8	1.0	1.1	1.2	1.3	1.4	1.5	1.7	2.1
4.00	0.20	0.100	0.6	0.8	0.9	1.0	1.1	1.2	1.3	1.5	1.9
4.25	0.01	0.005	1.3	1.5	1.6	1.6	1.7	1.8	2.0	2.2	2.6
4.25	0.02	0.010	1.1	1.3	1.4	1.5	1.6	1.7	1.8	2.0	2.5
4.25	0.05	0.025	1.0	1.2	1.2	1.3	1.4	1.5	1.6	1.8	2.2
4.25	0.10	0.050	0.8	1.0	1.1	1.1	1.2	1.3	1.5	1.7	2.0
4.25	0.20	0.100	0.6	0.8	0.9	1.0	1.0	1.1	1.3	1.5	1.8

Table 10.2

Table 10.2 Continued.

△	α 2-sided	1-sided	0.50	0.65	0.70	0.75	0.80	0.85	0.90	0.95	0.99
						Power $1-\beta$					

80 subjects entered per unit time

△	2-sided	1-sided	0.50	0.65	0.70	0.75	0.80	0.85	0.90	0.95	0.99
1.05	0.01	0.005	140.8	185.7	203.4	223.4	246.8	274.1	312.5	374.2	504.8
1.05	0.02	0.010	115.2	155.9	172.2	190.6	212.3	239.0	273.4	331.2	454.7
1.05	0.05	0.025	82.2	117.0	131.1	147.3	166.3	190.1	222.2	272.9	385.9
1.05	0.10	0.050	58.3	88.1	100.3	114.5	131.3	152.5	181.4	228.8	331.2
1.05	0.20	0.100	36.0	59.8	70.0	81.8	96.2	114.3	139.5	181.4	273.4
1.10	0.01	0.005	38.0	49.8	54.4	59.7	65.8	73.3	83.4	99.6	133.8
1.10	0.02	0.010	31.3	42.0	46.2	51.1	56.8	63.8	73.2	88.3	120.7
1.10	0.05	0.025	22.7	31.8	35.5	39.7	44.7	50.9	59.4	73.0	102.6
1.10	0.10	0.050	16.4	24.2	27.4	31.1	35.5	41.1	48.7	61.1	88.3
1.10	0.20	0.100	10.6	16.8	19.5	22.6	26.3	31.1	37.7	48.7	73.2
1.15	0.01	0.005	18.5	24.0	26.2	28.6	31.5	35.0	39.6	47.2	63.1
1.15	0.02	0.010	15.4	20.4	22.4	24.6	27.2	30.5	34.9	41.9	57.0
1.15	0.05	0.025	11.4	15.6	17.4	19.3	21.6	24.5	28.5	34.8	48.6
1.15	0.10	0.050	8.5	12.1	13.6	15.3	17.4	20.0	23.5	29.3	41.9
1.15	0.20	0.100	5.7	8.7	9.9	11.4	13.1	15.3	18.4	23.5	34.9
1.20	0.01	0.005	11.6	14.8	16.1	17.5	19.2	21.2	24.0	28.4	37.7
1.20	0.02	0.010	9.7	12.7	13.8	15.1	16.7	18.6	21.2	25.3	34.2
1.20	0.05	0.025	7.4	9.9	10.9	12.0	13.4	15.1	17.4	21.1	29.2
1.20	0.10	0.050	5.6	7.8	8.7	9.7	10.9	12.4	14.5	17.9	25.3
1.20	0.20	0.100	3.9	5.7	6.5	7.3	8.4	9.7	11.5	14.5	21.2
1.25	0.01	0.005	8.3	10.4	11.3	12.3	13.4	14.7	16.6	19.5	25.8
1.25	0.02	0.010	7.1	9.0	9.8	10.7	11.7	13.0	14.7	17.5	23.4
1.25	0.05	0.025	5.4	7.1	7.8	8.6	9.5	10.7	12.2	14.7	20.1
1.25	0.10	0.050	4.2	5.7	6.3	7.0	7.8	8.9	10.2	12.5	17.5
1.25	0.20	0.100	3.0	4.3	4.8	5.4	6.1	7.0	8.2	10.2	14.7
1.30	0.01	0.005	6.5	8.0	8.7	9.4	10.2	11.2	12.5	14.6	19.1
1.30	0.02	0.010	5.6	7.0	7.6	8.2	9.0	9.9	11.1	13.1	17.4
1.30	0.05	0.025	4.3	5.6	6.1	6.7	7.4	8.2	9.3	11.1	15.0
1.30	0.10	0.050	3.4	4.6	5.0	5.5	6.1	6.9	7.9	9.5	13.1
1.30	0.20	0.100	2.5	3.5	3.9	4.3	4.9	5.5	6.4	7.9	11.1
1.35	0.01	0.005	5.3	6.6	7.0	7.6	8.2	9.0	10.0	11.6	15.1
1.35	0.02	0.010	4.6	5.7	6.2	6.7	7.3	8.0	8.9	10.5	13.8
1.35	0.05	0.025	3.6	4.7	5.1	5.5	6.0	6.7	7.5	8.9	11.9
1.35	0.10	0.050	2.9	3.8	4.2	4.6	5.1	5.7	6.4	7.7	10.5
1.35	0.20	0.100	2.1	3.0	3.3	3.6	4.1	4.6	5.3	6.4	8.9
1.40	0.01	0.005	4.6	5.6	6.0	6.4	6.9	7.5	8.3	9.6	12.4
1.40	0.02	0.010	4.0	4.9	5.3	5.7	6.1	6.7	7.5	8.7	11.3
1.40	0.05	0.025	3.2	4.0	4.3	4.7	5.1	5.7	6.4	7.5	9.9
1.40	0.10	0.050	2.5	3.3	3.6	4.0	4.3	4.8	5.5	6.5	8.7
1.40	0.20	0.100	1.9	2.6	2.9	3.2	3.5	3.9	4.5	5.5	7.5
1.45	0.01	0.005	4.0	4.9	5.2	5.6	6.0	6.5	7.2	8.3	10.5
1.45	0.02	0.010	3.5	4.3	4.6	5.0	5.4	5.8	6.5	7.5	9.7
1.45	0.05	0.025	2.8	3.5	3.8	4.1	4.5	4.9	5.5	6.5	8.5
1.45	0.10	0.050	2.3	2.9	3.2	3.5	3.8	4.2	4.8	5.7	7.5
1.45	0.20	0.100	1.7	2.3	2.5	2.8	3.1	3.5	4.0	4.8	6.5
1.50	0.01	0.005	3.6	4.3	4.6	4.9	5.3	5.7	6.3	7.2	9.2
1.50	0.02	0.010	3.2	3.9	4.1	4.4	4.8	5.2	5.7	6.6	8.4
1.50	0.05	0.025	2.5	3.2	3.4	3.7	4.0	4.4	4.9	5.7	7.4
1.50	0.10	0.050	2.1	2.7	2.9	3.1	3.4	3.8	4.3	5.0	6.6
1.50	0.20	0.100	1.5	2.1	2.3	2.5	2.8	3.1	3.6	4.3	5.7
1.55	0.01	0.005	3.3	3.9	4.2	4.5	4.8	5.2	5.7	6.5	8.1
1.55	0.02	0.010	2.9	3.5	3.7	4.0	4.3	4.7	5.2	5.9	7.5
1.55	0.05	0.025	2.3	2.9	3.1	3.4	3.7	4.0	4.4	5.2	6.6
1.55	0.10	0.050	1.9	2.4	2.6	2.9	3.1	3.5	3.9	4.5	5.9
1.55	0.20	0.100	1.4	1.9	2.1	2.3	2.6	2.9	3.3	3.9	5.2

Table 10.2

Table 10.2 Continued.

△	α 2-sided	α 1-sided	0.50	0.65	0.70	0.75	0.80	0.85	0.90	0.95	0.99
						Power $1 - \beta$					

80 subjects entered per unit time

△	2-sided	1-sided	0.50	0.65	0.70	0.75	0.80	0.85	0.90	0.95	0.99
1.60	0.01	0.005	3.0	3.6	3.8	4.1	4.4	4.7	5.2	5.9	7.4
1.60	0.02	0.010	2.7	3.2	3.4	3.7	4.0	4.3	4.7	5.4	6.8
1.60	0.05	0.025	2.2	2.7	2.9	3.1	3.4	3.7	4.1	4.7	6.0
1.60	0.10	0.050	1.8	2.3	2.4	2.7	2.9	3.2	3.6	4.2	5.4
1.60	0.20	0.100	1.3	1.8	2.0	2.2	2.4	2.7	3.0	3.6	4.7
1.65	0.01	0.005	2.8	3.4	3.6	3.8	4.1	4.4	4.8	5.4	6.7
1.65	0.02	0.010	2.5	3.0	3.2	3.4	3.7	4.0	4.4	5.0	6.2
1.65	0.05	0.025	2.0	2.5	2.7	2.9	3.1	3.4	3.8	4.4	5.5
1.65	0.10	0.050	1.7	2.1	2.3	2.5	2.7	3.0	3.3	3.9	5.0
1.65	0.20	0.100	1.3	1.7	1.8	2.0	2.2	2.5	2.8	3.3	4.4
1.70	0.01	0.005	2.6	3.2	3.3	3.5	3.8	4.1	4.4	5.0	6.2
1.70	0.02	0.010	2.3	2.8	3.0	3.2	3.4	3.7	4.1	4.6	5.8
1.70	0.05	0.025	1.9	2.4	2.5	2.7	2.9	3.2	3.5	4.1	5.1
1.70	0.10	0.050	1.6	2.0	2.2	2.3	2.5	2.8	3.1	3.6	4.6
1.70	0.20	0.100	1.2	1.6	1.7	1.9	2.1	2.3	2.6	3.1	4.1
1.75	0.01	0.005	2.5	3.0	3.2	3.3	3.6	3.8	4.2	4.7	5.8
1.75	0.02	0.010	2.2	2.7	2.8	3.0	3.2	3.5	3.8	4.3	5.4
1.75	0.05	0.025	1.8	2.2	2.4	2.6	2.8	3.0	3.3	3.8	4.8
1.75	0.10	0.050	1.5	1.9	2.0	2.2	2.4	2.6	2.9	3.4	4.3
1.75	0.20	0.100	1.1	1.5	1.6	1.8	2.0	2.2	2.5	2.9	3.8
1.80	0.01	0.005	2.4	2.8	3.0	3.2	3.4	3.6	3.9	4.4	5.4
1.80	0.02	0.010	2.1	2.5	2.7	2.9	3.1	3.3	3.6	4.1	5.1
1.80	0.05	0.025	1.7	2.1	2.3	2.4	2.6	2.9	3.2	3.6	4.5
1.80	0.10	0.050	1.4	1.8	1.9	2.1	2.3	2.5	2.8	3.2	4.1
1.80	0.20	0.100	1.1	1.4	1.6	1.7	1.9	2.1	2.4	2.8	3.6
1.85	0.01	0.005	2.3	2.7	2.8	3.0	3.2	3.4	3.7	4.2	5.1
1.85	0.02	0.010	2.0	2.4	2.6	2.7	2.9	3.1	3.4	3.9	4.8
1.85	0.05	0.025	1.7	2.0	2.2	2.3	2.5	2.7	3.0	3.4	4.3
1.85	0.10	0.050	1.4	1.7	1.9	2.0	2.2	2.4	2.6	3.1	3.9
1.85	0.20	0.100	1.0	1.4	1.5	1.7	1.8	2.0	2.3	2.6	3.4
1.90	0.01	0.005	2.2	2.6	2.7	2.9	3.1	3.3	3.6	4.0	4.9
1.90	0.02	0.010	1.9	2.3	2.5	2.6	2.8	3.0	3.3	3.7	4.6
1.90	0.05	0.025	1.6	2.0	2.1	2.2	2.4	2.6	2.9	3.3	4.1
1.90	0.10	0.050	1.3	1.7	1.8	1.9	2.1	2.3	2.5	2.9	3.7
1.90	0.20	0.100	1.0	1.3	1.5	1.6	1.7	1.9	2.2	2.5	3.3
1.95	0.01	0.005	2.1	2.5	2.6	2.8	2.9	3.1	3.4	3.8	4.7
1.95	0.02	0.010	1.9	2.2	2.4	2.5	2.7	2.9	3.1	3.6	4.4
1.95	0.05	0.025	1.5	1.9	2.0	2.2	2.3	2.5	2.8	3.1	3.9
1.95	0.10	0.050	1.3	1.6	1.7	1.9	2.0	2.2	2.4	2.8	3.6
1.95	0.20	0.100	1.0	1.3	1.4	1.5	1.7	1.9	2.1	2.4	3.1
2.00	0.01	0.005	2.0	2.4	2.5	2.7	2.8	3.0	3.3	3.7	4.5
2.00	0.02	0.010	1.8	2.1	2.3	2.4	2.6	2.8	3.0	3.4	4.2
2.00	0.05	0.025	1.5	1.8	1.9	2.1	2.2	2.4	2.7	3.0	3.8
2.00	0.10	0.050	1.2	1.5	1.7	1.8	1.9	2.1	2.3	2.7	3.4
2.00	0.20	0.100	0.9	1.2	1.4	1.5	1.6	1.8	2.0	2.3	3.0
2.25	0.01	0.005	1.8	2.1	2.2	2.3	2.4	2.6	2.8	3.1	3.8
2.25	0.02	0.010	1.6	1.9	2.0	2.1	2.2	2.4	2.6	2.9	3.5
2.25	0.05	0.025	1.3	1.6	1.7	1.8	1.9	2.1	2.3	2.6	3.2
2.25	0.10	0.050	1.1	1.3	1.5	1.6	1.7	1.8	2.0	2.3	2.9
2.25	0.20	0.100	0.8	1.1	1.2	1.3	1.4	1.6	1.7	2.0	2.6
2.50	0.01	0.005	1.6	1.9	2.0	2.1	2.2	2.3	2.5	2.8	3.4
2.50	0.02	0.010	1.4	1.7	1.8	1.9	2.0	2.2	2.3	2.6	3.2
2.50	0.05	0.025	1.2	1.4	1.5	1.6	1.7	1.9	2.1	2.3	2.9
2.50	0.10	0.050	1.0	1.2	1.3	1.4	1.5	1.7	1.8	2.1	2.6
2.50	0.20	0.100	0.7	1.0	1.1	1.2	1.3	1.4	1.6	1.8	2.3

Table 10.2

Table 10.2 Continued.

△	α 2-sided	α 1-sided	0.50	0.65	0.70	0.75	0.80	0.85	0.90	0.95	0.99
					Power 1 − β						

					80 subjects entered per unit time						
2.75	0.01	0.005	1.5	1.7	1.8	1.9	2.0	2.2	2.3	2.6	3.1
2.75	0.02	0.010	1.3	1.6	1.7	1.8	1.9	2.0	2.2	2.4	2.9
2.75	0.05	0.025	1.1	1.3	1.4	1.5	1.6	1.7	1.9	2.2	2.6
2.75	0.10	0.050	0.9	1.1	1.2	1.3	1.4	1.5	1.7	1.9	2.4
2.75	0.20	0.100	0.7	0.9	1.0	1.1	1.2	1.3	1.5	1.7	2.2
3.00	0.01	0.005	1.4	1.6	1.7	1.8	1.9	2.0	2.2	2.4	2.9
3.00	0.02	0.010	1.2	1.5	1.6	1.6	1.8	1.9	2.0	2.3	2.7
3.00	0.05	0.025	1.0	1.3	1.3	1.4	1.5	1.6	1.8	2.0	2.5
3.00	0.10	0.050	0.9	1.1	1.2	1.2	1.3	1.5	1.6	1.8	2.3
3.00	0.20	0.100	0.7	0.9	0.9	1.0	1.1	1.2	1.4	1.6	2.0
3.25	0.01	0.005	1.3	1.5	1.6	1.7	1.8	1.9	2.1	2.3	2.7
3.25	0.02	0.010	1.2	1.4	1.5	1.6	1.7	1.8	1.9	2.2	2.6
3.25	0.05	0.025	1.0	1.2	1.3	1.4	1.5	1.6	1.7	1.9	2.3
3.25	0.10	0.050	0.8	1.0	1.1	1.2	1.3	1.4	1.5	1.7	2.2
3.25	0.20	0.100	0.6	0.8	0.9	1.0	1.1	1.2	1.3	1.5	1.9
3.50	0.01	0.005	1.3	1.5	1.6	1.6	1.7	1.9	2.0	2.2	2.6
3.50	0.02	0.010	1.1	1.4	1.4	1.5	1.6	1.7	1.9	2.1	2.5
3.50	0.05	0.025	1.0	1.2	1.2	1.3	1.4	1.5	1.6	1.9	2.3
3.50	0.10	0.050	0.8	1.0	1.1	1.1	1.2	1.3	1.5	1.7	2.1
3.50	0.20	0.100	0.6	0.8	0.9	1.0	1.0	1.1	1.3	1.5	1.9
3.75	0.01	0.005	1.2	1.4	1.5	1.6	1.7	1.8	1.9	2.1	2.5
3.75	0.02	0.010	1.1	1.3	1.4	1.5	1.6	1.7	1.8	2.0	2.4
3.75	0.05	0.025	0.9	1.1	1.2	1.3	1.4	1.5	1.6	1.8	2.2
3.75	0.10	0.050	0.8	1.0	1.0	1.1	1.2	1.3	1.4	1.6	2.0
3.75	0.20	0.100	0.6	0.8	0.9	0.9	1.0	1.1	1.2	1.4	1.8
4.00	0.01	0.005	1.2	1.4	1.5	1.6	1.6	1.7	1.9	2.1	2.5
4.00	0.02	0.010	1.1	1.3	1.4	1.4	1.5	1.6	1.7	1.9	2.3
4.00	0.05	0.025	0.9	1.1	1.2	1.2	1.3	1.4	1.6	1.7	2.1
4.00	0.10	0.050	0.8	0.9	1.0	1.1	1.2	1.3	1.4	1.6	1.9
4.00	0.20	0.100	0.6	0.8	0.8	0.9	1.0	1.1	1.2	1.4	1.7
4.25	0.01	0.005	1.2	1.4	1.4	1.5	1.6	1.7	1.8	2.0	2.4
4.25	0.02	0.010	1.1	1.3	1.3	1.4	1.5	1.6	1.7	1.9	2.3
4.25	0.05	0.025	0.9	1.1	1.1	1.2	1.3	1.4	1.5	1.7	2.1
4.25	0.10	0.050	0.7	0.9	1.0	1.1	1.1	1.2	1.4	1.5	1.9
4.25	0.20	0.100	0.6	0.7	0.8	0.9	1.0	1.1	1.2	1.4	1.7

					90 subjects entered per unit time						
1.05	0.01	0.005	125.4	165.2	180.9	198.7	219.5	245.1	277.8	332.6	448.7
1.05	0.02	0.010	102.5	138.8	153.2	169.6	188.9	212.6	244.5	294.4	404.2
1.05	0.05	0.025	73.2	104.2	116.7	131.1	148.0	169.1	197.7	244.1	343.0
1.05	0.10	0.050	52.0	78.4	89.3	101.9	116.9	135.7	161.4	203.5	294.4
1.05	0.20	0.100	32.1	53.4	62.4	72.9	85.6	101.8	124.1	161.4	244.5
1.10	0.01	0.005	34.0	44.4	48.5	53.2	58.7	65.4	74.3	88.7	119.1
1.10	0.02	0.010	28.0	37.5	41.3	45.6	50.6	56.8	65.2	78.7	107.4
1.10	0.05	0.025	20.3	28.4	31.7	35.5	39.9	45.4	52.9	65.1	91.4
1.10	0.10	0.050	14.8	21.7	24.5	27.8	31.8	36.7	43.4	54.5	78.7
1.10	0.20	0.100	9.6	15.1	17.5	20.2	23.6	27.8	33.7	43.4	65.2
1.15	0.01	0.005	16.7	21.5	23.4	25.6	28.1	31.2	35.4	42.1	56.2
1.15	0.02	0.010	13.9	18.3	20.0	22.0	24.4	27.3	31.2	37.4	50.8
1.15	0.05	0.025	10.3	14.1	15.6	17.3	19.4	22.0	25.5	31.1	43.4
1.15	0.10	0.050	7.7	10.9	12.3	13.8	15.6	17.9	21.0	26.2	37.4
1.15	0.20	0.100	5.2	7.9	9.0	10.3	11.8	13.8	16.5	21.0	31.2
1.20	0.01	0.005	10.5	13.3	14.5	15.7	17.2	19.0	21.5	25.4	33.7
1.20	0.02	0.010	8.8	11.4	12.5	13.6	15.0	16.7	19.0	22.7	30.5
1.20	0.05	0.025	6.7	9.0	9.9	10.9	12.1	13.6	15.6	19.0	26.2
1.20	0.10	0.050	5.2	7.1	7.9	8.8	9.9	11.2	13.1	16.1	22.7
1.20	0.20	0.100	3.6	5.3	5.9	6.7	7.6	8.8	10.4	13.1	19.0

Table 10.2

Table 10.2 Continued.

$\triangle$	α 2-sided	1-sided	0.50	0.65	0.70	Power $1 - \beta$ 0.75	0.80	0.85	0.90	0.95	0.99
						90 subjects entered per unit time					
1.25	0.01	0.005	7.5	9.5	10.2	11.1	12.1	13.3	14.9	17.5	23.1
1.25	0.02	0.010	6.4	8.2	8.9	9.7	10.6	11.7	13.3	15.7	21.0
1.25	0.05	0.025	5.0	6.5	7.1	7.8	8.6	9.7	11.0	13.2	18.0
1.25	0.10	0.050	3.9	5.3	5.8	6.4	7.1	8.0	9.3	11.3	15.7
1.25	0.20	0.100	2.8	4.0	4.4	5.0	5.6	6.4	7.5	9.3	13.3
1.30	0.01	0.005	5.9	7.3	7.9	8.5	9.2	10.1	11.3	13.2	17.2
1.30	0.02	0.010	5.1	6.4	6.9	7.5	8.2	9.0	10.1	11.9	15.7
1.30	0.05	0.025	4.0	5.2	5.6	6.1	6.7	7.5	8.5	10.1	13.5
1.30	0.10	0.050	3.2	4.2	4.6	5.1	5.6	6.3	7.2	8.7	11.9
1.30	0.20	0.100	2.3	3.2	3.6	4.0	4.5	5.1	5.9	7.2	10.1
1.35	0.01	0.005	4.9	6.0	6.4	6.9	7.5	8.2	9.1	10.5	13.6
1.35	0.02	0.010	4.2	5.3	5.7	6.1	6.6	7.3	8.1	9.5	12.4
1.35	0.05	0.025	3.4	4.3	4.7	5.1	5.5	6.1	6.9	8.1	10.8
1.35	0.10	0.050	2.7	3.5	3.9	4.2	4.7	5.2	5.9	7.0	9.5
1.35	0.20	0.100	2.0	2.7	3.0	3.4	3.7	4.2	4.9	5.9	8.1
1.40	0.01	0.005	4.2	5.1	5.5	5.9	6.3	6.9	7.6	8.8	11.2
1.40	0.02	0.010	3.7	4.5	4.8	5.2	5.6	6.2	6.8	7.9	10.3
1.40	0.05	0.025	2.9	3.7	4.0	4.3	4.7	5.2	5.8	6.8	9.0
1.40	0.10	0.050	2.4	3.1	3.3	3.7	4.0	4.4	5.0	6.0	7.9
1.40	0.20	0.100	1.8	2.4	2.7	2.9	3.3	3.6	4.2	5.0	6.8
1.45	0.01	0.005	3.7	4.5	4.8	5.1	5.5	5.9	6.6	7.5	9.5
1.45	0.02	0.010	3.2	4.0	4.3	4.6	4.9	5.4	5.9	6.9	8.8
1.45	0.05	0.025	2.6	3.3	3.5	3.8	4.2	4.6	5.1	5.9	7.7
1.45	0.10	0.050	2.1	2.7	3.0	3.2	3.5	3.9	4.4	5.2	6.9
1.45	0.20	0.100	1.6	2.1	2.4	2.6	2.9	3.2	3.7	4.4	5.9
1.50	0.01	0.005	3.3	4.0	4.3	4.5	4.9	5.3	5.8	6.6	8.3
1.50	0.02	0.010	2.9	3.6	3.8	4.1	4.4	4.8	5.3	6.1	7.7
1.50	0.05	0.025	2.4	3.0	3.2	3.4	3.7	4.1	4.5	5.3	6.8
1.50	0.10	0.050	1.9	2.5	2.7	2.9	3.2	3.5	3.9	4.6	6.1
1.50	0.20	0.100	1.4	2.0	2.1	2.4	2.6	2.9	3.3	3.9	5.3
1.55	0.01	0.005	3.1	3.6	3.9	4.1	4.4	4.8	5.2	5.9	7.4
1.55	0.02	0.010	2.7	3.3	3.5	3.7	4.0	4.3	4.8	5.4	6.9
1.55	0.05	0.025	2.2	2.7	2.9	3.1	3.4	3.7	4.1	4.7	6.1
1.55	0.10	0.050	1.8	2.3	2.5	2.7	2.9	3.2	3.6	4.2	5.4
1.55	0.20	0.100	1.3	1.8	2.0	2.2	2.4	2.7	3.0	3.6	4.8
1.60	0.01	0.005	2.8	3.4	3.6	3.8	4.0	4.4	4.8	5.4	6.7
1.60	0.02	0.010	2.5	3.0	3.2	3.4	3.7	4.0	4.4	5.0	6.2
1.60	0.05	0.025	2.0	2.5	2.7	2.9	3.1	3.4	3.8	4.3	5.5
1.60	0.10	0.050	1.6	2.1	2.3	2.5	2.7	3.0	3.3	3.8	5.0
1.60	0.20	0.100	1.2	1.7	1.8	2.0	2.2	2.5	2.8	3.3	4.4
1.65	0.01	0.005	2.6	3.1	3.3	3.5	3.8	4.0	4.4	5.0	6.2
1.65	0.02	0.010	2.3	2.8	3.0	3.2	3.4	3.7	4.0	4.6	5.7
1.65	0.05	0.025	1.9	2.3	2.5	2.7	2.9	3.2	3.5	4.0	5.1
1.65	0.10	0.050	1.5	2.0	2.1	2.3	2.5	2.8	3.1	3.6	4.6
1.65	0.20	0.100	1.2	1.6	1.7	1.9	2.1	2.3	2.6	3.1	4.0
1.70	0.01	0.005	2.5	2.9	3.1	3.3	3.5	3.8	4.1	4.6	5.7
1.70	0.02	0.010	2.2	2.6	2.8	3.0	3.2	3.4	3.8	4.3	5.3
1.70	0.05	0.025	1.8	2.2	2.4	2.5	2.7	3.0	3.3	3.8	4.7
1.70	0.10	0.050	1.5	1.9	2.0	2.2	2.4	2.6	2.9	3.3	4.3
1.70	0.20	0.100	1.1	1.5	1.6	1.8	2.0	2.2	2.4	2.9	3.8
1.75	0.01	0.005	2.3	2.8	2.9	3.1	3.3	3.5	3.9	4.3	5.3
1.75	0.02	0.010	2.1	2.5	2.6	2.8	3.0	3.2	3.5	4.0	5.0
1.75	0.05	0.025	1.7	2.1	2.2	2.4	2.6	2.8	3.1	3.5	4.4
1.75	0.10	0.050	1.4	1.8	1.9	2.1	2.2	2.4	2.7	3.2	4.0
1.75	0.20	0.100	1.1	1.4	1.5	1.7	1.9	2.1	2.3	2.7	3.5

Table 10.2

Table 10.2 Continued.

△	α 2-sided	1-sided	0.50	0.65	0.70	Power 1 − β 0.75	0.80	0.85	0.90	0.95	0.99
						90 subjects entered per unit time					
1.80	0.01	0.005	2.2	2.6	2.8	2.9	3.1	3.4	3.7	4.1	5.0
1.80	0.02	0.010	2.0	2.4	2.5	2.7	2.8	3.1	3.4	3.8	4.7
1.80	0.05	0.025	1.6	2.0	2.1	2.3	2.5	2.7	2.9	3.4	4.2
1.80	0.10	0.050	1.3	1.7	1.8	2.0	2.1	2.3	2.6	3.0	3.8
1.80	0.20	0.100	1.0	1.3	1.5	1.6	1.8	2.0	2.2	2.6	3.4
1.85	0.01	0.005	2.1	2.5	2.6	2.8	3.0	3.2	3.5	3.9	4.8
1.85	0.02	0.010	1.9	2.3	2.4	2.5	2.7	2.9	3.2	3.6	4.4
1.85	0.05	0.025	1.5	1.9	2.0	2.2	2.3	2.5	2.8	3.2	4.0
1.85	0.10	0.050	1.3	1.6	1.7	1.9	2.0	2.2	2.5	2.8	3.6
1.85	0.20	0.100	1.0	1.3	1.4	1.5	1.7	1.9	2.1	2.5	3.2
1.90	0.01	0.005	2.0	2.4	2.5	2.7	2.8	3.1	3.3	3.7	4.5
1.90	0.02	0.010	1.8	2.2	2.3	2.4	2.6	2.8	3.0	3.4	4.2
1.90	0.05	0.025	1.5	1.8	1.9	2.1	2.2	2.4	2.7	3.0	3.8
1.90	0.10	0.050	1.2	1.5	1.7	1.8	2.0	2.1	2.4	2.7	3.4
1.90	0.20	0.100	0.9	1.2	1.4	1.5	1.6	1.8	2.0	2.4	3.0
1.95	0.01	0.005	2.0	2.3	2.4	2.6	2.7	2.9	3.2	3.6	4.3
1.95	0.02	0.010	1.7	2.1	2.2	2.3	2.5	2.7	2.9	3.3	4.0
1.95	0.05	0.025	1.4	1.8	1.9	2.0	2.2	2.3	2.6	2.9	3.6
1.95	0.10	0.050	1.2	1.5	1.6	1.7	1.9	2.0	2.3	2.6	3.3
1.95	0.20	0.100	0.9	1.2	1.3	1.4	1.6	1.7	1.9	2.3	2.9
2.00	0.01	0.005	1.9	2.2	2.3	2.5	2.6	2.8	3.1	3.4	4.1
2.00	0.02	0.010	1.7	2.0	2.1	2.3	2.4	2.6	2.8	3.2	3.9
2.00	0.05	0.025	1.4	1.7	1.8	1.9	2.1	2.3	2.5	2.8	3.5
2.00	0.10	0.050	1.1	1.4	1.6	1.7	1.8	2.0	2.2	2.5	3.2
2.00	0.20	0.100	0.9	1.2	1.3	1.4	1.5	1.7	1.9	2.2	2.8
2.25	0.01	0.005	1.6	1.9	2.0	2.1	2.3	2.4	2.6	2.9	3.5
2.25	0.02	0.010	1.5	1.7	1.8	2.0	2.1	2.2	2.4	2.7	3.3
2.25	0.05	0.025	1.2	1.5	1.6	1.7	1.8	2.0	2.1	2.4	3.0
2.25	0.10	0.050	1.0	1.3	1.4	1.5	1.6	1.7	1.9	2.2	2.7
2.25	0.20	0.100	0.8	1.0	1.1	1.2	1.3	1.5	1.6	1.9	2.4
2.50	0.01	0.005	1.5	1.7	1.8	1.9	2.0	2.2	2.4	2.6	3.1
2.50	0.02	0.010	1.3	1.6	1.7	1.8	1.9	2.0	2.2	2.4	2.9
2.50	0.05	0.025	1.1	1.3	1.4	1.5	1.6	1.8	1.9	2.2	2.7
2.50	0.10	0.050	0.9	1.1	1.2	1.3	1.4	1.6	1.7	2.0	2.4
2.50	0.20	0.100	0.7	0.9	1.0	1.1	1.2	1.3	1.5	1.7	2.2
2.75	0.01	0.005	1.4	1.6	1.7	1.8	1.9	2.0	2.2	2.4	2.9
2.75	0.02	0.010	1.2	1.5	1.5	1.6	1.7	1.9	2.0	2.2	2.7
2.75	0.05	0.025	1.0	1.2	1.3	1.4	1.5	1.6	1.8	2.0	2.5
2.75	0.10	0.050	0.9	1.1	1.1	1.2	1.3	1.4	1.6	1.8	2.2
2.75	0.20	0.100	0.7	0.9	0.9	1.0	1.1	1.2	1.4	1.6	2.0
3.00	0.01	0.005	1.3	1.5	1.6	1.7	1.8	1.9	2.0	2.3	2.7
3.00	0.02	0.010	1.2	1.4	1.5	1.5	1.6	1.8	1.9	2.1	2.5
3.00	0.05	0.025	1.0	1.2	1.3	1.3	1.4	1.5	1.7	1.9	2.3
3.00	0.10	0.050	0.8	1.0	1.1	1.2	1.3	1.4	1.5	1.7	2.1
3.00	0.20	0.100	0.6	0.8	0.9	1.0	1.1	1.2	1.3	1.5	1.9
3.25	0.01	0.005	1.2	1.4	1.5	1.6	1.7	1.8	2.0	2.2	2.6
3.25	0.02	0.010	1.1	1.3	1.4	1.5	1.6	1.7	1.8	2.0	2.4
3.25	0.05	0.025	0.9	1.1	1.2	1.3	1.4	1.5	1.6	1.8	2.2
3.25	0.10	0.050	0.8	1.0	1.0	1.1	1.2	1.3	1.4	1.6	2.0
3.25	0.20	0.100	0.6	0.8	0.9	0.9	1.0	1.1	1.2	1.4	1.8
3.50	0.01	0.005	1.2	1.4	1.5	1.5	1.6	1.7	1.9	2.1	2.5
3.50	0.02	0.010	1.1	1.3	1.3	1.4	1.5	1.6	1.7	1.9	2.3
3.50	0.05	0.025	0.9	1.1	1.2	1.2	1.3	1.4	1.5	1.7	2.1
3.50	0.10	0.050	0.7	0.9	1.0	1.1	1.2	1.3	1.4	1.6	1.9
3.50	0.20	0.100	0.6	0.8	0.8	0.9	1.0	1.1	1.2	1.4	1.7

Table 10.2

Table 10.2 Continued.

△	α 2-sided	1-sided	0.50	0.65	0.70	Power 1 − β 0.75	0.80	0.85	0.90	0.95	0.99
						90 subjects entered per unit time					
3.75	0.01	0.005	1.2	1.4	1.4	1.5	1.6	1.7	1.8	2.0	2.4
3.75	0.02	0.010	1.0	1.2	1.3	1.4	1.5	1.6	1.7	1.9	2.2
3.75	0.05	0.025	0.9	1.1	1.1	1.2	1.3	1.4	1.5	1.7	2.0
3.75	0.10	0.050	0.7	0.9	1.0	1.0	1.1	1.2	1.3	1.5	1.9
3.75	0.20	0.100	0.6	0.7	0.8	0.9	1.0	1.0	1.2	1.3	1.7
4.00	0.01	0.005	1.1	1.3	1.4	1.5	1.5	1.6	1.8	1.9	2.3
4.00	0.02	0.010	1.0	1.2	1.3	1.3	1.4	1.5	1.6	1.8	2.2
4.00	0.05	0.025	0.9	1.0	1.1	1.2	1.2	1.3	1.5	1.6	2.0
4.00	0.10	0.050	0.7	0.9	1.0	1.0	1.1	1.2	1.3	1.5	1.8
4.00	0.20	0.100	0.5	0.7	0.8	0.8	0.9	1.0	1.1	1.3	1.6
4.25	0.01	0.005	1.1	1.3	1.4	1.4	1.5	1.6	1.7	1.9	2.3
4.25	0.02	0.010	1.0	1.2	1.2	1.3	1.4	1.5	1.6	1.8	2.1
4.25	0.05	0.025	0.8	1.0	1.1	1.1	1.2	1.3	1.4	1.6	1.9
4.25	0.10	0.050	0.7	0.9	0.9	1.0	1.1	1.2	1.3	1.4	1.8
4.25	0.20	0.100	0.5	0.7	0.8	0.8	0.9	1.0	1.1	1.3	1.6
						100 subjects entered per unit time					
1.05	0.01	0.005	113.0	148.8	163.0	179.0	197.7	220.7	251.5	299.3	403.8
1.05	0.02	0.010	92.4	125.0	138.0	152.8	170.1	191.5	220.2	265.0	363.8
1.05	0.05	0.025	66.0	93.9	105.2	118.1	133.4	152.3	178.0	219.8	308.7
1.05	0.10	0.050	46.9	70.7	80.6	91.9	105.4	122.3	145.4	183.3	265.0
1.05	0.20	0.100	29.1	48.2	56.3	65.8	77.2	91.8	111.9	145.4	220.2
1.10	0.01	0.005	30.7	40.1	43.8	48.0	52.9	59.0	67.0	80.0	107.3
1.10	0.02	0.010	25.3	33.9	37.3	41.2	45.7	51.3	58.8	71.0	96.8
1.10	0.05	0.025	18.4	25.7	28.7	32.1	36.1	41.1	47.8	58.7	82.4
1.10	0.10	0.050	13.4	19.7	22.2	25.2	28.7	33.2	39.2	49.2	71.0
1.10	0.20	0.100	8.8	13.8	15.9	18.4	21.4	25.2	30.4	39.2	58.8
1.15	0.01	0.005	15.1	19.5	21.2	23.2	25.5	28.3	32.0	38.0	50.8
1.15	0.02	0.010	12.6	16.6	18.2	20.0	22.1	24.7	28.2	33.9	45.9
1.15	0.05	0.025	9.4	12.8	14.2	15.8	17.6	19.9	23.1	28.2	????
1.15	0.10	0.050	7.1	10.0	11.2	12.6	14.2	16.3	19.1	23.7	33.9
1.15	0.20	0.100	4.9	7.2	8.2	9.4	10.8	12.6	15.0	19.1	28.2
1.20	0.01	0.005	9.6	12.2	13.2	14.3	15.7	17.3	19.5	23.0	30.5
1.20	0.02	0.010	8.1	10.4	11.4	12.4	13.7	15.2	17.3	20.6	27.6
1.20	0.05	0.025	6.2	8.2	9.0	10.0	11.0	12.4	14.2	17.2	23.7
1.20	0.10	0.050	4.8	6.5	7.2	8.1	9.0	10.3	11.9	14.6	20.6
1.20	0.20	0.100	3.4	4.9	5.5	6.2	7.0	8.1	9.5	11.9	17.3
1.25	0.01	0.005	6.9	8.7	9.4	10.1	11.0	12.1	13.6	16.0	20.9
1.25	0.02	0.010	5.9	7.5	8.2	8.9	9.7	10.7	12.1	14.3	19.0
1.25	0.05	0.025	4.6	6.0	6.6	7.2	7.9	8.9	10.1	12.1	16.4
1.25	0.10	0.050	3.6	4.9	5.4	5.9	6.6	7.4	8.5	10.3	14.3
1.25	0.20	0.100	2.6	3.7	4.1	4.6	5.2	5.9	6.9	8.5	12.1
1.30	0.01	0.005	5.5	6.7	7.2	7.8	8.5	9.3	10.3	12.0	15.7
1.30	0.02	0.010	4.7	5.9	6.4	6.9	7.5	8.2	9.3	10.9	14.3
1.30	0.05	0.025	3.7	4.8	5.2	5.7	6.2	6.9	7.8	9.2	12.4
1.30	0.10	0.050	3.0	3.9	4.3	4.7	5.2	5.8	6.6	8.0	10.9
1.30	0.20	0.100	2.2	3.0	3.3	3.7	4.2	4.7	5.4	6.6	9.3
1.35	0.01	0.005	4.5	5.5	5.9	6.4	6.9	7.5	8.3	9.6	12.4
1.35	0.02	0.010	4.0	4.9	5.3	5.7	6.1	6.7	7.5	8.7	11.3
1.35	0.05	0.025	3.2	4.0	4.3	4.7	5.1	5.6	6.3	7.5	9.9
1.35	0.10	0.050	2.5	3.3	3.6	3.9	4.3	4.8	5.5	6.5	8.7
1.35	0.20	0.100	1.9	2.6	2.8	3.1	3.5	3.9	4.5	5.5	7.5
1.40	0.01	0.005	3.9	4.7	5.1	5.4	5.8	6.3	7.0	8.0	10.3
1.40	0.02	0.010	3.4	4.2	4.5	4.8	5.2	5.7	6.3	7.3	9.4
1.40	0.05	0.025	2.8	3.5	3.7	4.0	4.4	4.8	5.4	6.3	8.2
1.40	0.10	0.050	2.2	2.9	3.1	3.4	3.7	4.1	4.7	5.5	7.3
1.40	0.20	0.100	1.6	2.2	2.5	2.7	3.0	3.4	3.9	4.7	6.3

Table 10.2

Table 10.2 Continued.

△	α 2-sided	1-sided	0.50	0.65	0.70	0.75	0.80	0.85	0.90	0.95	0.99
						Power $1-\beta$					

100 subjects entered per unit time

△	2-sided	1-sided	0.50	0.65	0.70	0.75	0.80	0.85	0.90	0.95	0.99
1.45	0.01	0.005	3.5	4.2	4.4	4.7	5.1	5.5	6.1	6.9	8.8
1.45	0.02	0.010	3.0	3.7	4.0	4.2	4.6	5.0	5.5	6.3	8.1
1.45	0.05	0.025	2.4	3.1	3.3	3.6	3.9	4.2	4.7	5.5	7.1
1.45	0.10	0.050	2.0	2.6	2.8	3.0	3.3	3.6	4.1	4.8	6.3
1.45	0.20	0.100	1.5	2.0	2.2	2.4	2.7	3.0	3.4	4.1	5.5
1.50	0.01	0.005	3.1	3.7	4.0	4.2	4.5	4.9	5.4	6.1	7.7
1.50	0.02	0.010	2.7	3.3	3.6	3.8	4.1	4.4	4.9	5.6	7.1
1.50	0.05	0.025	2.2	2.8	3.0	3.2	3.5	3.8	4.2	4.9	6.3
1.50	0.10	0.050	1.8	2.3	2.5	2.7	3.0	3.3	3.7	4.3	5.6
1.50	0.20	0.100	1.4	1.8	2.0	2.2	2.4	2.7	3.1	3.7	4.9
1.55	0.01	0.005	2.9	3.4	3.6	3.8	4.1	4.4	4.8	5.5	6.9
1.55	0.02	0.010	2.5	3.0	3.2	3.5	3.7	4.0	4.4	5.1	6.3
1.55	0.05	0.025	2.0	2.5	2.7	2.9	3.2	3.5	3.8	4.4	5.6
1.55	0.10	0.050	1.7	2.1	2.3	2.5	2.7	3.0	3.4	3.9	5.1
1.55	0.20	0.100	1.3	1.7	1.9	2.0	2.2	2.5	2.8	3.4	4.4
1.60	0.01	0.005	2.6	3.1	3.3	3.5	3.8	4.1	4.4	5.0	6.2
1.60	0.02	0.010	2.3	2.8	3.0	3.2	3.4	3.7	4.1	4.6	5.8
1.60	0.05	0.025	1.9	2.3	2.5	2.7	2.9	3.2	3.5	4.1	5.1
1.60	0.10	0.050	1.6	2.0	2.1	2.3	2.5	2.8	3.1	3.6	4.6
1.60	0.20	0.100	1.2	1.6	1.7	1.9	2.1	2.3	2.6	3.1	4.1
1.65	0.01	0.005	2.5	2.9	3.1	3.3	3.5	3.8	4.1	4.6	5.7
1.65	0.02	0.010	2.2	2.6	2.8	3.0	3.2	3.4	3.8	4.3	5.3
1.65	0.05	0.025	1.8	2.2	2.4	2.5	2.7	3.0	3.3	3.8	4.7
1.65	0.10	0.050	1.5	1.9	2.0	2.2	2.4	2.6	2.9	3.3	4.3
1.65	0.20	0.100	1.1	1.5	1.6	1.8	2.0	2.2	2.4	2.9	3.8
1.70	0.01	0.005	2.3	2.7	2.9	3.1	3.3	3.5	3.8	4.3	5.3
1.70	0.02	0.010	2.0	2.5	2.6	2.8	3.0	3.2	3.5	4.0	4.9
1.70	0.05	0.025	1.7	2.1	2.2	2.4	2.6	2.8	3.1	3.5	4.4
1.70	0.10	0.050	1.4	1.7	1.9	2.0	2.2	2.4	2.7	3.1	4.0
1.70	0.20	0.100	1.0	1.4	1.5	1.7	1.8	2.0	2.3	2.7	3.5
1.75	0.01	0.005	2.2	2.6	2.7	2.9	3.1	3.3	3.6	4.1	5.0
1.75	0.02	0.010	1.9	2.3	2.5	2.6	2.8	3.0	3.3	3.7	4.6
1.75	0.05	0.025	1.6	2.0	2.1	2.2	2.4	2.6	2.9	3.3	4.1
1.75	0.10	0.050	1.3	1.7	1.8	1.9	2.1	2.3	2.6	2.9	3.7
1.75	0.20	0.100	1.0	1.3	1.5	1.6	1.7	1.9	2.2	2.6	3.3
1.80	0.01	0.005	2.1	2.5	2.6	2.8	2.9	3.1	3.4	3.8	4.7
1.80	0.02	0.010	1.8	2.2	2.3	2.5	2.7	2.9	3.1	3.5	4.4
1.80	0.05	0.025	1.5	1.9	2.0	2.1	2.3	2.5	2.7	3.1	3.9
1.80	0.10	0.050	1.3	1.6	1.7	1.8	2.0	2.2	2.4	2.8	3.5
1.80	0.20	0.100	1.0	1.3	1.4	1.5	1.7	1.8	2.1	2.4	3.1
1.85	0.01	0.005	2.0	2.3	2.5	2.6	2.8	3.0	3.2	3.6	4.4
1.85	0.02	0.010	1.8	2.1	2.2	2.4	2.5	2.7	3.0	3.4	4.1
1.85	0.05	0.025	1.5	1.8	1.9	2.0	2.2	2.4	2.6	3.0	3.7
1.85	0.10	0.050	1.2	1.5	1.6	1.8	1.9	2.1	2.3	2.7	3.4
1.85	0.20	0.100	0.9	1.2	1.3	1.5	1.6	1.8	2.0	2.3	3.0
1.90	0.01	0.005	1.9	2.2	2.4	2.5	2.7	2.9	3.1	3.5	4.2
1.90	0.02	0.010	1.7	2.0	2.2	2.3	2.4	2.6	2.9	3.2	3.9
1.90	0.05	0.025	1.4	1.7	1.8	2.0	2.1	2.3	2.5	2.9	3.5
1.90	0.10	0.050	1.2	1.5	1.6	1.7	1.8	2.0	2.2	2.6	3.2
1.90	0.20	0.100	0.9	1.2	1.3	1.4	1.5	1.7	1.9	2.2	2.9
1.95	0.01	0.005	1.8	2.2	2.3	2.4	2.6	2.7	3.0	3.3	4.0
1.95	0.02	0.010	1.6	2.0	2.1	2.2	2.3	2.5	2.7	3.1	3.8
1.95	0.05	0.025	1.4	1.7	1.8	1.9	2.0	2.2	2.4	2.7	3.4
1.95	0.10	0.050	1.1	1.4	1.5	1.6	1.8	1.9	2.1	2.5	3.1
1.95	0.20	0.100	0.9	1.1	1.2	1.4	1.5	1.6	1.8	2.1	2.7

Table 10.2

Table 10.2 Continued.

$\triangle$	α 2-sided	1-sided	Power $1 - \beta$ 0.50	0.65	0.70	0.75	0.80	0.85	0.90	0.95	0.99

100 subjects entered per unit time

$\triangle$	2-sided	1-sided	0.50	0.65	0.70	0.75	0.80	0.85	0.90	0.95	0.99
2.00	0.01	0.005	1.8	2.1	2.2	2.3	2.5	2.6	2.9	3.2	3.9
2.00	0.02	0.010	1.6	1.9	2.0	2.1	2.3	2.4	2.6	3.0	3.6
2.00	0.05	0.025	1.3	1.6	1.7	1.8	2.0	2.1	2.3	2.6	3.3
2.00	0.10	0.050	1.1	1.4	1.5	1.6	1.7	1.9	2.1	2.4	3.0
2.00	0.20	0.100	0.8	1.1	1.2	1.3	1.4	1.6	1.8	2.1	2.6
2.25	0.01	0.005	1.6	1.8	1.9	2.0	2.1	2.3	2.5	2.7	3.3
2.25	0.02	0.010	1.4	1.6	1.7	1.8	2.0	2.1	2.3	2.6	3.1
2.25	0.05	0.025	1.1	1.4	1.5	1.6	1.7	1.8	2.0	2.3	2.8
2.25	0.10	0.050	0.9	1.2	1.3	1.4	1.5	1.6	1.8	2.0	2.6
2.25	0.20	0.100	0.7	1.0	1.0	1.1	1.2	1.4	1.5	1.8	2.3
2.50	0.01	0.005	1.4	1.6	1.7	1.8	1.9	2.1	2.2	2.5	2.9
2.50	0.02	0.010	1.3	1.5	1.6	1.7	1.8	1.9	2.1	2.3	2.8
2.50	0.05	0.025	1.0	1.3	1.3	1.4	1.5	1.7	1.8	2.0	2.5
2.50	0.10	0.050	0.9	1.1	1.2	1.2	1.3	1.5	1.6	1.8	2.3
2.50	0.20	0.100	0.7	0.9	1.0	1.0	1.1	1.2	1.4	1.6	2.1
2.75	0.01	0.005	1.3	1.5	1.6	1.7	1.8	1.9	2.0	2.3	2.7
2.75	0.02	0.010	1.2	1.4	1.5	1.5	1.6	1.8	1.9	2.1	2.5
2.75	0.05	0.025	1.0	1.2	1.3	1.3	1.4	1.5	1.7	1.9	2.3
2.75	0.10	0.050	0.8	1.0	1.1	1.2	1.3	1.4	1.5	1.7	2.1
2.75	0.20	0.100	0.6	0.8	0.9	1.0	1.1	1.2	1.3	1.5	1.9
3.00	0.01	0.005	1.2	1.4	1.5	1.6	1.7	1.8	1.9	2.1	2.5
3.00	0.02	0.010	1.1	1.3	1.4	1.5	1.5	1.6	1.8	2.0	2.4
3.00	0.05	0.025	0.9	1.1	1.2	1.3	1.3	1.5	1.6	1.8	2.2
3.00	0.10	0.050	0.8	1.0	1.0	1.1	1.2	1.3	1.4	1.6	2.0
3.00	0.20	0.100	0.6	0.8	0.8	0.9	1.0	1.1	1.2	1.4	1.8
3.25	0.01	0.005	1.2	1.4	1.4	1.5	1.6	1.7	1.8	2.0	2.4
3.25	0.02	0.010	1.1	1.2	1.3	1.4	1.5	1.6	1.7	1.9	2.3
3.25	0.05	0.025	0.9	1.1	1.1	1.2	1.3	1.4	1.5	1.7	2.1
3.25	0.10	0.050	0.7	0.9	1.0	1.1	1.1	1.2	1.3	1.5	1.9
3.25	0.20	0.100	0.6	0.7	0.8	0.9	1.0	1.1	1.2	1.3	1.7
3.50	0.01	0.005	1.1	1.3	1.4	1.5	1.5	1.6	1.8	1.9	2.3
3.50	0.02	0.010	1.0	1.2	1.3	1.3	1.4	1.5	1.6	1.8	2.2
3.50	0.05	0.025	0.9	1.0	1.1	1.2	1.2	1.3	1.5	1.6	2.0
3.50	0.10	0.050	0.7	0.9	0.9	1.0	1.1	1.2	1.3	1.5	1.8
3.50	0.20	0.100	0.5	0.7	0.8	0.9	0.9	1.0	1.1	1.3	1.6
3.75	0.01	0.005	1.1	1.3	1.3	1.4	1.5	1.6	1.7	1.9	2.2
3.75	0.02	0.010	1.0	1.2	1.2	1.3	1.4	1.5	1.6	1.8	2.1
3.75	0.05	0.025	0.8	1.0	1.1	1.1	1.2	1.3	1.4	1.6	1.9
3.75	0.10	0.050	0.7	0.9	0.9	1.0	1.1	1.1	1.3	1.4	1.8
3.75	0.20	0.100	0.5	0.7	0.8	0.8	0.9	1.0	1.1	1.3	1.6
4.00	0.01	0.005	1.1	1.2	1.3	1.4	1.5	1.5	1.7	1.8	2.2
4.00	0.02	0.010	1.0	1.1	1.2	1.3	1.3	1.4	1.5	1.7	2.0
4.00	0.05	0.025	0.8	1.0	1.0	1.1	1.2	1.3	1.4	1.5	1.9
4.00	0.10	0.050	0.7	0.8	0.9	1.0	1.0	1.1	1.2	1.4	1.7
4.00	0.20	0.100	0.5	0.7	0.7	0.8	0.9	1.0	1.1	1.2	1.5
4.25	0.01	0.005	1.1	1.2	1.3	1.4	1.4	1.5	1.6	1.8	2.1
4.25	0.02	0.010	1.0	1.1	1.2	1.2	1.3	1.4	1.5	1.7	2.0
4.25	0.05	0.025	0.8	1.0	1.0	1.1	1.2	1.2	1.3	1.5	1.8
4.25	0.10	0.050	0.7	0.8	0.9	0.9	1.0	1.1	1.2	1.4	1.7
4.25	0.20	0.100	0.5	0.7	0.7	0.8	0.9	0.9	1.0	1.2	1.5

Table 10.2

Table 10.2 Continued.

△	α 2-sided	α 1-sided	Power 1 − β 0.50	0.65	0.70	0.75	0.80	0.85	0.90	0.95	0.99
			150 subjects entered per unit time								
1.05	0.01	0.005	75.8	99.7	109.1	119.8	132.3	147.7	168.2	201.0	269.2
1.05	0.02	0.010	62.1	83.9	92.5	102.4	113.9	128.2	147.3	178.1	244.0
1.05	0.05	0.025	44.5	63.1	70.6	79.2	89.4	102.1	119.2	147.0	207.3
1.05	0.10	0.050	31.8	47.7	54.2	61.7	70.7	82.0	97.4	122.7	178.1
1.05	0.20	0.100	19.9	32.6	38.0	44.3	52.0	61.7	75.1	97.4	147.3
1.10	0.01	0.005	21.0	27.3	29.7	32.5	35.8	39.8	45.2	53.8	72.1
1.10	0.02	0.010	17.4	23.1	25.4	28.0	31.0	34.7	39.7	47.8	65.1
1.10	0.05	0.025	12.8	17.7	19.6	21.9	24.6	27.9	32.4	39.7	55.5
1.10	0.10	0.050	9.5	13.6	15.3	17.3	19.7	22.6	26.7	33.3	47.8
1.10	0.20	0.100	6.3	9.7	11.1	12.7	14.8	17.3	20.8	26.7	39.7
1.15	0.01	0.005	10.6	13.5	14.7	16.0	17.5	19.4	21.9	25.9	34.4
1.15	0.02	0.010	8.9	11.6	12.7	13.9	15.3	17.0	19.3	23.1	31.1
1.15	0.05	0.025	6.8	9.1	10.0	11.0	12.3	13.8	15.9	19.3	26.6
1.15	0.10	0.050	5.2	7.2	8.0	8.9	10.0	11.4	13.2	16.3	23.1
1.15	0.20	0.100	3.6	5.3	6.0	6.8	7.7	8.9	10.5	13.2	19.3
1.20	0.01	0.005	6.9	8.6	9.3	10.1	11.0	12.1	13.5	15.9	20.9
1.20	0.02	0.010	5.9	7.5	8.1	8.8	9.7	10.7	12.0	14.3	19.0
1.20	0.05	0.025	4.6	6.0	6.5	7.1	7.9	8.8	10.0	12.0	16.3
1.20	0.10	0.050	3.6	4.8	5.3	5.9	6.5	7.3	8.5	10.3	14.3
1.20	0.20	0.100	2.6	3.7	4.1	4.6	5.1	5.9	6.8	8.5	12.0
1.25	0.01	0.005	5.1	6.3	6.8	7.3	7.9	8.6	9.6	11.2	14.5
1.25	0.02	0.010	4.4	5.5	5.9	6.4	7.0	7.7	8.6	10.1	13.2
1.25	0.05	0.025	3.5	4.5	4.9	5.3	5.8	6.4	7.2	8.6	11.5
1.25	0.10	0.050	2.8	3.7	4.0	4.4	4.9	5.4	6.2	7.4	10.1
1.25	0.20	0.100	2.0	2.8	3.1	3.5	3.9	4.4	5.1	6.2	8.6
1.30	0.01	0.005	4.1	5.0	5.3	5.7	6.2	6.7	7.4	8.6	11.0
1.30	0.02	0.010	3.6	4.4	4.7	5.1	5.5	6.0	6.7	7.8	10.1
1.30	0.05	0.025	2.9	3.6	3.9	4.2	4.6	5.1	5.7	6.7	8.8
1.30	0.10	0.050	2.3	3.0	3.3	3.6	3.9	4.3	4.9	5.8	7.8
1.30	0.20	0.100	1.7	2.3	2.6	2.9	3.2	3.6	4.1	4.9	6.7
1.35	0.01	0.005	3.5	4.2	4.4	4.7	5.1	5.5	6.1	7.0	8.8
1.35	0.02	0.010	3.0	3.7	4.0	4.2	4.6	5.0	5.5	6.4	8.1
1.35	0.05	0.025	2.4	3.1	3.3	3.6	3.9	4.2	4.7	5.5	7.1
1.35	0.10	0.050	2.0	2.5	2.8	3.0	3.3	3.6	4.1	4.8	6.4
1.35	0.20	0.100	1.5	2.0	2.2	2.4	2.7	3.0	3.4	4.1	5.5
1.40	0.01	0.005	3.0	3.6	3.8	4.1	4.4	4.7	5.2	5.9	7.4
1.40	0.02	0.010	2.6	3.2	3.4	3.7	3.9	4.3	4.7	5.4	6.8
1.40	0.05	0.025	2.1	2.7	2.9	3.1	3.3	3.7	4.1	4.7	6.0
1.40	0.10	0.050	1.7	2.2	2.4	2.6	2.9	3.2	3.5	4.1	5.4
1.40	0.20	0.100	1.3	1.8	1.9	2.1	2.4	2.6	3.0	3.5	4.7
1.45	0.01	0.005	2.7	3.2	3.4	3.6	3.8	4.1	4.5	5.1	6.4
1.45	0.02	0.010	2.4	2.9	3.0	3.2	3.5	3.8	4.1	4.7	5.9
1.45	0.05	0.025	1.9	2.4	2.6	2.8	3.0	3.2	3.6	4.1	5.3
1.45	0.10	0.050	1.6	2.0	2.2	2.3	2.6	2.8	3.1	3.7	4.7
1.45	0.20	0.100	1.2	1.6	1.7	1.9	2.1	2.3	2.7	3.1	4.1
1.50	0.01	0.005	2.4	2.9	3.1	3.2	3.5	3.7	4.1	4.6	5.7
1.50	0.02	0.010	2.1	2.6	2.7	2.9	3.1	3.4	3.7	4.2	5.3
1.50	0.05	0.025	1.7	2.2	2.3	2.5	2.7	2.9	3.2	3.7	4.7
1.50	0.10	0.050	1.4	1.8	2.0	2.1	2.3	2.5	2.8	3.3	4.2
1.50	0.20	0.100	1.1	1.5	1.6	1.7	1.9	2.1	2.4	2.8	3.7
1.55	0.01	0.005	2.2	2.6	2.8	3.0	3.2	3.4	3.7	4.2	5.1
1.55	0.02	0.010	2.0	2.4	2.5	2.7	2.9	3.1	3.4	3.8	4.7
1.55	0.05	0.025	1.6	2.0	2.1	2.3	2.5	2.7	3.0	3.4	4.2
1.55	0.10	0.050	1.3	1.7	1.8	2.0	2.1	2.3	2.6	3.0	3.8
1.55	0.20	0.100	1.0	1.3	1.5	1.6	1.8	2.0	2.2	2.6	3.4

Table 10.2

Table 10.2 Continued.

△	α 2-sided	1-sided	Power 1 − β 0.50	0.65	0.70	0.75	0.80	0.85	0.90	0.95	0.99
\multicolumn — 150 subjects entered per unit time											

△	2-sided	1-sided	0.50	0.65	0.70	0.75	0.80	0.85	0.90	0.95	0.99
1.60	0.01	0.005	2.1	2.4	2.6	2.7	2.9	3.1	3.4	3.8	4.7
1.60	0.02	0.010	1.8	2.2	2.3	2.5	2.6	2.9	3.1	3.5	4.3
1.60	0.05	0.025	1.5	1.8	2.0	2.1	2.3	2.5	2.7	3.1	3.9
1.60	0.10	0.050	1.2	1.6	1.7	1.8	2.0	2.2	2.4	2.8	3.5
1.60	0.20	0.100	0.9	1.3	1.4	1.5	1.6	1.8	2.1	2.4	3.1
1.65	0.01	0.005	1.9	2.3	2.4	2.5	2.7	2.9	3.2	3.5	4.3
1.65	0.02	0.010	1.7	2.1	2.2	2.3	2.5	2.7	2.9	3.3	4.0
1.65	0.05	0.025	1.4	1.7	1.9	2.0	2.1	2.3	2.5	2.9	3.6
1.65	0.10	0.050	1.2	1.5	1.6	1.7	1.9	2.0	2.2	2.6	3.3
1.65	0.20	0.100	0.9	1.2	1.3	1.4	1.5	1.7	1.9	2.2	2.9
1.70	0.01	0.005	1.8	2.1	2.3	2.4	2.5	2.7	3.0	3.3	4.0
1.70	0.02	0.010	1.6	1.9	2.0	2.2	2.3	2.5	2.7	3.1	3.8
1.70	0.05	0.025	1.3	1.6	1.7	1.9	2.0	2.2	2.4	2.7	3.4
1.70	0.10	0.050	1.1	1.4	1.5	1.6	1.7	1.9	2.1	2.4	3.1
1.70	0.20	0.100	0.8	1.1	1.2	1.3	1.5	1.6	1.8	2.1	2.7
1.75	0.01	0.005	1.7	2.0	2.1	2.3	2.4	2.6	2.8	3.1	3.8
1.75	0.02	0.010	1.5	1.8	1.9	2.1	2.2	2.4	2.6	2.9	3.5
1.75	0.05	0.025	1.3	1.6	1.7	1.8	1.9	2.1	2.3	2.6	3.2
1.75	0.10	0.050	1.0	1.3	1.4	1.5	1.7	1.8	2.0	2.3	2.9
1.75	0.20	0.100	0.8	1.1	1.2	1.3	1.4	1.5	1.7	2.0	2.6
1.80	0.01	0.005	1.6	1.9	2.0	2.2	2.3	2.4	2.6	3.0	3.6
1.80	0.02	0.010	1.5	1.7	1.9	2.0	2.1	2.2	2.4	2.7	3.4
1.80	0.05	0.025	1.2	1.5	1.6	1.7	1.8	2.0	2.2	2.4	3.0
1.80	0.10	0.050	1.0	1.3	1.4	1.5	1.6	1.7	1.9	2.2	2.7
1.80	0.20	0.100	0.8	1.0	1.1	1.2	1.3	1.5	1.6	1.9	2.4
1.85	0.01	0.005	1.6	1.9	2.0	2.1	2.2	2.3	2.5	2.8	3.4
1.85	0.02	0.010	1.4	1.7	1.8	1.9	2.0	2.1	2.3	2.6	3.2
1.85	0.05	0.025	1.2	1.4	1.5	1.6	1.7	1.9	2.1	2.3	2.9
1.85	0.10	0.050	1.0	1.2	1.3	1.4	1.5	1.7	1.8	2.1	2.6
1.85	0.20	0.100	0.7	1.0	1.1	1.2	1.3	1.4	1.6	1.8	2.3
1.90	0.01	0.005	1.5	1.8	1.9	2.0	2.1	2.2	2.4	2.7	3.2
1.90	0.02	0.010	1.4	1.6	1.7	1.8	1.9	2.1	2.2	2.5	3.0
1.90	0.05	0.025	1.1	1.4	1.5	1.6	1.7	1.8	2.0	2.2	2.8
1.90	0.10	0.050	0.9	1.2	1.3	1.4	1.5	1.6	1.8	2.0	2.5
1.90	0.20	0.100	0.7	0.9	1.0	1.1	1.2	1.4	1.5	1.8	2.2
1.95	0.01	0.005	1.5	1.7	1.8	1.9	2.0	2.2	2.3	2.6	3.1
1.95	0.02	0.010	1.3	1.6	1.6	1.7	1.9	2.0	2.2	2.4	2.9
1.95	0.05	0.025	1.1	1.3	1.4	1.5	1.6	1.7	1.9	2.1	2.6
1.95	0.10	0.050	0.9	1.1	1.2	1.3	1.4	1.5	1.7	1.9	2.4
1.95	0.20	0.100	0.7	0.9	1.0	1.1	1.2	1.3	1.5	1.7	2.2
2.00	0.01	0.005	1.4	1.7	1.7	1.8	2.0	2.1	2.2	2.5	3.0
2.00	0.02	0.010	1.3	1.5	1.6	1.7	1.8	1.9	2.1	2.3	2.8
2.00	0.05	0.025	1.0	1.3	1.4	1.5	1.6	1.7	1.8	2.1	2.5
2.00	0.10	0.050	0.9	1.1	1.2	1.3	1.4	1.5	1.6	1.9	2.3
2.00	0.20	0.100	0.7	0.9	1.0	1.0	1.1	1.3	1.4	1.6	2.1
2.25	0.01	0.005	1.2	1.4	1.5	1.6	1.7	1.8	1.9	2.2	2.6
2.25	0.02	0.010	1.1	1.3	1.4	1.5	1.6	1.7	1.8	2.0	2.4
2.25	0.05	0.025	0.9	1.1	1.2	1.3	1.4	1.5	1.6	1.8	2.2
2.25	0.10	0.050	0.8	1.0	1.0	1.1	1.2	1.3	1.4	1.6	2.0
2.25	0.20	0.100	0.6	0.8	0.8	0.9	1.0	1.1	1.2	1.4	1.8
2.50	0.01	0.005	1.1	1.3	1.4	1.4	1.5	1.6	1.8	1.9	2.3
2.50	0.02	0.010	1.0	1.2	1.3	1.3	1.4	1.5	1.6	1.8	2.2
2.50	0.05	0.025	0.8	1.0	1.1	1.2	1.2	1.3	1.4	1.6	2.0
2.50	0.10	0.050	0.7	0.9	0.9	1.0	1.1	1.2	1.3	1.5	1.8
2.50	0.20	0.100	0.5	0.7	0.8	0.8	0.9	1.0	1.1	1.3	1.6

Table 10.2

164

Table 10.2 Continued.

$\triangle$	α 2-sided	1-sided	0.50	0.65	0.70	Power $1 - \beta$ 0.75	0.80	0.85	0.90	0.95	0.99
						150 subjects entered per unit time					
2.75	0.01	0.005	1.0	1.2	1.3	1.3	1.4	1.5	1.6	1.8	2.1
2.75	0.02	0.010	0.9	1.1	1.2	1.2	1.3	1.4	1.5	1.7	2.0
2.75	0.05	0.025	0.8	0.9	1.0	1.1	1.1	1.2	1.3	1.5	1.8
2.75	0.10	0.050	0.7	0.8	0.9	0.9	1.0	1.1	1.2	1.4	1.7
2.75	0.20	0.100	0.5	0.7	0.7	0.8	0.9	0.9	1.0	1.2	1.5
3.00	0.01	0.005	1.0	1.1	1.2	1.3	1.3	1.4	1.5	1.7	2.0
3.00	0.02	0.010	0.9	1.0	1.1	1.2	1.2	1.3	1.4	1.6	1.9
3.00	0.05	0.025	0.7	0.9	1.0	1.0	1.1	1.2	1.3	1.4	1.7
3.00	0.10	0.050	0.6	0.8	0.8	0.9	1.0	1.0	1.1	1.3	1.6
3.00	0.20	0.100	0.5	0.6	0.7	0.7	0.8	0.9	1.0	1.1	1.4
3.25	0.01	0.005	0.9	1.1	1.2	1.2	1.3	1.4	1.5	1.6	1.9
3.25	0.02	0.010	0.9	1.0	1.1	1.1	1.2	1.3	1.4	1.5	1.8
3.25	0.05	0.025	0.7	0.9	0.9	1.0	1.0	1.1	1.2	1.4	1.6
3.25	0.10	0.050	0.6	0.7	0.8	0.8	0.9	1.0	1.1	1.2	1.5
3.25	0.20	0.100	0.5	0.6	0.6	0.7	0.8	0.8	0.9	1.1	1.4
3.50	0.01	0.005	0.9	1.1	1.1	1.2	1.2	1.3	1.4	1.6	1.8
3.50	0.02	0.010	0.8	1.0	1.0	1.1	1.1	1.2	1.3	1.5	1.7
3.50	0.05	0.025	0.7	0.8	0.9	0.9	1.0	1.1	1.2	1.3	1.6
3.50	0.10	0.050	0.6	0.7	0.8	0.8	0.9	1.0	1.0	1.2	1.5
3.50	0.20	0.100	0.4	0.6	0.6	0.7	0.7	0.8	0.9	1.0	1.3
3.75	0.01	0.005	0.9	1.0	1.1	1.1	1.2	1.3	1.4	1.5	1.8
3.75	0.02	0.010	0.8	0.9	1.0	1.0	1.1	1.2	1.3	1.4	1.7
3.75	0.05	0.025	0.7	0.8	0.9	0.9	1.0	1.0	1.1	1.3	1.5
3.75	0.10	0.050	0.6	0.7	0.7	0.8	0.9	0.9	1.0	1.2	1.4
3.75	0.20	0.100	0.4	0.6	0.6	0.7	0.7	0.8	0.9	1.0	1.3
4.00	0.01	0.005	0.9	1.0	1.1	1.1	1.2	1.2	1.3	1.5	1.7
4.00	0.02	0.010	0.8	0.9	1.0	1.0	1.1	1.2	1.2	1.4	1.6
4.00	0.05	0.025	0.6	0.8	0.8	0.9	1.0	1.0	1.1	1.2	1.5
4.00	0.10	0.050	0.5	0.7	0.7	0.8	0.8	0.9	1.0	1.1	1.4
4.00	0.20	0.100	0.4	0.5	0.6	0.6	0.7	0.8	0.9	1.0	1.2
4.25	0.01	0.005	0.8	1.0	1.0	1.1	1.1	1.2	1.3	1.4	1.7
4.25	0.02	0.010	0.8	0.9	1.0	1.0	1.1	1.1	1.2	1.3	1.6
4.25	0.05	0.025	0.6	0.8	0.8	0.9	0.9	1.0	1.1	1.2	1.5
4.25	0.10	0.050	0.5	0.7	0.7	0.8	0.8	0.9	1.0	1.1	1.3
4.25	0.20	0.100	0.4	0.5	0.6	0.6	0.7	0.8	0.8	1.0	1.2
						200 subjects entered per unit time					
1.05	0.01	0.005	57.2	75.1	82.2	90.2	99.6	111.1	126.5	151.1	203.4
1.05	0.02	0.010	46.9	63.3	69.8	77.1	85.8	96.5	110.8	134.0	183.4
1.05	0.05	0.025	33.8	47.7	53.3	59.8	67.4	76.9	89.8	110.7	155.8
1.05	0.10	0.050	24.2	36.1	41.0	46.7	53.4	61.9	73.4	92.4	134.0
1.05	0.20	0.100	15.3	24.8	28.9	33.6	39.4	46.6	56.7	73.4	110.8
1.10	0.01	0.005	16.1	20.8	22.7	24.8	27.2	30.2	34.3	40.7	54.4
1.10	0.02	0.010	13.4	17.7	19.4	21.3	23.6	26.4	30.2	36.2	49.2
1.10	0.05	0.025	10.0	13.6	15.1	16.8	18.8	21.3	24.7	30.1	42.0
1.10	0.10	0.050	7.5	10.6	11.9	13.4	15.1	17.3	20.4	25.3	36.2
1.10	0.20	0.100	5.1	7.6	8.7	9.9	11.4	13.3	16.0	20.4	30.2
1.15	0.01	0.005	8.4	10.5	11.4	12.4	13.5	14.9	16.8	19.8	26.2
1.15	0.02	0.010	7.1	9.1	9.9	10.8	11.8	13.1	14.9	17.7	23.7
1.15	0.05	0.025	5.4	7.2	7.9	8.7	9.6	10.8	12.3	14.9	20.4
1.15	0.10	0.050	4.2	5.7	6.3	7.0	7.9	8.9	10.3	12.6	17.7
1.15	0.20	0.100	3.0	4.3	4.8	5.4	6.1	7.0	8.3	10.3	14.9
1.20	0.01	0.005	5.5	6.9	7.4	7.9	8.6	9.5	10.6	12.3	16.1
1.20	0.02	0.010	4.8	6.0	6.5	7.0	7.6	8.4	9.4	11.1	14.6
1.20	0.05	0.025	3.8	4.8	5.3	5.7	6.3	7.0	7.9	9.4	12.7
1.20	0.10	0.050	3.0	3.9	4.3	4.8	5.3	5.9	6.7	8.1	11.1
1.20	0.20	0.100	2.2	3.0	3.4	3.7	4.2	4.7	5.5	6.7	9.4

Table 10.2 Continued.

△	α 2-sided	1-sided	0.50	0.65	0.70	0.75	0.80	0.85	0.90	0.95	0.99

Power $1 - \beta$ column headers: 0.50, 0.65, 0.70, 0.75, 0.80, 0.85, 0.90, 0.95, 0.99

200 subjects entered per unit time

△	2-sided	1-sided	0.50	0.65	0.70	0.75	0.80	0.85	0.90	0.95	0.99
1.25	0.01	0.005	4.2	5.1	5.4	5.8	6.3	6.9	7.6	8.8	11.3
1.25	0.02	0.010	3.6	4.5	4.8	5.2	5.6	6.1	6.8	8.0	10.3
1.25	0.05	0.025	2.9	3.7	4.0	4.3	4.7	5.2	5.8	6.8	9.0
1.25	0.10	0.050	2.3	3.0	3.3	3.6	4.0	4.4	5.0	5.9	8.0
1.25	0.20	0.100	1.7	2.4	2.6	2.9	3.2	3.6	4.1	5.0	6.8
1.30	0.01	0.005	3.4	4.1	4.3	4.6	5.0	5.4	6.0	6.8	8.7
1.30	0.02	0.010	3.0	3.6	3.9	4.2	4.5	4.9	5.4	6.2	8.0
1.30	0.05	0.025	2.4	3.0	3.2	3.5	3.8	4.1	4.6	5.4	7.0
1.30	0.10	0.050	1.9	2.5	2.7	3.0	3.2	3.6	4.0	4.7	6.2
1.30	0.20	0.100	1.4	2.0	2.2	2.4	2.6	3.0	3.4	4.0	5.4
1.35	0.01	0.005	2.9	3.4	3.6	3.9	4.2	4.5	4.9	5.6	7.0
1.35	0.02	0.010	2.5	3.1	3.3	3.5	3.8	4.1	4.5	5.1	6.5
1.35	0.05	0.025	2.0	2.5	2.7	3.0	3.2	3.5	3.9	4.5	5.7
1.35	0.10	0.050	1.7	2.1	2.3	2.5	2.7	3.0	3.4	3.9	5.1
1.35	0.20	0.100	1.3	1.7	1.9	2.0	2.3	2.5	2.9	3.4	4.5
1.40	0.01	0.005	2.5	3.0	3.2	3.4	3.6	3.9	4.2	4.8	6.0
1.40	0.02	0.010	2.2	2.7	2.8	3.0	3.3	3.5	3.9	4.4	5.5
1.40	0.05	0.025	1.8	2.2	2.4	2.6	2.8	3.0	3.4	3.9	4.9
1.40	0.10	0.050	1.5	1.9	2.0	2.2	2.4	2.6	2.9	3.4	4.4
1.40	0.20	0.100	1.1	1.5	1.6	1.8	2.0	2.2	2.5	2.9	3.9
1.45	0.01	0.005	2.2	2.7	2.8	3.0	3.2	3.4	3.7	4.2	5.2
1.45	0.02	0.010	2.0	2.4	2.5	2.7	2.9	3.1	3.4	3.9	4.8
1.45	0.05	0.025	1.6	2.0	2.1	2.3	2.5	2.7	3.0	3.4	4.3
1.45	0.10	0.050	1.3	1.7	1.8	2.0	2.1	2.4	2.6	3.0	3.9
1.45	0.20	0.100	1.0	1.4	1.5	1.6	1.8	2.0	2.2	2.6	3.4
1.50	0.01	0.005	2.0	2.4	2.5	2.7	2.9	3.1	3.4	3.8	4.6
1.50	0.02	0.010	1.8	2.2	2.3	2.4	2.6	2.8	3.1	3.5	4.3
1.50	0.05	0.025	1.5	1.8	1.9	2.1	2.3	2.4	2.7	3.1	3.9
1.50	0.10	0.050	1.2	1.5	1.7	1.8	2.0	2.1	2.4	2.7	3.5
1.50	0.20	0.100	0.9	1.2	1.4	1.5	1.6	1.8	2.0	2.4	3.1
1.55	0.01	0.005	1.9	2.2	2.3	2.5	2.6	2.8	3.1	3.4	4.2
1.55	0.02	0.010	1.7	2.0	2.1	2.2	2.4	2.6	2.8	3.2	3.9
1.55	0.05	0.025	1.4	1.7	1.8	1.9	2.1	2.2	2.5	2.8	3.5
1.55	0.10	0.050	1.1	1.4	1.5	1.7	1.8	2.0	2.2	2.5	3.2
1.55	0.20	0.100	0.9	1.1	1.3	1.4	1.5	1.7	1.9	2.2	2.8
1.60	0.01	0.005	1.7	2.1	2.2	2.3	2.4	2.6	2.8	3.2	3.8
1.60	0.02	0.010	1.6	1.9	2.0	2.1	2.2	2.4	2.6	2.9	3.6
1.60	0.05	0.025	1.3	1.6	1.7	1.8	1.9	2.1	2.3	2.6	3.2
1.60	0.10	0.050	1.1	1.3	1.4	1.5	1.7	1.8	2.0	2.3	2.9
1.60	0.20	0.100	0.8	1.1	1.2	1.3	1.4	1.5	1.7	2.0	2.6
1.65	0.01	0.005	1.6	1.9	2.0	2.1	2.3	2.4	2.6	2.9	3.6
1.65	0.02	0.010	1.5	1.7	1.8	2.0	2.1	2.2	2.4	2.7	3.3
1.65	0.05	0.025	1.2	1.5	1.6	1.7	1.8	1.9	2.1	2.4	3.0
1.65	0.10	0.050	1.0	1.3	1.3	1.5	1.6	1.7	1.9	2.2	2.7
1.65	0.20	0.100	0.8	1.0	1.1	1.2	1.3	1.5	1.6	1.9	2.4
1.70	0.01	0.005	1.5	1.8	1.9	2.0	2.1	2.3	2.5	2.8	3.3
1.70	0.02	0.010	1.4	1.6	1.7	1.8	2.0	2.1	2.3	2.6	3.1
1.70	0.05	0.025	1.1	1.4	1.5	1.6	1.7	1.8	2.0	2.3	2.8
1.70	0.10	0.050	0.9	1.2	1.3	1.4	1.5	1.6	1.8	2.0	2.6
1.70	0.20	0.100	0.7	1.0	1.0	1.1	1.2	1.4	1.5	1.8	2.3
1.75	0.01	0.005	1.5	1.7	1.8	1.9	2.0	2.2	2.3	2.6	3.2
1.75	0.02	0.010	1.3	1.6	1.6	1.7	1.9	2.0	2.2	2.4	2.9
1.75	0.05	0.025	1.1	1.3	1.4	1.5	1.6	1.7	1.9	2.2	2.7
1.75	0.10	0.050	0.9	1.1	1.2	1.3	1.4	1.5	1.7	1.9	2.4
1.75	0.20	0.100	0.7	0.9	1.0	1.1	1.2	1.3	1.5	1.7	2.2

Table 10.2

Table 10.2 Continued.

△	α 2-sided	1-sided	0.50	0.65	0.70	0.75	0.80	0.85	0.90	0.95	0.99
						Power 1 − β					
200 subjects entered per unit time											
1.80	0.01	0.005	1.4	1.6	1.7	1.8	1.9	2.1	2.2	2.5	3.0
1.80	0.02	0.010	1.3	1.5	1.6	1.7	1.8	1.9	2.1	2.3	2.8
1.80	0.05	0.025	1.0	1.3	1.3	1.4	1.5	1.7	1.8	2.1	2.5
1.80	0.10	0.050	0.9	1.1	1.2	1.3	1.3	1.5	1.6	1.8	2.3
1.80	0.20	0.100	0.7	0.9	0.9	1.0	1.1	1.3	1.4	1.6	2.1
1.85	0.01	0.005	1.3	1.6	1.7	1.7	1.8	2.0	2.1	2.4	2.8
1.85	0.02	0.010	1.2	1.4	1.5	1.6	1.7	1.8	2.0	2.2	2.7
1.85	0.05	0.025	1.0	1.2	1.3	1.4	1.5	1.6	1.7	2.0	2.4
1.85	0.10	0.050	0.8	1.0	1.1	1.2	1.3	1.4	1.6	1.8	2.2
1.85	0.20	0.100	0.6	0.8	0.9	1.0	1.1	1.2	1.3	1.6	2.0
1.90	0.01	0.005	1.3	1.5	1.6	1.7	1.8	1.9	2.0	2.3	2.7
1.90	0.02	0.010	1.2	1.4	1.4	1.5	1.6	1.7	1.9	2.1	2.6
1.90	0.05	0.025	1.0	1.2	1.2	1.3	1.4	1.5	1.7	1.9	2.3
1.90	0.10	0.050	0.8	1.0	1.1	1.1	1.2	1.4	1.5	1.7	2.1
1.90	0.20	0.100	0.6	0.8	0.9	1.0	1.0	1.1	1.3	1.5	1.9
1.95	0.01	0.005	1.3	1.5	1.5	1.6	1.7	1.8	2.0	2.2	2.6
1.95	0.02	0.010	1.1	1.3	1.4	1.5	1.6	1.7	1.8	2.0	2.5
1.95	0.05	0.025	0.9	1.1	1.2	1.3	1.4	1.5	1.6	1.8	2.2
1.95	0.10	0.050	0.8	1.0	1.0	1.1	1.2	1.3	1.4	1.6	2.0
1.95	0.20	0.100	0.6	0.8	0.8	0.9	1.0	1.1	1.2	1.4	1.8
2.00	0.01	0.005	1.2	1.4	1.5	1.6	1.7	1.8	1.9	2.1	2.5
2.00	0.02	0.010	1.1	1.3	1.4	1.4	1.5	1.6	1.8	2.0	2.4
2.00	0.05	0.025	0.9	1.1	1.2	1.2	1.3	1.4	1.6	1.8	2.1
2.00	0.10	0.050	0.7	0.9	1.0	1.1	1.2	1.3	1.4	1.6	2.0
2.00	0.20	0.100	0.6	0.8	0.8	0.9	1.0	1.1	1.2	1.4	1.8
2.25	0.01	0.005	1.1	1.2	1.3	1.4	1.4	1.5	1.7	1.8	2.2
2.25	0.02	0.010	0.9	1.1	1.2	1.2	1.3	1.4	1.5	1.7	2.0
2.25	0.05	0.025	0.8	1.0	1.0	1.1	1.2	1.2	1.4	1.5	1.9
2.25	0.10	0.050	0.7	0.8	0.3	0.9	1.0	1.1	1.2	1.4	1.7
2.25	0.20	0.100	0.5	0.7	0.7	0.8	0.9	0.9	1.1	1.2	1.5
2.50	0.01	0.005	1.0	1.1	1.2	1.2	1.3	1.4	1.5	1.7	2.0
2.50	0.02	0.010	0.9	1.0	1.1	1.1	1.2	1.3	1.4	1.6	1.8
2.50	0.05	0.025	0.7	0.9	0.9	1.0	1.1	1.1	1.2	1.4	1.7
2.50	0.10	0.050	0.6	0.7	0.8	0.9	0.9	1.0	1.1	1.3	1.6
2.50	0.20	0.100	0.5	0.6	0.7	0.7	0.8	0.9	1.0	1.1	1.4
2.75	0.01	0.005	0.9	1.0	1.1	1.2	1.2	1.3	1.4	1.5	1.8
2.75	0.02	0.010	0.8	0.9	1.0	1.1	1.1	1.2	1.3	1.4	1.7
2.75	0.05	0.025	0.7	0.8	0.9	0.9	1.0	1.1	1.1	1.3	1.6
2.75	0.10	0.050	0.6	0.7	0.8	0.8	0.9	0.9	1.0	1.2	1.4
2.75	0.20	0.100	0.4	0.6	0.6	0.7	0.7	0.8	0.9	1.0	1.3
3.00	0.01	0.005	0.9	1.0	1.0	1.1	1.1	1.2	1.3	1.4	1.7
3.00	0.02	0.010	0.8	0.9	0.9	1.0	1.1	1.1	1.2	1.4	1.6
3.00	0.05	0.025	0.6	0.8	0.8	0.9	0.9	1.0	1.1	1.2	1.5
3.00	0.10	0.050	0.5	0.7	0.7	0.8	0.8	0.9	1.0	1.1	1.4
3.00	0.20	0.100	0.4	0.5	0.6	0.6	0.7	0.8	0.8	1.0	1.2
3.25	0.01	0.005	0.8	0.9	1.0	1.0	1.1	1.2	1.3	1.4	1.6
3.25	0.02	0.010	0.7	0.9	0.9	1.0	1.0	1.1	1.2	1.3	1.5
3.25	0.05	0.025	0.6	0.7	0.8	0.8	0.9	1.0	1.0	1.2	1.4
3.25	0.10	0.050	0.5	0.6	0.7	0.7	0.8	0.9	0.9	1.1	1.3
3.25	0.20	0.100	0.4	0.5	0.6	0.6	0.7	0.7	0.8	0.9	1.2
3.50	0.01	0.005	0.8	0.9	1.0	1.0	1.1	1.1	1.2	1.3	1.6
3.50	0.02	0.010	0.7	0.8	0.9	0.9	1.0	1.0	1.1	1.2	1.5
3.50	0.05	0.025	0.6	0.7	0.8	0.8	0.9	0.9	1.0	1.1	1.4
3.50	0.10	0.050	0.5	0.6	0.7	0.7	0.8	0.8	0.9	1.0	1.2
3.50	0.20	0.100	0.4	0.5	0.5	0.6	0.6	0.7	0.8	0.9	1.1

Table 10.2

Table 10.2 Continued.

△	α 2-sided	1-sided	Power 1 − β 0.50	0.65	0.70	0.75	0.80	0.85	0.90	0.95	0.99
3.75	0.01	0.005	0.8	0.9	0.9	1.0	1.0	1.1	1.2	1.3	1.5
3.75	0.02	0.010	0.7	0.8	0.9	0.9	1.0	1.0	1.1	1.2	1.4
3.75	0.05	0.025	0.6	0.7	0.7	0.8	0.8	0.9	1.0	1.1	1.3
3.75	0.10	0.050	0.5	0.6	0.6	0.7	0.7	0.8	0.9	1.0	1.2
3.75	0.20	0.100	0.4	0.5	0.5	0.6	0.6	0.7	0.8	0.9	1.1
4.00	0.01	0.005	0.7	0.9	0.9	1.0	1.0	1.1	1.1	1.3	1.5
4.00	0.02	0.010	0.7	0.8	0.8	0.9	0.9	1.0	1.1	1.2	1.4
4.00	0.05	0.025	0.6	0.7	0.7	0.8	0.8	0.9	1.0	1.1	1.3
4.00	0.10	0.050	0.5	0.6	0.6	0.7	0.7	0.8	0.9	1.0	1.2
4.00	0.20	0.100	0.4	0.5	0.5	0.6	0.6	0.7	0.7	0.9	1.1
4.25	0.01	0.005	0.7	0.8	0.9	0.9	1.0	1.0	1.1	1.2	1.4
4.25	0.02	0.010	0.7	0.8	0.8	0.9	0.9	1.0	1.0	1.2	1.4
4.25	0.05	0.025	0.5	0.7	0.7	0.7	0.8	0.9	0.9	1.0	1.3
4.25	0.10	0.050	0.5	0.6	0.6	0.7	0.7	0.8	0.8	1.0	1.2
4.25	0.20	0.100	0.4	0.5	0.5	0.5	0.6	0.7	0.7	0.8	1.0

Table 10.2

Chapter 11
Phase II Trials (Gehan's Method)

11.1 INTRODUCTION

In the development of new drugs for possible use as anticancer agents preliminary studies of efficacy are made in a small group of patients. Such a trial is described by Kleeberg, Mulder, Rumke *et al* (1982) who tested N – (Phosphonacetyl)–L–Aspartate (PALA) on a group of 29 patients with measurable advanced malignant melanoma. These trials are known as Phase II trials. In such trials, although details are often omitted in the published report, a minimum requirement of efficacy is often set and patients are recruited in two stages. If no responses are observed in the first stage, e.g. no tumour regressions in the above example, patients are not recruited for the second stage. On the other hand if one or more responses are observed then recruitment to the second stage depends on their number.

The objective of a Phase II trial is to decide if a particular therapeutic regimen is effective enough to warrant further study.

11.2 THEORY AND FORMULAE

If we assume that the probability an individual patient will respond to a particular treatment is π, and this is constant for all patients, then the probability of r successive patients failing on the drug is

$$\beta = (1-\pi)^r.$$

$$(11.1)$$

By specifying the error of rejection β and the minimum efficacy π equation (11.1) can be solved to give

$$n_1 = \log\beta/\log(1-\pi),$$

$$(11.2)$$

where n_1 is the number of patients to be recruited to the first stage of the Phase II trial. Assume r_1 responses are then observed in these n_1 patients.

If $r_1 > 0$ then n_2 patients are recruited to the second stage of such a trial to give a total of $N = n_1 + n_2$ patients in all. The value of n_2 is chosen to give a specified precision ε, for the final estimate of the efficacy π. This implies that

$$\sqrt{\frac{p(1-p)}{N}} = \varepsilon,$$

$$(11.3)$$

where $p = (r_1 + r_2)/(n_1 + n_2)$ is the estimate of the efficacy of the drug based on N patients, r_1 is the number of successes in the first stage from n_1 patients and r_2 the number of successes in the second stage from n_2 patients.

Rearranging equation (11.3) we obtain the required number of patients for the second stage as

$$n_2 = \frac{p(1-p)}{\varepsilon^2} - n_1.$$

$$(11.4)$$

However, at the end of the first stage, we do not know p, only $p_1 = r_1/n_1$ the proportion of successes in the first stage. Thus to estimate n_2 we must use p_1 rather than p since the latter is not available to us.

Now n_1 is usually small so that the resulting p_1 will be rather imprecise. As a consequence Gehan (1961) suggests not using p_1 in (11.4) to estimate p but instead π_U the one-sided upper 75% confidence limit for π. It can be shown that π_U is the solution of

$$\sum_{s=0}^{r1} \binom{n_1}{s} \pi_U^s (1 - \pi_U)^{n_i - s} \; ; = 0.25 \; . \tag{11.5}$$

The final estimate for n_2 is therefore

$$n_2 = \frac{\pi_U(1 - \pi_U)}{\varepsilon^2} - n_1 \tag{11.60}$$

and depends rather critically on the number of successes r_1 observed in the first stage of the trial.

11.3 BIBLIOGRAPHY

Gehan (1961) gives the theory and formulae for equations (11.2) and (11.6) and corresponding tables. A method of obtaining an approximate solution to equation (11.5), which we use in calculating our tables, is given by Fujino (1980). Lee (1980) and Herson (1984) give more complete discussions of Phase II trials.

11.4 DESCRIPTION OF THE TABLES

Table 11.1

Table 11.1 gives the probability of r successive treatment failures obtained from equation (11.1) for given efficacy π ranging from 0.1 by steps of 0.05 to 0.95. The last value of r tabulated for each π is that for which the value of β is first less than or equal to 0.01.

Table 11.2

Table 11.2 gives the initial sample size n_1 required for anticipated therapeutic effectiveness π ranging from 0.05 to 0.90 for power $1 - \beta$ ranging from 0.8 to 0.999. It should be recognized that the solutions to equation (11.2) are rounded up to the nearest whole number.

Table 11.3

Table 11.3 gives the number of additional patients n_2 required in the second stage of a Phase II trial, for the same range of π as Table 11.2 for $1 - \beta = 0.90$ and 0.95. The values for the precision, ε, are 0.05 and 0.10.

11.5 USE OF THE TABLES

Table 11.1

Example 11.1

If the probability of a treatment success is 0.3, what is the probability that six consecutively treated patients will all fail to respond?

Table 11.1 with $\pi = 0.3$ and $r = 6$ gives $\beta = 0.1176$. Thus the probability that none will respond is approximately 0.12.

Example 11.2

If the probability of a treatment success is 0.3 how many patients ought to be recruited if it is desired that the probability that none respond is to be less than or equal to 0.05?

Again, referring to Table 11.1, with $\pi = 0.3$, one searches for the first value of r for which $\beta \leqslant 0.05$. In fact for $r = 9$, $\beta = 0.0404$. Thus, if we recruit 9 patients to the trial with a true response rate of 30%, we have a probability of less than 0.05 that none of the patients will respond.

Table 11.2

Example 11.3

How many patients must be recruited to the first stage of a Phase II trial to ensure that the chance of rejecting a drug of efficacy at least 20% is less than 0.05?

Here $\pi = 0.2$, $1 - \beta = 0.95$ and Table 11.2 gives $n_1 = 14$. With $n_1 = 14$ and using either equation (11.1) or Table 11.1 with $\pi = 0.2$ the probability of no responses as more precisely 0.956.

Example 11.4

Olweny, Katongole-Mbidde, Toya and Kyalwazi (1975) recruited 14 patients with hepatocellular carcinoma to a Phase II trial of Adriamycin. If they had regarded a minimum efficacy for complete tumour regression as 15%, how many patients should have been recruited to such a trial had they specified a chance of 0.01 of rejecting the drug with such efficacy?

Here $\pi = 0.15$, $1 - \beta = 0.90$ and Table 11.2 gives $n_1 = 15$.

Table 11.3

Example 11.5

The first stage of the Phase II trial described in Example 11.3 gave 3 successes. If the total precision of the trial is set at 10%, how many patients should be recruited to the second stage?

Here $\pi = 0.2$, $1 - \beta = 0.95$, $r_1 = 3$ and $\varepsilon = 0.1$ then Table 11.3 indicates that a further $n_2 = 9$ patients should be recruited to the second stage. This gives a total recruitment of $n_1 + n_2 = 14 + 9 = 23$ patients.

Example 11.6

If the first stage of the Phase II trial described in Example 11.4 gave 2 patients with complete tumour regression, how many patients should be recruited to the second stage if the precision is set at 5%?

Here $\pi = 0.15$, $1 - \beta = 0.90$, $r_1 = 2$, $\varepsilon = 0.05$ and Table 11.3 indicates that a further $n_2 = 59$ patients should be recruited. Thus a total of $N = n_1 + n_2 = 15 + 59 = 74$ patients would eventually be recruited to the trial.

11.6 REFERENCES

Fujino Y. (1980) Approximate binomial confidence limits. *Biometrika*, **67**, 677–681.
Gehan E.A. (1961) The determination of the number of patients required in a preliminary and follow–up trial of a new chemotherapeutic agent. *J. Chron. Dis.*, **13**, 346–353.

Herson. J. (1984) Statistical aspects in the design and analysis of Phase II clinical trials. In: *Cancer Clinical Trials: Methods and Practice.* ed. Buyse M.E., Staquet M.J. & Sylvestor R.J. Oxford Medical Publications, Oxford.

Kleeberg U.R., Mulder J.H., Rumke P., Thomas D. & Rozencweig M. (1982). N–(Phosphonacetyl)–L–Asparate (PALA) in advanced malignant melanoma. A phase II trial of the EORTC malignant melanoma co–operative group. *Eur. J. Cancer Clin. Oncol.* **18,** 723–726.

Lee E.T. (1980) *Statistical Methods for Survival Data Analysis.* Lifetime Learning Publications, Belmont, California.

Olweny C.L.M., Katongole–Mbidde E., Toya T. & Kyalwazi S.K. (1975) Phase II Studies with Adriamycin. In *Adriamycin Review* Part IV, 291–302, ed. Staquet M., Tagnon H., Kenis Y., Bonadonna G., Carter S.K., Sokal G. Trouet A., Ghione M., Praga C., Lenaz L. & Karim O.S. European Press Medikon, Ghent, Belgium.

Table 11.1 Probability of a given number of successive treatment failures.

Consecutive patients	Probability of patient response, π																	
r	0.10	0.15	0.20	0.25	0.30	0.35	0.40	0.45	0.50	0.55	0.60	0.65	0.70	0.75	0.80	0.85	0.90	0.95
1	0.9000	0.8500	0.8000	0.7500	0.7000	0.6500	0.6000	0.5500	0.5000	0.4500	0.4000	0.3500	0.3000	0.2500	0.2000	0.1500	0.1000	0.0500
2	0.8100	0.7225	0.6400	0.5625	0.4900	0.4225	0.3600	0.3025	0.2500	0.2025	0.1600	0.1225	0.0900	0.0625	0.0400	0.0225	0.0100	0.0025
3	0.7290	0.6141	0.5120	0.4219	0.3430	0.2746	0.2160	0.1664	0.1250	0.0911	0.0640	0.0429	0.0270	0.0156	0.0080	0.0034		
4	0.6561	0.5220	0.4096	0.3164	0.2401	0.1785	0.1296	0.0915	0.0625	0.0410	0.0256	0.0150	0.0081	0.0039				
5	0.5905	0.4337	0.3277	0.2373	0.1681	0.1160	0.0778	0.0503	0.0313	0.0185	0.0102	0.0053						
6	0.5314	0.3771	0.2621	0.1780	0.1176	0.0754	0.0467	0.0277	0.0156	0.0083	0.0041							
7	0.4783	0.3206	0.2097	0.1335	0.0824	0.0490	0.0280	0.0152	0.0078									
8	0.4305	0.2725	0.1678	0.1001	0.0576	0.0319	0.0168	0.0084										
9	0.3874	0.2316	0.1342	0.0751	0.0404	0.0207	0.0101											
10	0.3487	0.1969	0.1074	0.0563	0.0282	0.0135	0.0060											
11	0.3138	0.1673	0.0859	0.0422	0.0198	0.0088												
12	0.2824	0.1422	0.0687	0.0317	0.0138													
13	0.2542	0.1209	0.0550	0.0238	0.0097													
14	0.2288	0.1028	0.0440	0.0178														
15	0.2059	0.0874	0.0352	0.0134														
16	0.1853	0.0743	0.0281	0.0100														
17	0.1668	0.0631	0.0225	0.0075														
18	0.1501	0.0536	0.0180															
19	0.1351	0.0456	0.0144															
20	0.1216	0.0388	0.0115															
21	0.1094	0.0329	0.0092															
22	0.0985	0.0280																
23	0.0886	0.0238																
24	0.0798	0.0202																
25	0.0718	0.0172																
26	0.0646	0.0146																
27	0.0581	0.0124																
28	0.0523	0.0106																
29	0.0471	0.0090																
30	0.0424																	
31	0.0382																	
32	0.0343																	
33	0.0309																	
34	0.0278																	
35	0.0250																	
36	0.0225																	
37	0.0203																	
38	0.0182																	
39	0.0164																	
40	0.0148																	
41	0.0133																	
42	0.0120																	
43	0.0109																	
44	0.0097																	
45																		

Table 11.1

Table 11.2 Number of patients, n_1, required for the first stage of a Phase II trial.

Therapeutic effectiveness	Power $1 - \beta$					
π	0.80	0.85	0.90	0.95	0.99	0.999
0.05	32	37	45	59	90	135
0.06	27	31	38	49	75	112
0.07	23	27	32	42	64	96
0.08	20	23	28	36	56	83
0.09	18	21	25	32	49	74
0.10	16	19	22	44	66	
0.11	14	17	20	26	40	60
0.12	13	15	19	24	37	55
0.13	12	14	17	22	34	50
0.14	11	13	16	20	31	46
0.15	10	12	15	19	29	43
0.16	10	11	14	18	27	40
0.17	9	11	13	17	25	38
0.18	9	10	12	16	24	35
0.19	8	10	11	15	22	33
0.20	8	9	11	14	21	31
0.25	6	7	9	11	17	25
0.30	5	6	7	9	13	20
0.35	4	5	6	7	11	17
0.40	4	4	5	6	10	14
0.45	3	4	4	6	8	12
0.50	3	3	4	5	7	10
0.60	2	3	3	4	6	8
0.70	2	2	2	3	4	6
0.80	1	2	2	2	3	5
0.90	1	1	1	2	2	3

Table 11.2

Table 11.3. Additional number of patients required for the second stage of a Phase II trial.

Therapeutic effectiveness	Number of patients in preliminary trial	Number of treatment successes in preliminary trial, r_1					
π	n_1	1	2	3	4	5	6
Specified precision $\varepsilon = 0.05$, power $1-\beta = 0.90$							
0.05	45	0	0	0	2	9	16
0.06	38	0	0	8	16	24	31
0.07	32	0	10	21	30	38	45
0.08	28	6	19	30	40	48	55
0.09	25	13	27	38	48	57	63
0.10	22	20	35	47	57	65	71
0.11	20	25	41	54	64	71	76
0.12	19	28	45	57	67	74	79
0.13	17	35	52	64	74	80	83
0.14	16	38	55	68	77	82	84
0.15	15	42	59	72	80	85	85
0.16	14	46	64	76	83	86	86
0.17	13	50	68	80	86	87	87
0.18	12	55	73	84	88	88	88
0.19	11	60	77	87	89	89	89
0.20	11	60	77	87	89	89	89
0.25	9	71	87	91	91	91	91
0.30	7	83	93	93	93	93	93
0.35	6	90	93	93	93	93	93
0.40	5	95	95	95	95	95	–
0.45	4	96	96	96	96	–	
0.50	4	96	96	96	96	–	
0.60	3	85	85	85	–		
0.70	2	45	45	–			
0.80	2	45	45	–			

Table 11.3

174

Table 11.3 Continued.

Therapeutic effectiveness	Number of patients in preliminary trial	Number of treatment successes in preliminary trial, r_1					
π	n_1	1	2	3	4	5	6
Specified precision $\varepsilon = 0.10$, power $1-\beta = 0.90$							
0.05	45	0	0	0	0	0	0
0.06	38	0	0	0	0	0	0
0.07	32	0	0	0	0	0	0
0.08	28	0	0	0	0	0	0
0.09	25	0	0	0	0	0	0
0.10	22	0	0	0	0	0	0
0.11	20	0	0	0	1	3	4
0.12	19	0	0	0	1	3	4
0.13	17	0	1	4	6	8	8
0.14	16	0	2	5	8	9	9
0.15	15	0	4	7	9	10	10
0.16	14	1	6	9	11	11	11
0.17	13	3	8	11	12	12	12
0.18	12	5	10	12	13	13	13
0.19	11	7	11	14	14	14	14
0.20	11	7	11	14	14	14	14
0.25	9	11	15	16	16	16	16
0.30	7	16	18	18	18	18	18
0.35	6	18	19	19	19	19	19
0.40	5	20	20	20	20	20	—
0.45	4	21	21	21	21	—	—
0.50	4	21	21	21	21	—	
0.60	3	19	19	19	—		
0.70	2	10	10	—			
0.80	2	10	10	—			

Table 11.3

Table 11.3 Continued.

Therapeutic effectiveness	Number of patients in preliminary trial	Number of treatment successes in preliminary trial r_1					
π	n_1	1	2	3	4	5	6
Specified precision $\varepsilon = 0.05$, power $1-\beta = 0.95$							
0.05	59	0	0	0	0	0	0
0.06	49	0	0	0	0	2	8
0.07	42	0	0	0	8	15	22
0.08	36	0	2	12	21	28	35
0.09	32	0	10	21	30	38	45
0.10	29	4	17	28	38	46	53
0.11	26	11	24	36	46	54	61
0.12	24	15	29	41	51	60	66
0.13	22	20	35	47	58	65	71
0.14	20	25	41	54	64	71	77
0.15	19	29	45	58	67	75	79
0.16	18	31	48	61	71	77	81
0.17	17	35	52	65	74	80	83
0.18	16	38	56	68	77	83	84
0.19	15	42	60	72	81	85	85
0.20	14	46	64	76	84	86	86
0.25	11	60	78	87	89	89	89
0.30	9	71	87	91	91	91	91
0.35	7	84	93	93	93	93	93
0.40	6	90	93	93	93	93	93
0.45	6	90	93	93	93	93	93
0.50	5	95	95	95	95	95	–
0.60	4	95	95	95	95	–	–
0.70	3	82	82	82	–		
0.80	2	19	19	–			
0.90	2	19	19	–			

Table 11.3

Table 11.3 Continued.

Therapeutic effectiveness	Number of patients in preliminary trial	Number of treatment successes in preliminary trial r_1					
π	n_1	1	2	3	4	5	6
Specified precision $\varepsilon = 0.10$ power $1-\beta = 0.95$							
0.05	59	0	0	0	0	0	0
0.06	49	0	0	0	0	0	0
0.07	42	0	0	0	0	0	0
0.08	36	0	0	0	0	0	0
0.09	32	0	0	0	0	0	0
0.10	29	0	0	0	0	0	0
0.11	26	0	0	0	0	0	0
0.12	24	0	0	0	0	0	0
0.13	22	0	0	0	0	0	2
0.14	20	0	0	0	1	3	5
0.15	19	0	0	0	3	5	6
0.16	18	0	0	2	5	6	7
0.17	17	0	1	4	6	8	8
0.18	16	0	2	5	8	9	9
0.19	15	0	4	7	9	10	10
0.20	14	1	6	9	11	11	11
0.25	11	7	12	14	14	14	14
0.30	9	11	15	16	16	16	16
0.35	7	16	18	18	18	18	18
0.40	6	18	19	19	19	19	19
0.45	6	18	19	19	19	19	19
0.50	5	20	20	20	20	20	–
0.60	4	21	21	21	21	–	
0.70	3	19	19	19	–		
0.80	2	4	4	–			

Table 11.3

Chapter 12
Phase II Trials (Fleming's Single Stage Procedure)

12.1 INTRODUCTION

In considering the design of a Phase II trial of a new drug, the investigators will usually have some knowledge of the activity of other drugs for the same disease. The anticipated response to the new drug is therefore compared, at the planning stage, with the observed responses to other therapies. This may lead to the investigators prespecifying a response probability, which if the new drug does not achieve it, results in no further investigation. They might also have some idea of a response probability which if achieved or exceeded would certainly imply that the new drug has an efficacy worthy of further investigation, perhaps in a Phase III randomized trial. Such considerations led Fleming (1982) to propose a single stage procedure for Phase II trials.

12.2 THEORY AND FORMULAE

Sample Size

Suppose the investigators set the largest response proportion as π_o, which if true would clearly imply that the treatment regimen does not warrant further investigation. For example, for a new anti-tumour drug this may be set at 0.1. The investigators then judge that the smallest response proportion that would imply the treatment warrants further investigation is π_a. For a new anti-tumour drug this may be set at 0.2. This implies that the one-sided hypotheses to be tested are that $\pi \leqslant \pi_o$ versus $\pi \geqslant \pi_a$, where π is the actual probability of response.

In addition to specifying π_o and π_a it is necessary to specify α, the probability of rejecting the hypothesis $\pi \leqslant \pi_o$ when it is in fact true, together with β the probability of rejecting the hypothesis $\pi \geqslant \pi_a$ when that is true.

If N patients are recruited to the Phase II trial then the observed number of patient responses r will have a binomial distribution with parameter π. For N reasonably large and π not too small one can use a normal approximation to this binomial distribution.

This leads to the sample size required for Fleming's single-stage procedure as approximately

$$N = \frac{[z_{1-\alpha}\sqrt{\{\pi_o(1-\pi_o)\}} + z_{1-\beta}\sqrt{\{\pi_a(1-\pi_a)\}}]^2}{(\pi_a - \pi_o)^2}. \tag{12.1}$$

This expression is also given in equation (3.8).

Power

In contrast if one were asked for the power of a proposed trial of N patients then equation (12.1) could be inverted to give

$$z_{1-\beta} = \frac{(\pi_a - \pi_o)\sqrt{N} - z_{1-\alpha}\sqrt{\{\pi_o(1-\pi_o)\}}}{\sqrt{\{\pi_a(1-\pi_a)\}}} \tag{12.2}$$

12.3 BIBLIOGRAPHY

Fleming (*loc cit*) compares the single-stage procedure of equation (12.1) with multiple testing alternatives. For these latter designs patients are recruited to the trial a few at a time and their response to therapy assessed. More patients are then recruited to the trial only if the one-sided tests of hypothesis for either one of the preset probabilities π_o and π_a are rejected, otherwise recruitment is stopped. Early termination plans have also been discussed by Herson (1979) & Lee, Staquet, Simon, Catane and Muggia (1979).

12.4 DESCRIPTION OF THE TABLES

Table 12.1

Table 17.1 gives the number of subjects required to be recruited to a Phase II trial using Fleming's single-stage procedure. Equation (12.1) is evaluated for π_o from 0.05 to 0.9 in steps of 0.05, with π_a taking values greater than π_o in steps of 0.05, for one-sided α from 0.005 to 0.1 and power $1 - \beta$ ranging from 0.50 to 0.99.

12.5 USE OF THE TABLES

Table 12.1

Example 12.1

A certain standard treatment regimen is expected to yield responses in approximately 35% of cancer patients who are of the type to be entered into a Phase II study. Phase I studies indicate that the experimental regimen is somewhat more toxic than the standard treatment. How many patients should be recruited to the study if the one-sided test size is set at 5% and the power at 80%?

Since the new experimental treatment is more toxic than the standard it is unlikely to be of much potential interest if it has an efficacy much lower than the standard 35%, neither is it likely that it will investigated further unless its efficacy is somewhat greater than the standard. Such considerations may cause the investigor to set $\pi_o = 0.3$ and $\pi_a = 0.5$, then from Table 12.1 with $\alpha = 0.05$ and $1-\beta = 0.8$ one obtains $N = 35$ patients to be recruited to the trial.

Example 12.2

If in Example 12.1 π_a had been set at 0.6 rather than 0.5, how many patients would need to be recruited to such a trial?

Again $\pi_o = 0.3$, $\alpha = 0.05$, $1 - \beta = 0.8$ but now $\pi_a = 0.6$. Table 12.1 gives $N = 16$ patients for such a trial. The patient numbers are smaller here as the value for π_a has been raised over that of Example 12.1. Patient numbers are therefore very sensitive to changes in the investigator's preset probabilities.

Example 12.3

Van der Zee, van Rhoon, Wike-Hooley, Faithfull & Reinhold (1983) describe a Phase II trial of whole-body hyperthermia in 27 patients with various cancers. They observed complete responses in 5 patients and partial responses in 5 more. Suppose a further Phase II trial is planned but only in patients with lung adenocarcinoma in which 2 of the 3 patients with this disease had complete remission in the previous trial. The investigators set the lowest response probability of interest to be 0.15 and the highest as 0.50. They also require a test size $\alpha = 0.01$ and a power $1 - \beta = 0.90$, how many patients should be recruited to such a trial?

Here $\pi_o = 0.15$, $\pi_a = 0.50$, $\alpha = 0.01$, $1 - \beta = 0.9$ and Table 12.1 gives $N = 18$ patients for such a trial.

Example 12.4

An alternative method of delivering a drug used for the control of an infantile epileptic fit by means of a suppository rather than an injection is proposed. The standard procedure is known to be 90% effective but there are difficulties in giving injections to children while in a fit. It is considered that the suppository, which is easy to insert, must be at least 50% effective and would certainly be investigated further if it had an efficacy of 75% or more. How many children should be recruited to such a trial?

Here $\pi_o = 0.50$, $\pi_a = 0.75$ and suppose further the investigator sets $\alpha = 0.10$ and $1-\beta = 0.90$. Table 12.1 gives $N = 23$ children experiencing an infantile fit to be recruited to the trial.

Example 12.5

If the investigators planning the trial described in Example 12.4 were only able to recruit 16 children to their trial, what would be the corresponding power?

Here $N = 16$, $\pi_o = 0.50$, $\pi_a = 0.75$ and $\alpha = 0.10$. Table 2.2 gives $z_{1-\alpha} = 1.6449$ and equation (12.2) gives $z_{1-\beta} = 0.41$. From Table 2.1 gives $1-\beta = 0.6591$. Thus the power of the proposed Phase II trial is approximately 66%.

12.6 REFERENCES

Fleming T. R. (1982) One-sample multiple testing procedure for Phase II clinical trials. *Biometrics,* **38,** 143-151.

Herson J. (1979) Predictive probability early termination plans for Phase II clinical trials. *Biometrics,* **35,** 775-783.

Lee Y. J., Staquet M., Simon R., Catane R. & Muggia F. (1979) Two stage plans for patient accrual in Phase II cancer clinical trials. *Cancer Treatment Reports,* **63,** 1721-1726.

van Der Zee J., van Rhoon G. C., Wike-Hooley J. L., Faithfull N. S. & Reinhold H. S. (1983). Whole-body hyperthermia in cancer therapy: a report of a PhaseI-II study. *Eur. J. Cancer Clin. Oncol.* **19,** 1189-1200.

Table 12.1 Sample sizes for phase II trials: Flemings' single stage procedure.

π_a	α 1-sided	Power $1-\beta$								
		0.50	0.65	0.70	0.75	0.80	0.85	0.90	0.95	0.99

$\pi_o = 0.05$

π_a	α 1-sided	0.50	0.65	0.70	0.75	0.80	0.85	0.90	0.95	0.99
0.10	0.005	127	184	207	234	265	305	358	446	635
0.10	0.010	103	156	177	202	231	268	318	401	581
0.10	0.025	73	118	137	159	185	218	264	340	507
0.10	0.050	52	90	107	126	150	180	221	291	447
0.10	0.100	32	63	77	93	114	140	177	239	382
0.15	0.005	32	49	57	65	75	87	104	132	194
0.15	0.010	26	42	49	56	66	77	94	120	179
0.15	0.025	19	32	38	45	53	64	79	103	159
0.15	0.050	13	25	30	36	44	54	67	90	142
0.15	0.100	8	18	22	28	34	43	55	76	124
0.20	0.005	15	23	27	31	36	43	52	67	99
0.20	0.010	12	20	23	27	32	38	47	61	92
0.20	0.025	9	16	19	22	26	32	40	53	82
0.20	0.050	6	12	15	18	22	27	34	46	74
0.20	0.100	4	9	11	14	17	22	28	40	66
0.25	0.005	8	14	16	19	22	26	32	41	62
0.25	0.010	7	12	14	16	19	23	29	38	58
0.25	0.025	5	9	11	13	16	20	25	33	52
0.25	0.050	4	7	9	11	14	17	21	29	47
0.25	0.100	2	5	7	9	11	14	18	25	42
0.30	0.005	6	9	11	13	15	18	22	28	43
0.30	0.010	5	8	9	11	13	16	20	26	40
0.30	0.025	3	6	8	9	11	14	17	23	36
0.30	0.050	3	5	6	8	9	12	15	20	33
0.30	0.100	2	4	5	6	8	10	13	18	29
0.35	0.005	4	7	8	9	11	13	16	21	32
0.35	0.010	3	6	7	8	10	12	14	19	30
0.35	0.025	3	5	6	7	8	10	12	17	27
0.35	0.050	2	4	5	6	7	9	11	15	24
0.35	0.100	1	3	4	5	6	7	9	13	22
0.40	0.005	3	5	6	7	8	10	12	16	24
0.40	0.010	3	4	5	6	7	9	11	15	23
0.40	0.025	2	4	4	5	6	8	10	13	21
0.40	0.050	2	3	4	4	5	7	8	12	19
0.40	0.100	1	2	3	4	4	6	7	10	17
0.45	0.005	2	4	5	6	7	8	9	12	19
0.45	0.010	2	4	4	5	6	7	9	11	18
0.45	0.025	2	3	3	4	5	6	8	10	16
0.45	0.050	1	2	3	4	4	5	7	9	15
0.45	0.100	1	2	2	3	4	4	6	8	13
0.50	0.005	2	3	4	4	5	6	8	10	15
0.50	0.010	2	3	3	4	5	6	7	9	14
0.50	0.025	1	2	3	3	4	5	6	8	13
0.50	0.050	1	2	2	3	3	4	5	7	12
0.50	0.100	1	2	2	2	3	4	5	6	11
0.55	0.005	2	3	3	4	4	5	6	8	12
0.55	0.010	2	2	3	3	4	5	6	8	12
0.55	0.025	1	2	2	3	3	4	5	7	11
0.55	0.050	1	2	2	2	3	4	4	6	10
0.55	0.100	1	1	2	2	2	3	4	5	9
0.60	0.005	2	2	3	3	4	4	5	7	10
0.60	0.010	1	2	2	3	3	4	5	6	9
0.60	0.025	1	2	2	2	3	3	4	6	9
0.60	0.050	1	1	2	2	2	3	4	5	8
0.60	0.100	1	1	1	2	2	3	3	4	7

Table 12.1

Table 12.1 Continued.

π_a	α 1-sided	0.50	0.65	0.70	0.75	Power $1-\beta$ 0.80	0.85	0.90	0.95	0.99
						$\pi_o = $ **0.05**				
0.65	0.005	1	2	2	3	3	4	4	6	8
0.65	0.010	1	2	2	2	3	3	4	5	8
0.65	0.025	1	2	2	2	2	3	3	5	7
0.65	0.050	1	1	2	2	2	3	3	4	6
0.65	0.100	1	1	1	2	2	2	3	4	6
0.70	0.005	1	2	2	2	3	3	4	5	7
0.70	0.010	1	2	2	2	2	3	3	4	6
0.70	0.025	1	1	2	2	2	2	3	4	6
0.70	0.050	1	1	1	2	2	2	3	3	5
0.70	0.100	1	1	1	1	2	2	2	3	5
0.75	0.005	1	2	2	2	2	3	3	4	6
0.75	0.010	1	1	2	2	2	2	3	4	5
0.75	0.025	1	1	1	2	2	2	2	3	5
0.75	0.050	1	1	1	1	2	2	2	3	4
0.75	0.100	1	1	1	1	1	2	2	3	4
0.80	0.005	1	1	2	2	2	2	3	3	4
0.80	0.010	1	1	1	2	2	2	2	3	4
0.80	0.025	1	1	1	1	2	2	2	3	4
0.80	0.050	1	1	1	1	1	2	2	2	3
0.80	0.100	1	1	1	1	1	1	2	2	3
0.85	0.005	1	1	1	2	2	2	2	3	4
0.85	0.010	1	1	1	1	2	2	2	2	3
0.85	0.025	1	1	1	1	1	1	2	2	3
0.85	0.050	1	1	1	1	1	1	2	2	3
0.85	0.100	1	1	1	1	1	1	1	2	2
0.90	0.005	1	1	1	1	1	2	2	2	3
0.90	0.010	1	1	1	1	1	1	2	2	3
0.90	0.025	1	1	1	1	1	1	1	2	2
0.90	0.050	1	1	1	1	1	1	1	2	2
0.90	0.100	1	1	1	1	1	1	1	1	2
0.95	0.005	1	1	1	1	1	1	1	2	2
0.95	0.010	1	1	1	1	1	1	1	1	2
0.95	0.025	1	1	1	1	1	1	1	1	2
0.95	0.050	1	1	1	1	1	1	1	1	1
0.95	0.100	1	1	1	1	1	1	1	1	1
						$\pi_o = $ **0.05**				
0.15	0.005	239	332	369	411	461	523	606	740	1029
0.15	0.010	195	280	314	353	399	457	535	661	935
0.15	0.025	139	211	241	275	316	368	438	553	806
0.15	0.050	98	160	186	216	253	299	362	468	702
0.15	0.100	60	110	131	157	188	228	284	378	591
0.20	0.005	60	86	97	109	124	141	166	205	291
0.20	0.010	49	73	83	94	108	124	147	184	266
0.20	0.025	35	56	64	74	86	101	122	156	231
0.20	0.050	25	42	50	59	69	83	102	133	203
0.20	0.100	15	30	36	43	53	64	81	109	173
0.25	0.005	27	40	45	51	58	67	79	99	141
0.25	0.010	22	34	39	44	51	59	70	89	130
0.25	0.025	16	26	30	35	41	48	59	76	114
0.25	0.050	11	20	24	28	33	40	49	65	101
0.25	0.100	7	14	17	21	25	31	40	54	87
0.30	0.005	15	23	26	30	34	39	47	59	85
0.30	0.010	13	20	23	26	30	35	42	53	78
0.30	0.025	9	15	18	21	24	29	35	46	69
0.30	0.050	7	12	14	17	20	24	30	39	61
0.30	0.100	4	8	10	13	15	19	24	33	53

Table 12.1

Table 12.1 Continued.

π_a	α 1-sided	0.50	0.65	0.70	0.75	Power $1-\beta$ 0.80	0.85	0.90	0.95	0.99
						$\pi_o = 0.10$				
0.35	0.005	10	15	17	20	23	26	31	39	57
0.35	0.010	8	13	15	17	20	23	28	36	53
0.35	0.025	6	10	12	14	16	19	24	31	47
0.35	0.050	4	8	9	11	13	16	20	27	42
0.35	0.100	3	6	7	8	10	13	16	22	36
0.40	0.005	7	11	12	14	16	19	22	28	41
0.40	0.010	6	9	11	12	14	17	20	26	38
0.40	0.025	4	7	8	10	12	14	17	22	34
0.40	0.050	3	6	7	8	10	12	14	19	30
0.40	0.100	2	4	5	6	8	9	12	16	26
0.45	0.005	5	8	9	11	12	14	17	21	31
0.45	0.010	4	7	8	9	11	13	15	19	29
0.45	0.025	3	5	6	7	9	10	13	17	25
0.45	0.050	2	4	5	6	7	9	11	15	23
0.45	0.100	2	3	4	5	6	7	9	12	20
0.50	0.005	4	6	7	8	9	11	13	16	24
0.50	0.010	4	5	6	7	8	10	12	15	22
0.50	0.025	3	4	5	6	7	8	10	13	20
0.50	0.050	2	3	4	5	6	7	9	11	18
0.50	0.100	1	3	3	4	5	6	7	10	15
0.55	0.005	3	5	6	7	8	9	10	13	19
0.55	0.010	3	4	5	6	7	8	9	12	17
0.55	0.025	2	4	4	5	6	7	8	10	16
0.55	0.050	2	3	3	4	5	6	7	9	14
0.55	0.100	1	2	3	3	4	5	6	8	12
0.60	0.005	3	4	5	5	6	7	8	10	15
0.60	0.010	2	4	4	5	5	6	8	10	14
0.60	0.025	2	3	3	4	5	5	6	8	12
0.60	0.050	1	2	3	3	4	5	6	7	11
0.60	0.100	1	2	2	3	3	4	5	6	10
0.65	0.005	2	4	4	4	5	6	7	9	12
0.65	0.010	2	3	3	4	4	5	6	8	11
0.65	0.025	2	2	3	3	4	4	5	7	10
0.65	0.050	1	2	2	3	3	4	5	6	9
0.65	0.100	1	2	2	2	3	3	4	5	8
0.70	0.005	2	3	3	4	4	5	6	7	10
0.70	0.010	2	3	3	3	4	4	5	6	9
0.70	0.025	1	2	2	3	3	4	4	6	8
0.70	0.050	1	2	2	2	3	3	4	5	7
0.70	0.100	1	1	2	2	2	3	3	4	6
0.75	0.005	2	3	3	3	4	4	5	6	8
0.75	0.010	2	2	3	3	3	4	4	5	7
0.75	0.025	1	2	2	2	3	3	4	5	7
0.75	0.050	1	2	2	2	2	3	3	4	6
0.75	0.100	1	1	1	2	2	2	3	3	5
0.80	0.005	2	2	2	3	3	3	4	5	6
0.80	0.010	1	2	2	2	3	3	3	4	6
0.80	0.025	1	2	2	2	2	3	3	4	5
0.80	0.050	1	1	2	2	2	2	3	3	5
0.80	0.100	1	1	1	1	2	2	2	3	4
0.85	0.005	2	2	2	2	3	3	3	4	5
0.85	0.010	1	2	2	2	2	3	3	3	5
0.85	0.025	1	1	2	2	2	2	2	3	4
0.85	0.050	1	1	1	1	2	2	2	3	4
0.85	0.100	1	1	1	1	1	2	2	2	3

Table 12.1

Table 12.1 Continued.

π_a	α 1-sided	\multicolumn{9}{c}{Power $1-\beta$}								
		0.50	0.65	0.70	0.75	0.80	0.85	0.90	0.95	0.99
\multicolumn{11}{c}{$\pi_o = 0.10$}										
0.90	0.005	1	2	2	2	2	2	3	3	4
0.90	0.010	1	2	2	2	2	2	2	3	4
0.90	0.025	1	1	1	1	2	2	2	2	3
0.90	0.050	1	1	1	1	1	2	2	2	3
0.90	0.100	1	1	1	1	1	1	1	2	2
0.95	0.005	1	2	2	2	2	2	2	2	3
0.95	0.010	1	1	1	1	2	2	2	2	3
0.95	0.025	1	1	1	1	1	1	2	2	2
0.95	0.050	1	1	1	1	1	1	1	2	2
0.95	0.100	1	1	1	1	1	1	1	1	2
\multicolumn{11}{c}{$\pi_o = 0.15$}										
0.20	0.005	339	462	511	567	632	713	821	996	1370
0.20	0.010	277	388	433	485	546	621	722	887	1241
0.20	0.025	196	292	331	377	430	497	589	738	1064
0.20	0.050	138	220	255	294	342	402	484	621	922
0.20	0.100	84	150	179	212	253	305	377	498	771
0.25	0.005	85	119	132	147	165	188	218	267	372
0.25	0.010	70	100	112	127	143	164	192	239	338
0.25	0.025	49	76	86	99	114	132	158	200	292
0.25	0.050	35	57	67	78	91	108	131	169	255
0.25	0.100	21	39	47	57	68	83	103	137	215
0.30	0.005	38	54	60	68	76	87	101	125	176
0.30	0.010	31	46	51	58	66	76	90	112	160
0.30	0.025	22	35	40	46	53	62	74	94	139
0.30	0.050	16	26	31	36	43	51	62	80	122
0.30	0.100	10	18	22	27	32	39	49	66	104
0.35	0.005	22	31	35	39	44	50	59	73	103
0.35	0.010	18	26	30	34	38	44	52	66	95
0.35	0.025	13	20	23	27	31	36	43	56	82
0.35	0.050	9	15	18	21	25	30	36	48	72
0.35	0.100	6	11	13	16	19	23	29	39	62
0.40	0.005	14	20	23	26	29	33	39	48	68
0.40	0.010	12	17	19	22	25	29	35	43	63
0.40	0.025	8	13	15	17	20	24	29	37	55
0.40	0.050	6	10	12	14	16	20	24	32	48
0.40	0.100	4	7	9	10	13	15	19	26	41
0.45	0.005	10	14	16	18	20	23	27	34	48
0.45	0.010	8	12	14	16	18	21	24	31	44
0.45	0.025	6	9	11	12	14	17	20	26	39
0.45	0.050	4	7	8	10	12	14	17	22	34
0.45	0.100	3	5	6	7	9	11	14	19	29
0.50	0.005	7	11	12	13	15	17	20	25	36
0.50	0.010	6	9	10	12	13	15	18	23	33
0.50	0.025	4	7	8	9	11	13	15	19	29
0.50	0.050	3	5	6	7	9	10	13	17	26
0.50	0.100	2	4	5	6	7	8	10	14	22
0.55	0.005	6	8	9	10	12	13	16	19	27
0.55	0.010	5	7	8	9	10	12	14	17	25
0.55	0.025	4	5	6	7	8	10	12	15	22
0.55	0.050	3	4	5	6	7	8	10	13	20
0.55	0.100	2	3	4	4	5	6	8	11	17
0.60	0.005	5	7	7	8	9	11	12	15	21
0.60	0.010	4	6	6	7	8	9	11	14	20
0.60	0.025	3	4	5	6	7	8	9	12	17
0.60	0.050	2	3	4	5	5	6	8	10	15
0.60	0.100	2	3	3	4	4	5	6	8	13

Table 12.1

Table 12.1 Continued.

π_a	α 1-sided	0.50	0.65	0.70	0.75	0.80	0.85	0.90	0.95	0.99
					Power $1-\beta$					
					$\pi_o = 0.15$					
0.65	0.005	4	5	6	7	7	8	10	12	17
0.65	0.010	3	5	5	6	7	8	9	11	16
0.65	0.025	2	4	4	5	5	6	7	9	14
0.65	0.050	2	3	3	4	4	5	6	8	12
0.65	0.100	1	2	3	3	3	4	5	7	10
0.70	0.005	3	4	5	5	6	7	8	10	14
0.70	0.010	3	4	4	5	5	6	7	9	12
0.70	0.025	2	3	3	4	4	5	6	7	11
0.70	0.050	2	2	3	3	4	4	5	6	10
0.70	0.100	1	2	2	2	3	3	4	5	8
0.75	0.005	3	4	4	5	5	6	7	8	11
0.75	0.010	2	3	4	4	4	5	6	7	10
0.75	0.025	2	3	3	3	4	4	5	6	9
0.75	0.050	1	2	2	3	3	3	4	5	8
0.75	0.100	1	2	2	2	2	3	3	4	6
0.80	0.005	3	3	4	4	4	5	5	6	9
0.80	0.010	2	3	3	3	4	4	5	6	8
0.80	0.025	2	2	2	3	3	3	4	5	7
0.80	0.050	1	2	2	2	3	3	3	4	6
0.80	0.100	1	1	2	2	2	2	3	3	5
0.85	0.005	2	3	3	3	4	4	4	5	7
0.85	0.010	2	2	3	3	3	3	4	5	6
0.85	0.025	1	2	2	2	3	3	3	4	5
0.85	0.050	1	2	2	2	2	2	3	3	5
0.85	0.100	1	1	1	1	2	2	2	3	4
0.90	0.005	2	2	3	3	3	3	4	4	5
0.90	0.010	2	2	2	2	3	3	3	4	5
0.90	0.025	1	2	2	2	2	2	3	3	4
0.90	0.050	1	1	1	2	2	2	2	3	3
0.90	0.100	1	1	1	1	1	2	2	2	3
0.95	0.005	2	2	2	2	2	3	3	3	4
0.95	0.010	2	2	2	2	2	2	2	3	3
0.95	0.025	1	1	2	2	2	2	2	2	3
0.95	0.050	1	1	1	1	1	2	2	2	2
0.95	0.100	1	1	1	1	1	1	1	2	2
					$\pi_o = 0.20$					
0.25	0.005	425	574	633	700	779	876	1006	1215	1661
0.25	0.010	347	482	537	598	671	762	883	1080	1503
0.25	0.025	246	362	409	464	528	608	718	896	1284
0.25	0.050	174	273	314	362	419	490	589	751	1110
0.25	0.100	106	185	219	260	308	370	456	601	925
0.30	0.005	107	146	162	180	201	227	262	319	440
0.30	0.010	87	123	138	154	174	198	231	284	399
0.30	0.025	62	93	105	120	137	159	189	237	343
0.30	0.050	44	70	81	94	109	129	156	200	298
0.30	0.100	27	48	57	68	81	98	121	161	250
0.35	0.005	48	66	73	82	92	104	120	147	204
0.35	0.010	39	56	62	70	79	91	106	131	185
0.35	0.025	28	42	48	55	63	73	87	110	160
0.35	0.050	20	32	37	43	50	60	72	93	139
0.35	0.100	12	22	26	31	38	46	57	75	117
0.40	0.005	27	38	42	47	53	60	69	85	118
0.40	0.010	22	32	36	40	46	52	61	76	108
0.40	0.025	16	24	28	32	36	42	50	64	93
0.40	0.050	11	18	21	25	29	34	42	54	81
0.40	0.100	7	13	15	18	22	27	33	44	69

Table 12.1

Table 12.1 Continued.

π_a	α 1-sided	0.50	0.65	0.70	0.75	0.80	0.85	0.90	0.95	0.99
					Power $1-\beta$					
					$\pi_o = 0.20$					
0.45	0.005	17	24	27	30	34	39	45	55	77
0.45	0.010	14	21	23	26	30	34	40	49	70
0.45	0.025	10	16	18	21	24	28	33	42	61
0.45	0.050	7	12	14	16	19	23	27	35	53
0.45	0.100	5	8	10	12	14	17	22	29	45
0.50	0.005	12	17	19	21	24	27	32	39	54
0.50	0.010	10	15	16	18	21	24	28	35	49
0.50	0.025	7	11	13	14	17	19	23	29	43
0.50	0.050	5	9	10	12	13	16	19	25	37
0.50	0.100	3	6	7	9	10	12	15	20	32
0.55	0.005	9	13	14	16	18	20	23	28	40
0.55	0.010	8	11	12	14	15	18	21	25	36
0.55	0.025	6	8	9	11	12	14	17	21	31
0.55	0.050	4	6	7	9	10	12	14	18	27
0.55	0.100	3	5	5	6	8	9	11	15	23
0.60	0.005	7	10	11	12	14	15	18	22	30
0.60	0.010	6	8	9	10	12	13	16	19	27
0.60	0.025	4	6	7	8	9	11	13	16	24
0.60	0.050	3	5	6	7	8	9	11	14	21
0.60	0.100	2	4	4	5	6	7	9	11	18
0.65	0.005	6	8	9	10	11	12	14	17	23
0.65	0.010	5	7	7	8	9	11	12	15	21
0.65	0.025	4	5	6	7	7	9	10	13	18
0.65	0.050	3	4	5	5	6	7	8	11	16
0.65	0.100	2	3	3	4	5	6	7	9	13
0.70	0.005	5	6	7	8	9	10	11	13	18
0.70	0.010	4	5	6	7	7	8	10	12	16
0.70	0.025	3	4	5	5	6	7	8	10	14
0.70	0.050	2	3	4	4	5	6	7	8	12
0.70	0.100	2	2	3	3	4	4	5	7	10
0.75	0.005	4	5	6	6	7	8	9	11	14
0.75	0.010	3	4	5	5	6	7	8	9	13
0.75	0.025	3	3	4	4	5	6	6	8	11
0.75	0.050	2	3	3	3	4	5	5	7	10
0.75	0.100	1	2	2	3	3	4	4	5	8
0.80	0.005	3	4	5	5	6	6	7	8	11
0.80	0.010	3	4	4	5	5	6	6	8	10
0.80	0.025	2	3	3	4	4	4	5	6	9
0.80	0.050	2	2	3	3	3	4	4	5	8
0.80	0.100	1	2	2	2	3	3	3	4	6
0.85	0.005	3	4	4	4	5	5	6	7	9
0.85	0.010	3	3	3	4	4	5	5	6	8
0.85	0.025	2	3	3	3	3	4	4	5	7
0.85	0.050	2	2	2	2	3	3	3	4	6
0.85	0.100	1	2	2	2	2	2	3	3	5
0.90	0.005	3	3	3	4	4	4	5	5	7
0.90	0.010	2	3	3	3	3	4	4	5	6
0.90	0.025	2	2	2	2	3	3	3	4	5
0.90	0.050	1	2	2	2	2	2	3	3	4
0.90	0.100	1	1	1	2	2	2	2	3	3
0.95	0.005	2	3	3	3	3	3	4	4	5
0.95	0.010	2	2	2	3	3	3	3	3	4
0.95	0.025	2	2	2	2	2	2	3	3	3
0.95	0.050	1	1	2	2	2	2	2	2	3
0.95	0.100	1	1	1	1	1	1	2	2	2

Table 12.1

Table 12.1 Continued.

π_a	α 1-sided	\multicolumn{9}{c}{Power $1-\beta$}
		0.50	0.65	0.70	0.75	0.80	0.85	0.90	0.95	0.99
\multicolumn{11}{c}{$\pi_o = 0.25$}
0.30	0.005	498	668	736	812	902	1012	1160	1398	1904
0.30	0.010	406	561	623	694	777	879	1018	1241	1720
0.30	0.025	289	421	475	537	610	701	825	1028	1467
0.30	0.050	203	316	363	418	483	564	676	860	1265
0.30	0.100	124	215	253	299	354	425	522	686	1052
0.35	0.005	125	169	187	207	231	260	299	361	496
0.35	0.010	102	142	159	177	199	226	262	322	449
0.35	0.025	73	107	121	137	157	181	214	267	384
0.35	0.050	51	81	93	107	125	146	176	225	332
0.35	0.100	31	55	65	77	92	111	136	180	278
0.40	0.005	56	76	84	93	104	118	136	165	227
0.40	0.010	46	64	72	80	90	103	119	147	205
0.40	0.025	33	48	55	62	71	82	97	122	176
0.40	0.050	23	37	42	49	57	67	80	103	153
0.40	0.100	14	25	30	35	42	51	63	83	128
0.45	0.005	32	43	48	53	59	67	77	94	130
0.45	0.010	26	36	41	46	51	58	68	84	118
0.45	0.025	19	28	31	36	41	47	56	70	101
0.45	0.050	13	21	24	28	32	38	46	59	88
0.45	0.100	8	14	17	20	24	29	36	48	74
0.50	0.005	20	28	31	34	38	43	50	61	84
0.50	0.010	17	24	26	29	33	38	44	54	76
0.50	0.025	12	18	20	23	26	30	36	45	65
0.50	0.050	9	14	16	18	21	25	30	38	57
0.50	0.100	5	9	11	13	16	19	23	31	48
0.55	0.005	14	19	22	24	27	30	35	42	58
0.55	0.010	12	16	18	21	23	26	31	38	53
0.55	0.025	9	13	14	16	18	21	25	31	45
0.55	0.050	6	10	11	13	15	17	21	27	39
0.55	0.100	4	7	8	9	11	13	16	21	33
0.60	0.005	11	14	16	18	20	22	25	31	42
0.60	0.010	9	12	14	15	17	19	22	27	38
0.60	0.025	6	9	10	12	13	16	18	23	33
0.60	0.050	5	7	8	9	11	13	15	19	28
0.60	0.100	3	5	6	7	8	10	12	16	24
0.65	0.005	8	11	12	13	15	17	19	23	31
0.65	0.010	7	9	10	12	13	15	17	21	29
0.65	0.025	5	7	8	9	10	12	14	17	24
0.65	0.050	4	6	6	7	8	10	11	15	21
0.65	0.100	2	4	5	5	6	7	9	12	18
0.70	0.005	7	9	10	11	12	13	15	18	24
0.70	0.010	6	7	8	9	10	11	13	16	22
0.70	0.025	4	6	6	7	8	9	11	13	19
0.70	0.050	3	4	5	6	6	7	9	11	16
0.70	0.100	2	3	4	4	5	6	7	9	13
0.75	0.005	5	7	8	8	9	10	12	14	19
0.75	0.010	5	6	7	7	8	9	10	12	17
0.75	0.025	3	5	5	6	6	7	8	10	14
0.75	0.050	3	4	4	5	5	6	7	9	12
0.75	0.100	2	3	3	3	4	5	5	7	10
0.80	0.005	5	6	6	7	7	8	9	11	14
0.80	0.010	4	5	5	6	6	7	8	10	13
0.80	0.025	3	4	4	5	5	6	7	8	11
0.80	0.050	2	3	3	4	4	5	5	7	9
0.80	0.100	2	2	2	3	3	4	4	5	8

Table 12.1

Table 12.1 Continued.

π_a	α 1-sided	0.50	0.65	0.70	0.75	0.80	0.85	0.90	0.95	0.99
						Power $1-\beta$				

$\pi_o = 0.25$

π_a	α 1-sided	0.50	0.65	0.70	0.75	0.80	0.85	0.90	0.95	0.99
0.85	0.005	4	5	5	6	6	7	7	9	11
0.85	0.010	3	4	4	5	5	6	6	8	10
0.85	0.025	3	3	3	4	4	5	5	6	8
0.85	0.050	2	3	3	3	3	4	4	5	7
0.85	0.100	1	2	2	2	3	3	3	4	6
0.90	0.005	3	4	4	5	5	5	6	7	8
0.90	0.010	3	3	4	4	4	5	5	6	7
0.90	0.025	2	3	3	3	3	4	4	5	6
0.90	0.050	2	2	2	2	3	3	3	4	5
0.90	0.100	1	2	2	2	2	2	3	3	4
0.95	0.005	3	3	4	4	4	4	4	5	6
0.95	0.010	3	3	3	3	3	4	4	4	5
0.95	0.025	2	2	2	3	3	3	3	3	4
0.95	0.050	2	2	2	2	2	2	3	3	4
0.95	0.100	1	1	1	2	2	2	2	2	3

$\pi_o = 0.30$

π_a	α 1-sided	0.50	0.65	0.70	0.75	0.80	0.85	0.90	0.95	0.99
0.35	0.005	558	745	819	903	1001	1122	1285	1545	2098
0.35	0.010	455	625	693	771	862	974	1126	1370	1894
0.35	0.025	323	469	528	596	676	776	912	1133	1613
0.35	0.050	228	352	404	463	534	624	746	947	1389
0.35	0.100	138	238	281	331	392	468	575	753	1152
0.40	0.005	140	188	207	229	254	285	327	395	539
0.40	0.010	114	158	176	196	219	248	287	351	487
0.40	0.025	81	119	134	151	172	198	233	291	416
0.40	0.050	57	89	103	118	136	160	191	244	359
0.40	0.100	35	61	72	85	100	120	148	195	299
0.45	0.005	62	84	93	103	114	128	147	178	243
0.45	0.010	51	71	79	88	98	112	129	158	220
0.45	0.025	36	53	60	68	78	89	105	131	188
0.45	0.050	26	40	46	53	62	72	87	110	163
0.45	0.100	16	27	32	38	45	55	67	88	136
0.50	0.005	35	48	53	58	65	73	83	101	138
0.50	0.010	29	40	45	50	56	63	73	90	125
0.50	0.025	21	30	34	39	44	51	60	75	107
0.50	0.050	15	23	26	30	35	41	49	63	92
0.50	0.100	9	16	19	22	26	31	38	50	77
0.55	0.005	23	31	34	37	41	47	53	64	88
0.55	0.010	19	26	29	32	36	41	47	57	80
0.55	0.025	13	20	22	25	28	32	38	48	68
0.55	0.050	10	15	17	19	22	26	31	40	59
0.55	0.100	6	10	12	14	17	20	25	32	49
0.60	0.005	16	21	23	26	29	32	37	44	60
0.60	0.010	13	18	20	22	25	28	32	39	55
0.60	0.025	9	14	15	17	20	22	26	33	47
0.60	0.050	7	10	12	14	16	18	22	28	40
0.60	0.100	4	7	8	10	12	14	17	22	34
0.65	0.005	12	16	17	19	21	23	27	32	43
0.65	0.010	10	13	15	16	18	20	23	28	39
0.65	0.025	7	10	11	13	14	16	19	24	33
0.65	0.050	5	8	9	10	11	13	16	20	29
0.65	0.100	3	5	6	7	8	10	12	16	24
0.70	0.005	9	12	13	14	16	18	20	24	32
0.70	0.010	8	10	11	12	14	15	18	21	29
0.70	0.025	6	8	9	10	11	12	14	18	25
0.70	0.050	4	6	7	8	9	10	12	15	21
0.70	0.100	3	4	5	6	6	8	9	12	18

Table 12.1

Table 12.1 Continued.

π_a	α 1-sided	0.50	0.65	0.70	0.75	0.80	0.85	0.90	0.95	0.99
						Power $1-\beta$				

$\pi_o = \mathbf{0.30}$

π_a	α 1-sided	0.50	0.65	0.70	0.75	0.80	0.85	0.90	0.95	0.99
0.75	0.005	7	9	10	11	12	14	15	18	24
0.75	0.010	6	8	9	10	11	12	13	16	22
0.75	0.025	4	6	7	7	8	9	11	13	18
0.75	0.050	3	5	5	6	7	8	9	11	16
0.75	0.100	2	3	4	4	5	6	7	9	13
0.80	0.005	6	8	8	9	10	11	12	14	18
0.80	0.010	5	6	7	8	8	9	10	12	16
0.80	0.025	4	5	5	6	7	7	8	10	14
0.80	0.050	3	4	4	5	5	6	7	8	12
0.80	0.100	2	3	3	3	4	5	5	7	10
0.85	0.005	5	6	7	7	8	8	9	11	14
0.85	0.010	4	5	6	6	7	7	8	10	12
0.85	0.025	3	4	4	5	5	6	7	8	10
0.85	0.050	2	3	3	4	4	5	5	6	9
0.85	0.100	2	2	2	3	3	4	4	5	7
0.90	0.005	4	5	5	6	6	7	7	8	10
0.90	0.010	4	4	5	5	5	6	6	7	9
0.90	0.025	3	3	4	4	4	5	5	6	8
0.90	0.050	2	3	3	3	3	4	4	5	6
0.90	0.100	1	2	2	2	2	3	3	4	5
0.95	0.005	4	4	4	5	5	5	6	6	7
0.95	0.010	3	4	4	4	4	4	5	5	6
0.95	0.025	2	3	3	3	3	3	4	4	5
0.95	0.050	2	2	2	2	3	3	3	3	4
0.95	0.100	1	2	2	2	2	2	2	3	3

$\pi_o = \mathbf{0.35}$

π_a	α 1-sided	0.50	0.65	0.70	0.75	0.80	0.85	0.90	0.95	0.99
0.40	0.005	604	804	883	973	1078	1206	1379	1656	2244
0.40	0.010	493	675	747	830	927	1047	1208	1468	2024
0.40	0.025	350	506	569	641	726	833	977	1212	1722
0.40	0.050	247	379	434	498	573	669	798	1012	1482
0.40	0.100	150	257	302	355	420	501	615	804	1227
0.45	0.005	151	202	222	245	272	305	349	419	570
0.45	0.010	124	170	188	209	234	265	306	372	514
0.45	0.025	88	127	143	162	184	211	248	308	438
0.45	0.050	62	96	110	126	145	170	203	257	378
0.45	0.100	38	65	77	90	107	127	156	205	313
0.50	0.005	68	90	99	109	121	136	156	187	255
0.50	0.010	55	76	84	94	105	118	137	166	230
0.50	0.025	39	57	64	72	82	94	111	138	196
0.50	0.050	28	43	49	56	65	76	91	115	169
0.50	0.100	17	29	34	40	48	57	70	92	140
0.55	0.005	38	51	56	62	68	77	88	105	143
0.55	0.010	31	43	47	53	59	67	77	93	129
0.55	0.025	22	32	36	41	46	53	62	77	110
0.55	0.050	16	24	28	32	37	43	51	65	95
0.55	0.100	10	17	20	23	27	32	39	52	79
0.60	0.005	25	33	36	39	44	49	56	67	90
0.60	0.010	20	27	30	34	38	42	49	59	81
0.60	0.025	14	21	23	26	30	34	40	49	69
0.60	0.050	10	16	18	20	23	27	32	41	60
0.60	0.100	6	11	13	15	17	21	25	33	50
0.65	0.005	17	23	25	27	30	33	38	46	61
0.65	0.010	14	19	21	23	26	29	33	40	55
0.65	0.025	10	14	16	18	20	23	27	33	47
0.65	0.050	7	11	12	14	16	19	22	28	40
0.65	0.100	5	8	9	10	12	14	17	22	33

Table 12.1

Table 12.1 Continued.

π_a	α 1-sided	0.50	0.65	0.70	0.75	0.80	0.85	0.90	0.95	0.99
					Power $1-\beta$					

					$\pi_o = 0.35$					
0.70	0.005	13	17	18	20	22	24	27	33	43
0.70	0.010	11	14	15	17	19	21	24	29	39
0.70	0.025	8	11	12	13	15	17	19	24	33
0.70	0.050	6	8	9	10	12	13	16	20	28
0.70	0.100	4	6	6	7	9	10	12	16	23
0.75	0.005	10	13	14	15	16	18	20	24	32
0.75	0.010	8	11	12	13	14	16	18	21	29
0.75	0.025	6	8	9	10	11	12	14	17	24
0.75	0.050	4	6	7	8	9	10	12	15	21
0.75	0.100	3	4	5	6	6	8	9	11	17
0.80	0.005	8	10	11	12	13	14	15	18	24
0.80	0.010	7	8	9	10	11	12	13	16	21
0.80	0.025	5	6	7	8	8	9	11	13	18
0.80	0.050	4	5	5	6	7	8	9	11	15
0.80	0.100	2	3	4	4	5	6	7	8	12
0.85	0.005	7	8	9	9	10	11	12	14	17
0.85	0.010	5	7	7	8	8	9	10	12	16
0.85	0.025	4	5	6	6	7	7	8	10	13
0.85	0.050	3	4	4	5	5	6	7	8	11
0.85	0.100	2	3	3	3	4	4	5	6	9
0.90	0.005	5	6	7	7	8	8	9	10	13
0.90	0.010	5	5	6	6	7	7	8	9	11
0.90	0.025	3	4	4	5	5	6	6	7	9
0.90	0.050	3	3	3	4	4	4	5	6	8
0.90	0.100	2	2	2	3	3	3	4	5	6
0.95	0.005	5	5	6	6	6	6	7	7	9
0.95	0.010	4	4	5	5	5	5	6	6	8
0.95	0.025	3	3	4	4	4	4	5	5	6
0.95	0.050	2	3	3	3	3	3	4	4	5
0.95	0.100	2	2	2	2	2	2	3	3	4

					$\pi_o = 0.40$					
0.45	0.005	637	846	928	1021	1130	1264	1444	1731	2342
0.45	0.010	520	710	785	871	972	1096	1264	1534	2111
0.45	0.025	369	531	597	672	761	872	1022	1266	1794
0.45	0.050	260	399	456	522	600	699	834	1056	1542
0.45	0.100	158	269	316	372	439	523	641	837	1275
0.50	0.005	160	212	233	256	284	317	363	435	589
0.50	0.010	130	178	197	219	244	275	317	385	531
0.50	0.025	93	133	150	169	191	219	257	318	451
0.50	0.050	65	100	115	131	151	176	210	266	388
0.50	0.100	40	68	80	94	110	132	161	211	321
0.55	0.005	71	94	104	114	126	141	161	193	261
0.55	0.010	58	79	88	97	108	122	141	171	235
0.55	0.025	41	59	67	75	85	97	114	141	200
0.55	0.050	29	45	51	58	67	78	93	118	172
0.55	0.100	18	30	36	42	49	59	72	93	142
0.60	0.005	40	53	58	64	71	79	90	107	145
0.60	0.010	33	45	49	55	61	68	79	95	130
0.60	0.025	24	34	38	42	48	54	64	78	111
0.60	0.050	17	25	29	33	38	44	52	65	95
0.60	0.100	10	17	20	23	28	33	40	52	79
0.65	0.005	26	34	37	41	45	50	57	68	90
0.65	0.010	21	29	31	35	38	43	50	60	81
0.65	0.025	15	21	24	27	30	34	40	49	69
0.65	0.050	11	16	18	21	24	28	33	41	59
0.65	0.100	7	11	13	15	17	21	25	32	49

Table 12.1

Table 12.1 Continued.

π_a	α 1-sided	\multicolumn{9}{c}{Power $1-\beta$}								
		0.50	0.65	0.70	0.75	0.80	0.85	0.90	0.95	0.99
\multicolumn{11}{c}{$\pi_o = 0.40$}										
0.70	0.005	18	23	26	28	31	34	38	46	61
0.70	0.010	15	20	22	24	26	29	34	40	55
0.70	0.025	11	15	17	18	21	23	27	33	46
0.70	0.050	8	11	13	14	16	19	22	28	39
0.70	0.100	5	8	9	10	12	14	17	22	32
0.75	0.005	13	17	19	20	22	24	27	32	43
0.75	0.010	11	14	16	17	19	21	24	28	38
0.75	0.025	8	11	12	13	15	17	19	23	32
0.75	0.050	6	8	9	10	12	13	16	19	27
0.75	0.100	4	6	6	7	9	10	12	15	22
0.80	0.005	10	13	14	15	16	18	20	24	31
0.80	0.010	9	11	12	13	14	16	18	21	27
0.80	0.025	6	8	9	10	11	12	14	17	23
0.80	0.050	5	6	7	8	9	10	11	14	19
0.80	0.100	3	4	5	6	6	7	9	11	16
0.85	0.005	8	10	11	12	13	14	15	17	22
0.85	0.010	7	9	9	10	11	12	13	15	20
0.85	0.025	5	6	7	8	8	9	10	12	16
0.85	0.050	4	5	5	6	7	7	8	10	14
0.85	0.100	2	3	4	4	5	5	6	8	11
0.90	0.005	7	8	9	9	10	10	11	13	16
0.90	0.010	6	7	7	8	8	9	10	11	14
0.90	0.025	4	5	5	6	6	7	8	9	11
0.90	0.050	3	4	4	5	5	5	6	7	10
0.90	0.100	2	3	3	3	4	4	5	6	8
0.95	0.005	6	6	7	7	7	8	8	9	11
0.95	0.010	5	5	6	6	6	7	7	8	9
0.95	0.025	4	4	4	5	5	5	6	6	8
0.95	0.050	3	3	3	4	4	4	4	5	6
0.95	0.100	2	2	2	2	3	3	3	4	5
\multicolumn{11}{c}{$\pi_o = 0.45$}										
0.50	0.005	657	870	954	1049	1160	1296	1478	1771	2391
0.50	0.010	536	730	807	894	997	1124	1294	1568	2154
0.50	0.025	381	546	613	689	780	892	1045	1293	1829
0.50	0.050	268	409	467	535	615	715	852	1077	1571
0.50	0.100	163	276	324	381	449	535	654	853	1298
0.55	0.005	165	218	238	262	290	323	369	441	595
0.55	0.010	134	182	202	223	249	280	323	391	536
0.55	0.025	96	137	153	172	195	223	261	322	455
0.55	0.050	67	103	117	134	154	178	212	268	391
0.55	0.100	41	69	81	95	112	133	163	212	323
0.60	0.005	73	97	106	116	128	143	163	194	261
0.60	0.010	60	81	89	99	110	124	142	172	235
0.60	0.025	43	61	68	76	86	98	115	141	199
0.60	0.050	30	46	52	59	68	79	93	118	171
0.60	0.100	19	31	36	42	49	59	72	93	141
0.65	0.005	42	54	59	65	71	79	90	107	143
0.65	0.010	34	45	50	55	61	69	79	95	129
0.65	0.025	24	34	38	43	48	54	63	78	109
0.65	0.050	17	26	29	33	38	44	52	65	93
0.65	0.100	11	17	20	24	27	33	39	51	77
0.70	0.005	27	35	38	41	45	50	56	67	89
0.70	0.010	22	29	32	35	39	43	49	59	80
0.70	0.025	16	22	24	27	30	34	40	48	67
0.70	0.050	11	16	18	21	24	27	32	40	57
0.70	0.100	7	11	13	15	17	20	25	31	47

191

Table 12.1

Table 12.1 Continued.

π_a	α 1-sided	0.50	0.65	0.70	0.75	0.80	0.85	0.90	0.95	0.99
						Power $1-\beta$				

$\pi_o = \mathbf{0.45}$

π_a	α 1-sided	0.50	0.65	0.70	0.75	0.80	0.85	0.90	0.95	0.99
0.75	0.005	19	24	26	28	31	34	38	45	59
0.75	0.010	15	20	22	24	26	29	33	39	53
0.75	0.025	11	15	17	18	20	23	27	32	44
0.75	0.050	8	11	13	14	16	18	21	27	38
0.75	0.100	5	8	9	10	12	14	16	21	31
0.80	0.005	14	17	19	20	22	24	27	31	40
0.80	0.010	11	15	16	17	19	21	23	27	36
0.80	0.025	8	11	12	13	15	16	19	22	30
0.80	0.050	6	8	9	10	11	13	15	18	25
0.80	0.100	4	6	6	7	8	10	11	14	21
0.85	0.005	11	13	14	15	16	18	19	22	28
0.85	0.010	9	11	12	13	14	15	17	20	25
0.85	0.025	6	8	9	10	11	12	13	16	21
0.85	0.050	5	6	7	8	8	9	11	13	17
0.85	0.100	3	4	5	5	6	7	8	10	14
0.90	0.005	9	10	11	11	12	13	14	16	20
0.90	0.010	7	9	9	10	10	11	12	14	17
0.90	0.025	5	6	7	7	8	9	10	11	14
0.90	0.050	4	5	5	6	6	7	8	9	12
0.90	0.100	3	3	4	4	4	5	6	7	9
0.95	0.005	7	8	8	9	9	10	10	11	13
0.95	0.010	6	7	7	7	8	8	9	10	12
0.95	0.025	4	5	5	6	6	6	7	8	9
0.95	0.050	3	4	4	4	5	5	5	6	8
0.95	0.100	2	3	3	3	3	3	4	4	6

$\pi_o = \mathbf{0.50}$

π_a	α 1-sided	0.50	0.65	0.70	0.75	0.80	0.85	0.90	0.95	0.99
0.55	0.005	664	876	960	1055	1166	1302	1483	1775	2392
0.55	0.010	542	735	812	899	1001	1128	1298	1571	2154
0.55	0.025	385	550	616	693	783	895	1047	1294	1828
0.55	0.050	271	412	470	537	617	717	853	1077	1568
0.55	0.100	165	278	326	382	449	535	654	852	1294
0.60	0.005	166	219	239	262	290	323	368	439	590
0.60	0.010	136	183	202	224	249	280	321	388	531
0.60	0.025	97	137	153	172	194	222	259	319	450
0.60	0.050	68	103	117	133	153	177	211	266	385
0.60	0.100	42	69	81	95	111	132	161	210	317
0.65	0.005	74	97	106	116	127	142	161	191	256
0.65	0.010	61	81	89	98	109	123	140	169	230
0.65	0.025	43	61	68	76	85	97	113	139	195
0.65	0.050	31	45	52	59	67	78	92	115	166
0.65	0.100	19	31	36	42	49	58	70	91	137
0.70	0.005	42	54	59	64	71	78	88	105	139
0.70	0.010	34	45	50	55	60	68	77	92	125
0.70	0.025	25	34	38	42	47	53	62	76	105
0.70	0.050	17	25	29	33	37	43	50	63	90
0.70	0.100	11	17	20	23	27	32	38	49	73
0.75	0.005	27	34	37	40	44	49	55	65	85
0.75	0.010	22	29	31	34	38	42	48	57	76
0.75	0.025	16	22	24	26	29	33	38	46	64
0.75	0.050	11	16	18	20	23	26	31	38	54
0.75	0.100	7	11	13	14	17	19	23	30	44
0.80	0.005	19	24	25	27	30	33	37	43	55
0.80	0.010	16	20	21	23	25	28	32	37	49
0.80	0.025	11	15	16	18	20	22	25	30	41
0.80	0.050	8	11	12	14	15	18	20	25	35
0.80	0.100	5	8	9	10	11	13	15	19	28

Table 12.1

Table 12.1 Continued.

π_a	α 1-sided	0.50	0.65	0.70	0.75	0.80	0.85	0.90	0.95	0.99
					Power $1-\beta$					

$\pi_o = 0.50$

π_a	α 1-sided	0.50	0.65	0.70	0.75	0.80	0.85	0.90	0.95	0.99
0.85	0.005	14	17	18	20	21	23	25	29	37
0.85	0.010	12	14	15	17	18	20	22	26	33
0.85	0.025	8	11	12	13	14	15	17	21	27
0.85	0.050	6	8	9	10	11	12	14	17	23
0.85	0.100	4	5	6	7	8	9	10	13	18
0.90	0.005	11	13	14	14	15	16	18	20	25
0.90	0.010	9	11	11	12	13	14	15	18	22
0.90	0.025	7	8	9	9	10	11	12	14	18
0.90	0.050	5	6	6	7	8	9	10	11	15
0.90	0.100	3	4	4	5	5	6	7	9	12
0.95	0.005	9	10	10	11	11	12	13	14	16
0.95	0.010	7	8	9	9	9	10	11	12	14
0.95	0.025	5	6	6	7	7	8	8	9	11
0.95	0.050	4	5	5	5	5	6	6	7	9
0.95	0.100	3	3	3	4	4	4	5	5	7

$\pi_o = 0.55$

π_a	α 1-sided	0.50	0.65	0.70	0.75	0.80	0.85	0.90	0.95	0.99
0.60	0.005	657	865	947	1040	1148	1281	1459	1743	2345
0.60	0.010	536	725	801	886	986	1110	1275	1542	2111
0.60	0.025	381	542	608	682	770	880	1028	1269	1789
0.60	0.050	268	406	463	528	606	704	837	1056	1534
0.60	0.100	163	274	321	375	441	525	641	834	1264
0.65	0.005	165	215	235	258	284	316	359	427	572
0.65	0.010	134	180	199	219	243	273	313	378	514
0.65	0.025	96	135	151	169	190	216	252	310	435
0.65	0.050	67	101	115	130	149	173	205	257	372
0.65	0.100	41	68	79	93	108	129	156	203	306
0.70	0.005	73	95	103	113	124	138	156	185	245
0.70	0.010	60	80	87	96	106	119	136	163	220
0.70	0.025	43	59	66	74	83	94	109	133	186
0.70	0.050	30	44	50	57	65	75	88	110	158
0.70	0.100	19	30	35	40	47	56	67	87	129
0.75	0.005	42	53	57	62	68	75	85	100	131
0.75	0.010	34	44	48	53	58	65	74	88	118
0.75	0.025	24	33	37	41	45	51	59	72	99
0.75	0.050	17	25	28	31	35	41	48	59	84
0.75	0.100	11	17	19	22	26	30	36	46	68
0.80	0.005	27	33	36	39	42	47	52	61	79
0.80	0.010	22	28	30	33	36	40	45	53	70
0.80	0.025	16	21	23	25	28	31	36	43	59
0.80	0.050	11	16	17	19	22	25	29	35	49
0.80	0.100	7	11	12	14	16	18	22	27	40
0.85	0.005	19	23	24	26	28	31	34	39	50
0.85	0.010	15	19	21	22	24	26	29	34	44
0.85	0.025	11	14	16	17	19	21	23	28	37
0.85	0.050	8	11	12	13	14	16	19	22	31
0.85	0.100	5	7	8	9	10	12	14	17	24
0.90	0.005	14	16	17	18	20	21	23	26	32
0.90	0.010	11	14	15	16	17	18	20	23	29
0.90	0.025	8	10	11	12	13	14	16	18	23
0.90	0.050	6	8	8	9	10	11	12	15	19
0.90	0.100	4	5	6	6	7	8	9	11	15
0.95	0.005	11	12	13	13	14	15	16	17	20
0.95	0.010	9	10	11	11	12	12	13	15	18
0.95	0.025	6	8	8	8	9	10	10	12	14
0.95	0.050	5	6	6	6	7	7	8	9	11
0.95	0.100	3	4	4	4	5	5	6	7	9

Table 12.1

Table 12.1 Continued.

π_a	α 1-sided	Power $1-\beta$								
		0.50	0.65	0.70	0.75	0.80	0.85	0.90	0.95	0.99
					$\pi_o = $ **0.60**					
0.65	0.005	637	836	915	1004	1107	1234	1404	1676	2250
0.65	0.010	520	701	773	855	950	1069	1227	1482	2024
0.65	0.025	369	524	586	658	742	847	988	1218	1714
0.65	0.050	260	392	446	509	583	677	804	1012	1468
0.65	0.100	158	264	309	361	424	504	615	798	1208
0.70	0.005	160	207	226	247	272	302	342	407	542
0.70	0.010	130	174	191	210	233	261	299	359	487
0.70	0.025	93	130	145	162	182	206	240	294	411
0.70	0.050	65	97	110	125	142	165	195	244	351
0.70	0.100	40	65	76	88	103	122	148	191	287
0.75	0.005	71	91	99	108	118	131	147	174	229
0.75	0.010	58	76	84	92	101	113	128	153	205
0.75	0.025	41	57	63	70	78	89	103	125	173
0.75	0.050	29	43	48	54	61	70	83	103	147
0.75	0.100	18	29	33	38	44	5263		80	119
0.80	0.005	40	51	55	59	64	71	79	93	121
0.80	0.010	33	42	46	50	55	61	69	81	108
0.80	0.025	24	32	35	38	43	48	55	66	90
0.80	0.050	17	24	26	29	33	38	44	54	76
0.80	0.100	10	16	18	21	24	28	33	42	61
0.85	0.005	26	32	34	37	40	43	48	55	71
0.85	0.010	21	27	29	31	34	37	41	48	63
0.85	0.025	15	20	22	24	26	29	33	39	52
0.85	0.050	11	15	16	18	20	23	26	32	43
0.85	0.100	7	10	11	13	14	16	19	24	35
0.90	0.005	18	22	23	24	26	28	31	35	43
0.90	0.010	15	18	19	21	22	24	26	30	38
0.90	0.025	11	13	14	16	17	18	21	24	31
0.90	0.050	8	10	11	12	13	14	16	19	26
0.90	0.100	5	7	7	8	9	10	12	14	20
0.95	0.005	13	15	16	17	18	19	20	22	26
0.95	0.010	11	13	13	14	15	16	17	19	23
0.95	0.025	8	9	10	11	11	12	13	15	18
0.95	0.050	6	7	7	8	8	9	10	12	15
0.95	0.100	4	5	5	5	6	6	7	8	11
					$\pi_o = $ **0.65**					
0.70	0.005	604	790	864	946	1043	1161	1319	1572	2107
0.70	0.010	493	662	729	806	895	1005	1152	1389	1894
0.70	0.025	350	495	553	619	698	796	927	1141	1602
0.70	0.050	247	370	421	479	548	635	753	947	1370
0.70	0.100	150	249	291	339	398	472	575	746	1126
0.75	0.005	151	195	212	232	254	282	319	377	500
0.75	0.010	124	163	179	197	218	243	278	332	449
0.75	0.025	88	122	136	151	169	192	222	272	378
0.75	0.050	62	91	103	116	133	153	180	225	322
0.75	0.100	38	61	71	82	96	113	136	176	262
0.80	0.005	68	85	92	100	109	120	135	159	208
0.80	0.010	55	71	78	85	93	104	117	139	185
0.80	0.025	39	53	59	65	72	81	94	113	155
0.80	0.050	28	40	44	50	56	64	75	93	131
0.80	0.100	17	27	30	35	40	47	57	72	106
0.85	0.005	38	47	51	54	59	64	72	83	107
0.85	0.010	31	39	43	46	50	55	62	72	95
0.85	0.025	22	29	32	35	39	43	49	58	78
0.85	0.050	16	22	24	27	30	34	39	48	66
0.85	0.100	10	15	16	19	21	25	29	36	52

Table 12.1

Table 12.1 Continued.

π_a	α 1-sided	0.50	0.65	0.70	0.75	0.80	0.85	0.90	0.95	0.99
					Power $1-\beta$					
					$\pi_o = 0.65$					
0.90	0.005	25	29	31	33	36	38	42	48	60
0.90	0.010	20	25	26	28	30	33	36	42	53
0.90	0.025	14	18	20	21	23	25	28	33	43
0.90	0.050	10	13	15	16	18	20	22	27	36
0.90	0.100	6	9	10	11	12	14	16	20	28
0.95	0.005	17	20	21	22	23	24	26	28	34
0.95	0.010	14	16	17	18	19	20	22	24	30
0.95	0.025	10	12	13	14	14	15	17	19	24
0.95	0.050	7	9	9	10	11	12	13	15	19
0.95	0.100	5	6	6	7	8	8	9	11	14
					$\pi_o = 0.70$					
0.75	0.005	558	727	793	868	955	1062	1205	1433	1915
0.75	0.010	455	609	669	738	819	918	1052	1265	1720
0.75	0.025	323	454	507	567	638	726	845	1038	1453
0.75	0.050	228	340	385	438	501	579	686	860	1241
0.75	0.100	138	228	266	310	363	430	522	676	1018
0.80	0.005	140	179	194	211	231	255	287	338	446
0.80	0.010	114	149	163	179	197	220	250	298	399
0.80	0.025	81	111	123	137	153	173	200	243	335
0.80	0.050	57	83	93	105	119	137	161	200	284
0.80	0.100	35	55	64	74	86	101	121	156	231
0.85	0.005	62	78	84	90	98	107	120	139	180
0.85	0.010	51	65	70	76	84	92	104	122	160
0.85	0.025	36	48	53	58	64	72	82	99	133
0.85	0.050	26	36	40	44	50	57	66	80	112
0.85	0.100	16	24	27	31	36	41	49	62	90
0.90	0.005	35	42	45	48	52	56	62	71	89
0.90	0.010	29	35	38	41	44	48	53	61	78
0.90	0.025	21	26	28	31	34	37	42	49	64
0.90	0.050	15	19	21	23	26	29	33	39	53
0.90	0.100	9	13	14	16	18	21	24	30	42
0.95	0.005	23	26	27	29	30	32	35	38	46
0.95	0.010	19	22	23	24	25	27	29	33	40
0.95	0.025	13	16	17	18	19	21	23	26	32
0.95	0.050	10	12	13	13	15	16	18	20	26
0.95	0.100	6	8	8	9	10	11	13	15	20
					$\pi_o = 0.75$					
0.80	0.005	498	645	703	768	844	937	1061	1258	1675
0.80	0.010	406	540	593	653	723	809	925	1110	1503
0.80	0.025	289	403	449	501	563	639	742	908	1267
0.80	0.050	203	301	341	386	441	508	601	751	1080
0.80	0.100	124	202	234	273	318	376	456	589	883
0.85	0.005	125	157	170	184	201	221	248	290	379
0.85	0.010	102	132	143	156	172	190	215	255	338
0.85	0.025	73	98	108	119	133	149	171	207	283
0.85	0.050	51	73	81	91	103	118	137	169	239
0.85	0.100	31	48	56	64	74	86	103	131	192
0.90	0.005	56	68	72	78	84	91	100	116	147
0.90	0.010	46	57	61	66	71	78	87	101	130
0.90	0.025	33	42	45	50	54	60	68	81	107
0.90	0.050	23	31	34	38	42	47	54	65	89
0.90	0.100	14	20	23	26	29	34	40	49	70
0.95	0.005	32	36	38	40	43	45	49	55	66
0.95	0.010	26	30	32	34	36	39	42	47	58
0.95	0.025	19	22	24	25	27	29	32	37	46
0.95	0.050	13	16	18	19	21	23	25	29	38
0.95	0.100	8	11	12	13	14	16	18	21	29

Table 12.1

Table 12.1 Continued.

π_a	α 1-sided	0.50	0.65	0.70	0.75	Power $1-\beta$ 0.80	0.85	0.90	0.95	0.99
					$\pi_o = 0.80$					
0.85	0.005	425	546	594	647	709	785	886	1047	1386
0.85	0.010	347	457	500	549	607	677	771	922	1241
0.85	0.025	246	340	378	421	471	533	617	753	1043
0.85	0.050	174	254	286	324	368	423	498	621	887
0.85	0.100	106	170	196	228	265	312	377	484	722
0.90	0.005	107	132	142	152	165	180	201	233	299
0.90	0.010	87	110	119	129	140	155	173	203	266
0.90	0.025	62	81	89	98	108	120	137	164	220
0.90	0.050	44	60	67	75	83	94	109	133	184
0.90	0.100	27	40	45	52	59	68	81	102	147
0.95	0.005	48	56	59	62	66	71	77	86	106
0.95	0.010	39	46	49	52	56	60	66	74	92
0.95	0.025	28	34	36	39	42	46	51	59	75
0.95	0.050	20	25	27	29	32	35	40	46	61
0.95	0.100	12	16	18	20	22	25	28	34	47
					$\pi_o = 0.85$					
0.90	0.005	339	429	465	504	550	606	681	799	1047
0.90	0.010	277	359	391	427	470	522	591	702	935
0.90	0.025	196	266	294	326	363	409	471	570	782
0.90	0.050	138	198	222	250	283	323	378	468	661
0.90	0.100	84	132	152	175	202	237	284	362	535
0.95	0.005	85	101	107	114	122	132	144	164	204
0.95	0.010	70	84	90	96	103	112	124	142	179
0.95	0.025	49	62	67	72	79	86	96	113	146
0.95	0.050	35	46	50	54	60	67	76	90	120
0.95	0.100	21	30	33	37	42	47	55	67	94
					$\pi_o = 0.90$					
0.95	0.005	239	294	315	339	366	399	443	512	656
0.95	0.010	195	245	264	286	311	342	382	447	581
0.95	0.025	139	181	198	217	239	265	301	359	480
0.95	0.050	98	134	148	165	184	207	239	291	401
0.95	0.100	60	88	100	113	130	150	177	221	318

Table 12.1

196

Chapter 13
Randomization

13.1 INTRODUCTION

Having decided on the number of subjects to include in a clinical trial, an investigator is faced with the problem of deciding how to allocate his patients to different treatments. In this chapter we discuss various forms of random allocation. In general when subjects have been assigned to treatment at random, then at the start of the trial each subject is equally likely to receive each of the treatments. The advantage of randomization is that it tends to produce groups comparable with respect to known, as well as unknown risk factors and ensures that the statistical tests will have valid significance levels.

In the examples in this chapter we will assume that the aim is to achieve roughly equal numbers of patients in each treatment group. The simplest form of randomization is to toss an unbiased coin whenever a subject is eligible for randomization, and allocate one treatment 'heads' and the other treatment 'tails'. Roughly half the subjects will then be allocated to one treatment and half to the other. Another method would be to use a random digit table. Select a row (or column) and working along it one could allocate to one treatment if the digit were even and the other if it were odd. Alternatively one could use 0-4 and 5-9 as the two digit sets and allocate to one treatment if the digit is 0, 1, 2, 3, 4 and to the other treatment otherwise. The disadvantage of this type of procedure is that it can result in a substantial difference in the numbers of patients in the two groups.

On the other hand blocked randomization balances the patient numbers in the two groups. Given that the patients enter the trial sequentially over time a block of even size (e.g. 2, 4, 6, 8 or 10) is chosen. Within each block the order in which the treatments are administered is randomized, and the numbers chosen so that one half of the subjects are assigned to one treatment and the other half to the alternative treatment. This process is repeated until the required number of patients is recruited.

One problem with blocked randomization is that if the size of the block is known, and the study is not double-blind, then the assignment for the last subject entering a block is known beforehand. One way around this is to alter the block size occasionally. Another problem is that many of the statistical techniques involved in the analysis usually assume that the allocation has been by simple randomization. However, the loss of power obtained by analysing the data, ignoring blocking, is usually small.

It is possible that the way a subject responds to a treatment is affected by measurable prognostic factors such as age, smoking, sex or disease status. In any randomized trial it is desirable that the treatment groups are similar in such factors. When the treatment groups are similar the statistical methods used in the analysis are usually simple and efficient. One method of achieving a balance is to divide the subjects into groups or strata of similar factors and randomize within strata. For example, one might group subjects of the same sex, and by 10 year age groups, and randomize within the age and sex specific groups.

Random number tables can also be used to select a random sample from a prespecified population. For example, a consultant has a register of diabetics, and he may wish to randomly select a small group on whom he can try out a new therapy.

13.2 THEORY AND FORMULAE

The uniform random numbers given in Table 13.1 were generated by the Numerical Algorithm Group (1978) routines GO5CBF(ISEED) and GO5CAF(Z). The corresponding manual describes the method and implementation.

13.3 BIBLIOGRAPHY

Fleiss (1981, chapter 4), Friedman *et al.* (1982, chapter 5) and Pocock (1983, chapter 5) have sections on how to randomize treatments in clinical trials. Altman & Gore (1982) also have fairly detailed descriptions of the methods. They also discuss the mechanics of randomization, such as the use of envelopes, or of a telephone centre. Zelen (1974) discusses in more detail stratified randomization, and what is called minimization. The Rand Corporation (1955) provide very extensive tables of random numbers. The books by Moses & Oakford (1963) and by Shwartz, Flamant & Lellouch (1980) give tables to facilitate blocked randomization.

13.4 DESCRIPTION OF THE TABLE

Table 13.1

Table 13.1 gives 10,000 random numbers, grouped in sets of 5. Each digit 0-9 is in theory equally likely, and not predictable from any of the other digits.

13.5 USE OF THE TABLE

Table 13.1

Example 13.1

Suppose that a doctor wishes to allocate patients to two treatments, *A* and *B* and he requires equal numbers in each group. How does he achieve this by blocked randomization, with a block size of length 6?

Choose any row or column of Table 13.1, say the tenth row on the first page. Reading along the row we ignore numbers 0, 7, 8 and 9 and any repetitions, and select the first three numbers between 1 and 6 inclusive. These are 3, 5 and 2. Thus we allocate treatment *A* to the 2nd, 3rd and 5th patients recruited to the trial and treatment *B* to the 1st, 4th and 6th patients in the trial. Continuing along the row, the next three numbers to be selected are 5, 1 and 2. Thus in the second block of six patients, we would allocate treatment *A* to the 1st, 2nd and 5th, and treatment *B* to the 3rd, 4th and 6th. We may then decide to change the block size, and a suitable choice is one of size 4, 8 or 10. Of these three, the first to be encountered in the row of random digits, starting where we left off before, is 8. Thus we would use a block of size 8 to continue the randomization process.

Example 13.2

It is possible that the way a patient responds to treatment is affected by his age. How can an investigator ensure that the two treatments are applied to approximately equal numbers of patients in the different age groups?

One method of achieving this is to use stratified randomization. Within each age group or strata, one would use blocked randomization, with block size of, say, 6. The assignments within each strata are chosen as in Example 13.1. The randomization schedule is drawn up before the trial begins, and for the first 12 patients in each group might be as follows:

Strata	Age	Assignment	
1	−34	ABABBA	AABABB
2	35–44	ABBABB	ABBBAA
3	45–54	BABAAB	BBABAA
4	55–64	BABBAA	ABABAB
5	65–	ABBBAA	BAABBA

Thus the first patient aged between 45 and 54 would receive treatment *B*, the second *A* etc. The first patient over 65 would receive *A* and the second *B* etc. As the number of prognostic factors increases, the number of strata get very large, and so there may be few subjects within each stratum which can make analysis difficult. Thus only important variables should be chosen, and as few as possible included, as stratifying factors. Note that the analysis of the data after the trial should take stratification into account for maximum power (Green & Byar 1978).

Example 13.3

Suppose that a doctor has 121 patients with a particular disease. How do we choose a random sample of 10 of them?

Firstly we need to allocate distinct numbers to each patient, say 001 to 121. Then choose at random (haphazardly) any column in Table 13.1. Say we choose the second column on the third page. These contain five digits, so we choose the first three. Going down the column, we select numbers between 001 and 121 and ignore 000 and those between 122 and 999. The first distinct four numbers are 022, 088, 037, 116. We then start on the next column to get 047, 019, 045, 050, 024, 097 to complete the ten. If any number occurs twice we ignore the second occurence. Therefore the random sample consists of patients 019, 022, 024, 037, 045, 047, 050, 088, 097, and 116.

13.6 REFERENCES

Altman D. G. & Gore S. M. (1982) *Statistics in Practice.* British Medical Association, London.

Fleiss J. L. (1981) *Statistical Methods for Rates and Proportions,* 2nd edn. Wiley, New York.

Friedman L. M., Furberg C. D. & DeMets D. L. (1982) *Fundamentals of Clinical Trials.* John Wright PSG Inc, Boston.

Green S. B. & Byar D. P. (1978) The effect of stratified randomization on size and power of statistical tests in clinical trials. *J. Chron. Dis.* **31,** 445-454.

Moses L. E. & Oakford R. V. (1963) *Tables of Random Permutations.* Stanford University Press, Stanford.

Numerical Algorithms Group (1978) *Nag Manual.* Numerical Algorithms Group Ltd, Oxford.

Pocock S. J. (1983) *Clinical Trials : A Practical Approach.* Wiley, Chichester.

Rand Corporation (1955) *A Million Random Digits With 100,000 Normal Deviates.* The Free Press, New York.

Schwartz D., Flamant R. & Lellouch J. (1980) *Clinical Trials.* Academic Press, London.

Zelen M. (1974) The randomization and stratification of patients in clinical trials. *J. Chron. Dis.* **27,** 365-375.

Table 13.1 10,000 random numbers. Each digit is equally likely to occur.

19695	55904	50846	64105	42780	23965	36499	57893	05960	11026
59228	58296	70129	87139	36011	83256	99232	37848	20810	78885
83448	88340	60067	64903	53454	25872	79051	71984	34610	06655
65066	61315	57693	32754	99042	27714	98461	08470	68718	30391
92350	93445	92791	59767	06150	89188	69695	77993	65543	48868
54108	92466	37685	90961	85376	03107	36889	60444	15556	23016
48899	66837	96651	68584	59011	71585	05837	08242	74712	05065
89721	59836	81970	62542	57631	05800	08346	59428	85825	20406
72534	84860	25700	01781	80133	50991	68006	58908	62223	72602
10398	78868	90973	40785	10956	47322	91563	40402	41510	70893
89971	65487	51394	60186	39181	28395	18275	22633	08893	43145
47570	08194	02562	20992	56879	05693	88934	71911	94339	61254
55138	32828	10968	17700	52889	64544	75914	93440	42284	19973
03280	94391	50718	17890	31348	90965	73205	87428	43324	87974
26882	96710	82060	48647	79210	76937	46807	01830	29728	70792
49011	78358	61382	54542	97935	98840	11802	23473	90236	45036
20104	44627	66416	34023	76942	57201	84455	43747	68258	02495
27642	32036	08374	00271	87720	72927	90019	67482	08342	55381
80525	00890	49465	29132	05055	97384	78570	68724	69531	98881
91518	51593	80375	37083	95401	12059	69517	33806	64950	96553
66388	11317	10822	73959	72465	54346	12493	72173	28913	41596
46205	84091	80180	34859	11056	92598	49508	30201	61548	56777
21275	89005	91684	25166	18564	75675	85708	96208	15809	48300
24480	15944	80693	37474	82915	45320	90802	49606	40133	35479
44687	11327	53020	97344	49769	83601	53876	06976	75032	37790
96578	33353	33655	00745	86377	59013	41384	81927	28423	73355
87625	70921	90065	24953	98297	97193	93978	09708	11551	59465
79590	16311	12467	40424	94692	01263	09766	62160	13310	84709
89696	63651	31390	72481	50129	87939	93528	03994	15973	82497
86536	73618	82431	10758	39108	39660	86058	75053	10367	59170
96080	54600	27031	09924	87125	15802	08169	38793	60222	35181
33735	88263	28488	56953	87996	43406	45684	92652	82743	66102
63516	79446	21215	23801	61204	65571	84584	15527	82077	48161
29447	16001	55752	20575	48266	35948	45029	56181	28070	29340
43587	96847	16487	71134	85237	80685	78818	29010	99641	68880
11765	52250	97854	36846	80629	37448	57220	17197	65633	46420
49013	94619	36084	43133	90100	80163	72639	50685	34733	77402
06272	55941	20708	18167	86838	63064	31896	16520	54940	90397
77729	26181	83307	76772	50238	33086	25699	78854	13658	92510
97264	27329	93342	68374	83820	06987	46633	45154	44068	23559
60129	52945	65912	12553	47496	40812	31617	01428	42978	83522
59892	74599	21199	08936	07551	42225	25880	25180	49969	34068
22379	50705	72965	72480	26610	58623	46941	53870	31155	08724
70492	84591	89860	32141	00004	27272	95321	69730	12415	85988
50764	88881	88393	93807	88684	65361	64197	70538	38400	22641
00872	77751	92179	39287	90572	51004	91633	55146	69732	90057
64551	38464	64352	23309	32205	62738	72153	79366	55464	77543
79005	33902	93864	36982	49632	86301	23966	65465	36333	44269
94742	06961	27440	81104	05948	76330	25478	37578	82918	26005
19039	72676	09785	38374	13534	55197	26421	45426	87114	69106
82215	85502	32887	78827	59159	34959	04208	89240	26306	85042
60713	46311	68456	32114	73634	16246	37555	89905	44377	31408
58413	89629	80038	49094	43220	47587	96218	77776	40342	03022
95590	10880	02462	71482	60906	26636	02804	04242	51242	43610
01767	03398	86673	36557	25268	01658	23722	55682	24525	86880
44889	93365	74246	40710	92068	15549	94581	13980	09833	05810
26770	32389	81640	53985	00098	23194	18701	14761	77301	44517
80379	99904	93794	70757	60793	90617	07392	08027	81291	35894
36204	20514	60944	60708	77288	60685	59232	74416	84358	28293
73852	59198	24429	55655	54370	04792	45206	58105	12144	73418
73169	36101	13321	07692	28534	96953	06471	10601	89899	38511
18342	43666	63416	65789	45805	60271	14760	46543	55367	59661
34295	79821	77078	33512	14717	84165	18459	71940	03561	72497
08175	92677	94883	87566	80696	14551	99039	42261	05468	37101
93011	98954	72102	08436	99614	55277	30252	36988	06930	10279
15934	85343	93282	44512	16615	08556	42153	86613	67403	33250

Table 13.1

Table 13.1 Continued.

12462	79221	03901	99851	92706	35731	12381	86041	79993	00186
69881	42631	91267	18691	13093	22570	55410	61567	99173	44602
48618	96772	70471	53672	53697	35435	70783	59050	63968	51650
15354	86197	20078	65067	71007	42604	86189	83678	94570	27136
06677	83627	28704	93706	64690	49381	98883	32437	61866	99152
99650	50217	98285	77220	24786	30346	93515	04538	61108	18362
41643	81483	43753	87976	82309	62796	85718	79610	53662	90835
46042	57283	50575	45998	83731	63936	81532	96381	75521	00335
15753	39862	59101	28374	54933	23475	00305	61071	19028	69078
26551	62338	53496	08707	98921	62422	18370	56179	90178	36676
58219	62189	31108	48888	12079	59814	19573	83878	07180	53657
23624	47799	49984	04423	27156	02513	93062	25317	81044	53562
91179	22121	96029	50293	07036	06208	08265	43431	44979	75005
52834	18951	71105	85383	24726	59826	49549	17302	63035	81937
52648	46847	33441	60780	89062	06275	47503	73398	76265	77194
84293	72038	68513	79769	08636	60502	88726	78208	31568	13172
45448	60712	38768	67721	35791	36334	93773	77642	20906	71106
20385	23842	48641	37809	17077	35575	88178	86352	01649	55437
15133	51593	46847	53959	81658	71898	06774	76049	35865	25888
39640	93349	59399	99139	87691	40591	40552	18045	46976	65541
14646	58605	98510	32745	35993	93665	77443	36114	27118	25554
49407	41924	91831	61532	95788	05937	56740	10144	19618	37605
69291	52332	89452	94929	76149	46369	26117	71146	60013	69906
62282	53969	50723	54983	93655	60757	18447	48790	90826	75347
23813	00912	23834	02071	74847	53181	71074	96593	41486	87597
95214	65353	16013	96543	69957	36571	02070	43358	44897	76978
93836	55639	31480	08670	80718	41746	00450	15348	03165	44323
41030	27068	52877	90698	55304	62015	70334	75673	58447	22336
08154	84442	69776	17338	93758	51413	57837	94414	67077	91634
92292	80326	91158	94904	59502	37678	37875	26913	83445	76098
46609	89295	83948	05329	47532	77675	61423	56344	26675	36358
84332	05232	85219	11532	68557	29376	11777	91888	01673	13959
60555	16518	62555	25948	19252	30492	66047	84466	02650	67274
47689	69843	61625	15148	35514	43133	88570	45117	43185	94735
65844	26200	04727	71539	66628	45052	62018	11708	48742	69068
80207	33446	57784	87108	10268	49346	75788	31162	58934	03122
13487	91008	01900	25295	44107	60293	13586	73558	55974	19599
36935	78255	46321	00219	31860	35558	26337	00115	54765	45939
19275	02240	71456	14336	43476	33028	88959	23362	82496	24893
35560	83364	22998	90659	08751	27981	52173	76316	93111	87613
03752	98831	77638	49285	62066	74713	21565	29280	81909	72158
81837	30172	04508	89071	62855	86324	06358	57717	76757	40291
99874	56369	05077	28725	64343	81289	49368	89824	07492	75785
47327	21427	49874	61534	89894	62261	90839	66899	84778	67134
67357	08825	76458	45108	17996	63511	83568	09568	69705	23086
79015	89239	14360	09587	44822	26949	49496	12175	26485	47178
14452	17714	25188	69313	77634	05696	61094	40330	53859	35021
92069	81043	37848	57331	39935	28147	06764	97004	66593	77229
50340	99088	61253	68136	29007	44507	24617	60046	84033	60656
91846	82337	26193	35571	35295	08252	74503	59943	24560	22374
93810	03714	50406	65699	18450	23170	40117	29354	80548	99877
11021	42326	85160	26011	40475	72222	21810	75732	89003	33924
06601	71764	02476	87112	90229	72586	61097	49386	66405	04539
48180	22602	71882	93942	57603	71269	04742	44440	23190	59719
47066	65926	26776	24462	55726	24327	74372	92890	04719	48818
33014	36442	16280	20680	38425	68293	48522	79765	61859	61066
62911	93772	86922	49796	55054	59808	75728	53113	91425	53369
69614	89513	30567	63322	84566	81552	62261	43956	00809	76291
96720	56308	54281	68204	59116	64149	46824	37176	53914	82257
55533	81214	73228	17760	11676	75740	94467	20320	52812	80008
20703	54346	62142	52713	27941	00436	35069	77948	93729	47949
88658	68178	76500	77522	51609	04763	88457	98665	49788	11336
03542	71590	51769	27994	45943	36487	24892	93979	63866	67614
28223	63293	70889	50017	18334	88083	12121	50741	17813	03705
17106	37074	39760	21545	80909	97657	86957	18994	81207	19108
25740	66793	09757	01564	30465	87608	20769	44855	80493	51133
26370	61078	55210	00642	98188	97197	11861	30767	77736	05131

Table 13.1

Table 13.1 Continued.

31123	29673	20649	35233	87798	45201	30228	26523	74318	72434
87699	16140	76297	35674	87089	63901	34730	13078	72090	92885
67425	24448	02933	52511	84552	52655	34894	29903	27328	95652
89270	11691	60037	20257	80787	95167	83268	13543	29583	72597
72622	48817	52954	03996	95738	47909	71694	15877	73377	75924
51774	88584	86697	13066	91050	49653	26735	64438	06469	36393
46309	99600	06214	51683	92480	34427	89140	41336	45583	27144
87597	57079	91766	92488	00289	77560	01421	81779	51894	82562
31646	57226	92937	93150	11344	39827	90260	72427	40824	07898
65302	69663	91004	44870	45905	34879	65968	86806	74176	79923
01577	71625	26774	57246	91613	07768	66347	80626	04460	58894
55634	71993	51160	22769	99912	47556	50567	28290	06554	60537
85512	93091	80037	69521	18523	14252	63117	07781	97167	52038
18866	39027	02543	76350	42750	83412	40894	22312	97887	06172
99669	87471	68675	96181	77472	62430	99488	36200	66822	92639
37479	15164	18679	79672	32567	94563	07617	21455	33382	09452
75140	57111	13664	91913	48505	40279	41529	45898	09166	28364
48363	66897	76768	85250	72098	66045	30728	33255	79288	32083
91791	31661	55011	43282	56782	71436	03002	83329	63492	66845
88551	03397	21971	10386	44973	92009	20743	69340	60607	26066
67369	27720	80190	40937	01623	34579	49770	05988	47388	58258
28783	12182	66338	91126	72584	54498	42086	01052	40377	72815
26241	83273	69559	40407	14443	29333	20876	29742	78308	50317
70633	54293	72527	35586	23033	26232	34429	24709	95497	11799
72495	70350	06422	98385	64926	71132	05204	16406	72123	65689
53543	91004	30200	77643	51598	91984	39404	95223	29527	21376
00882	02757	73330	86303	23218	19630	44572	65682	55202	60759
51545	69430	38495	83615	06349	30060	71655	98526	30601	08552
39877	34702	65214	34285	33858	82516	74122	88707	86412	35586
83519	54717	72198	86225	30562	26218	04539	19830	14236	99645
97954	70429	70913	37197	79284	33633	22498	42847	38627	17456
26569	90111	51495	27926	40401	23847	18162	85104	61245	09032
77769	89332	48094	25531	65874	18868	05954	12283	80627	20588
85268	38217	98991	31437	35204	01519	77043	62658	01284	98569
53260	52912	79063	33321	77747	22875	97211	78917	60833	32255
54604	57645	33482	12661	48860	10637	62200	30712	59933	66417
62916	57207	22686	21926	33200	00815	21315	95212	53600	80392
87455	37236	24657	44118	47952	08362	50751	52254	48662	97475
86468	02668	33286	73051	42155	96041	92439	43992	61509	13650
83389	05151	28091	91606	39899	46819	75737	36525	39289	16893
81177	12852	10449	71745	85282	09278	54991	90531	45197	41057
76896	74853	60262	08234	64399	14813	83353	92370	98900	57090
85516	01299	84969	15722	24320	89159	71453	84571	24566	33429
91397	47504	93285	76648	41823	52157	41564	61029	89468	31131
43792	99547	93027	30631	84162	06013	73752	61913	07411	79109
26015	20069	16498	95605	93859	65766	49439	33416	82862	50132
95512	08645	78663	25642	13942	08259	76612	02191	61680	35572
73754	28310	35170	82301	92847	88930	34491	53589	96240	42101
75298	59567	50881	34360	51194	14804	72812	21401	29218	09194
17302	51444	72863	34562	62575	59959	20950	93893	15665	28176
32961	36514	02548	42566	56408	45937	73377	56684	57996	08531
69245	60897	96547	17409	82300	54104	62695	69007	29324	59983
85209	54546	55487	05893	81524	64709	27136	20761	09671	25773
54914	75687	23563	08768	36671	08650	98082	99973	21967	20539
16857	35800	23087	45451	94788	89862	96548	19590	37045	13242
53246	15725	82742	41114	16860	52046	27253	67720	53924	03789
11437	35241	63696	47282	92215	09297	51943	34873	90340	79842
40181	05047	59688	82995	37630	81849	41888	66944	50392	87537
67978	49190	87861	81777	63840	23173	78560	91506	45852	39808
47868	64597	50638	71060	19003	17992	29486	48374	46187	15423
62115	25292	90772	05793	47022	42500	51269	86569	51482	85869
03124	68454	33644	47290	05076	43508	43307	43186	36021	29520
77932	06677	80016	24248	77543	90836	08359	79568	02889	38079
90914	89869	21239	57381	57977	98249	39164	09391	68098	75886
14996	59701	50020	61721	86297	14405	76298	62622	21162	71429
13293	25710	54883	13406	55861	48735	47286	78343	53914	40631
13690	45378	65594	88797	60585	48315	03587	35589	55237	53994

Table 13.1

Chapter 14
Bibliography of Other Statistical Tables and Texts

14.1 INTRODUCTION

The discussion in this text has concentrated on the comparison of two groups. It is recognized that many studies will involve more than two groups and that in certain circumstances extensions of the tables described here to more than two groups would be valuable. One method of estimating numbers for situations in which there are more than two alternative groups is to take, for example as design criteria, two of the alternatives which are likely to result in the most similar consequence. For given test size and power a corresponding number of subjects per group could be obtained by multiplying this by the corresponding number of treatment alternatives. There are many situations, however, where this can be shown to be inefficient, for example if there are four treatments organised in a 2×2 factorial design. Some arguments for the use of the factorial designs in clinical studies are given by Peto (1978). Another example would be when the alternative groups could be ordered; one group could be a series of patients treated with a placebo (dose zero), a second group with a low dose of an active treatment and a third group with twice that particular dose. An appropriate design and hence appropriate patient numbers should take into account the trend in dose in these three treatment groups. There are also other types of trial, in particular sequential trials, which have their own specialist tables. The design and analysis of such trials and appropriate tables are given in Armitage (1975) and Whitehead (1983).

14.2 STATISTICAL TABLES FOR DESIGN

Non-Randomized Comparative Studies

Makuch & Simon (1980) give tables required for calculating sample sizes in non-randomized comparative studies which include a historical control.

Historical controls are frequently used in the clinical evaluation of a new therapy. Sample size tables are needed for such non-randomized studies since these determinations can differ markedly from the widely published sample size estimates appropriate for randomized comparative studies. A comprehensive array of sample size tables are provided to aid investigators in the planning of non-randomized comparative trials. Guidelines for the use of historical controls and the pooling of several historical control groups are discussed.

Table 1 in Makuch & Simon (1980) gives the number of patients needed in an experimental group for a given probability of obtaining a significant result (one-sided test) with significance level $\alpha = 0.05$ and power $1 - \beta = 0.8$ when 20, 30, 40 and 50 historical controls are used for comparison. The proportion of success for the historical control patients varies from 0.1 to 0.8 and the corresponding proportion of success for experimental patients increases from 0.2 to 0.9.

Table 2 is similar to Table 1 except that the number of patients in the control group is 75, 100, 125 and 150. Table 3 is similar to Tables 1 and 2 except that the number of patients in the control group is 200, 250, 300 and 500.

Planning the Duration of a Comparative clinical Trial with Loss to Follow-up and a Period of Continued Observation

Rubinstein, Gail & Santner (1981) give tables to assist in designing clinical trials which are to be analyzed with the Mantel-Haenszel (logrank) test for comparing survival curves. The results allow for loss to follow-up during the trial and permit the planner reduce the total number of patients required by introducing a period of continued observation after the end of patient accrual.

The calculation of the required trial length is based on the assumption of exponential survival time so that the ratio of the maximum likelihood estimators of the hazard rates may be used to test the null hypothesis that the hazard ratio is equal to one. A method is developed to compute the approximate trial length required to assure a desired statistical power for given significance level, hazard ratio, accrual rate, loss to follow-up rate, and length of the period of continued observation.

Table 1 in Rubinstein, Gail & Santner (1981) gives the total required trial time assuming no loss to follow-up, for yearly patient entry rates varying from 20 to 240, with length of continued observation varying from 0 to 2 years and values of the ratio of median survival of control group to experimental group (hazard ratio) of 1.25, 1.5, 2 and 3. The one-sided significance level $\alpha = 0.05$ and power $1 - \beta = 0.8$ and 0.9.

Tables 2, 3 and 4 are similar to Table 1 but with the fraction of the loss to follow-up rate to the control hazard rate of ¼, ½ or 1 respectively.

Adjusting for Losses to Follow-Up in Sample Size Determination for Cohort Studies

Palta & McHugh (1979) argue that it is important to anticipate losses to follow-up and to adjust the sample size calculations accordingly, at the design stage of a study. They point out that the end-point and the occurrence of loss to follow-up are competing events and calculate appropriate sample sizes based on competing risk theory.

Table 1 of Palta & McHugh (1979) gives sample sizes for differing survival rates in the control and experimental groups ranging from 0.005 to 0.995 and net loss rates in either group ranging from 0 to 10% under five alternative assumptions concerning the probabilities of survival. Two-sided $\alpha = 0.05$ and power $1 - \beta = 0.90$ only are chosen.

Palta & McHugh (1980) extend Palta & McHugh (1979) to include losses due to non-compliance.

Table 2 of Palta & McHugh (1980) gives sample sizes for non compliance values of 0, 0.05, 0.1, 0.2 and 0.3 and loss rates of 0, 0.05, 0.1, 0.2 and 0.3. The control group probabilities range from 0.05 to 0.95. The excess of experimental group probability over this control value ranges from 0.01 to 0.1. Again two-sided $\alpha = 0.05$ and power $1 - \beta = 0.90$ are chosen.

Determination of Accrual Rate

Palta (1982) considers the determination of the necessary accrual rate to a trial of a given fixed duration. She points out that earlier tables given by Pasternack (1972) are misleading.

Table 1 of Palta (1982) gives the size of annual cohort to be entered per group (control and experimental) to determine the statistically significant difference in five year survival rates, one-sided $\alpha = 0.05$ and power $1 - \beta = 0.8, 0.9$ and 0.95. The anticipated five year survival rate of the control group ranges from 0.05 to 0.90 with an anticipated increase in the five year survival rate of the experimental group, ranging from 0.05 to 0.40.

Table 2 is similar to Table 1 except that one-sided $\alpha = 0.01$.

Sample Size Requirements for comparing Time to Failure Among Several Groups

Makuch & Simon (1982) give sample size requirements for the k (≥ 2) group comparative clinical trial in which time to failure is the measure of treatment efficacy (see also Le, 1983). Time to failure is assumed to have an exponential distribution and so these results generalize those of George & Desu (1973) discussed in Chapter 10. The authors show that the heuristic use of sample size formulae for comparing two treatment groups is not adequate for obtaining sufficient power and properly accounting for the multiple comparisons possible with 3 or more treatment groups.

Table 1 of Makuch & Simon (1982) gives values of the non-centrality parameter required to ensure power $1 - \beta$ of 0.8, 0.9 and 0.95 for testing the null hypothesis at level $\alpha = 0.05$ using a χ^2 statistic on $k - 1$ degrees of freedom, with k the number of treatment groups varying from 2 to 6. It should be noted that for $k = 2$ this table corresponds to Table 2.3 of this book.

Table 2 gives the number of patients required to detect a significant difference among k treatment groups for given largest ratio of the mean failure times amongst all possible treatment group pairings ranging from 1.2 to 2.5.

Two Stage Plans in Phase II Trials

Lee, Staquet, Simon, Catane & Muggia (1979) considered two stage patient accrual plans for Phase II trials of anti-cancer drugs, the purpose of the trial being to identify the therapeutic level of drugs as either above or below a reference response rate taken as 20%. Using such a plan, decisions concerning the future disposition of drugs can be made.

Table 1 of Lee, Staquet, Simon et al. (1979) gives the false negative probability of a conclusion of less than 20% efficacy together with false-positive probability of a conclusion of greater than 20% efficacy in the case of fixed-sampling patient accrual plan. The number of patients is 14, 25 and 40.

Table 2 gives an optimal two stage patient accrual plan which assumes that the maximum number of patients available is specified in advance. The design admits a number of patients to the first stage of the trial which is then closed if at least a certain number of responders are observed. If the number of responders is less than this but greater than a certain minimum then the remaining patients are recruited to the trial. The table gives the number of responders required to conclude that the drug is either ineffective, of uncertain efficacy, or at least effective at 20%.

Table 4 gives a two stage patient accrual plan with the initial stage size fixed at 15 patients as well as the maximum number of patients available being fixed in advance.

Analysis and Interpretation of Response Rate in Phase II Trials

Lee, Catane, Rozencweig, et al. (1979) describe a statistical method of interpreting data from retrospective analysis of studies evaluating the activity of anti-cancer agents against individual tumor types.

Table 1 of Lee, Catane, Rozencweig et al. (1979) gives maximum and minimum sample sizes necessary to accept or reject a new treatment at a chosen rate of effectiveness with false positive and false negative rates α and β not greater than 0.05. The rates of effectiveness are set at 10, 15, 20, 30 and 50% and the observed number of responses from 0 to 60.

Table 2 is similar to Table 1 except that the false positive rate, α, is now not greater than 0.5. Table 3 is also similar but $\alpha = 0.1$ and 0.2 while $\beta = 0.1$.

Size of Sample Required Using Analysis of Variance

Chapter 7 described methods of determining sample sizes when the results from a clinical trial are quantitative and the analysis entailed the use of the 't' test. If a trial consists of k (≥ 2) groups, then an analysis of variance may be required, resulting in an F test. Cohen (1977, Chapter 8) describes sample size tables for this type of trial. He defines an effect size f as the expected ratio of the between to the within groups standard deviation. The tables give sample sizes for trials with values of f between 0.05 and 0.80 in steps of 0.05, and degrees of freedom ($k-1$) from 1 to 24. The number of groups to be compared is one more than the degrees of freedom.

Lesser (1982) gives similar tables.

14.3 COLLECTIONS OF TABLES USEFUL IN ANALYSIS

Many statistical texts include tables of common use in the analysis of clinical trials. However, collections of tables of particular use at the analysis stage include:

Fisher R. A. & Yates F. (1963) *Statistical Tables for Biological, Medical and Agricultural Research.* Oliver and Boyd, Edinburgh.

Lindley D. V. & Scott W. F. (1984) *New Cambridge Elementary Statistical Tables.* Cambridge University Press, Cambridge.

Pearson E. S. & Hartley H. O. (1958) *Biometrika Tables for Statisticians Vol. I* Cambridge University Press, Cambridge.

Pearson E. S. & Hartley H. O. (1972) *Biometrika Tables for Statisticians Vol. II* Cambridge University Press, Cambridge.

14.4 RECOMMENDED READING ON CLINICAL TRIALS

Armitage P. (1971) *Statistical Methods in Medical Research,* Blackwell Scientific Publications, Oxford.
    A basic medical statistics text useful both in design and analysis of clinical trials.

Bulpitt C. J. (1983) *Randomized Controlled Clinical Trials.* Martinus Nijhoff. The Hague.
    A practical guide to the problems of planning and executing randomized trials

Buyse M. E., Staquet M. J. & Sylvester R. J. (1984) *Cancer Clinical Trials: Methods and Practice.* Oxford University Press. Oxford.
    A practical handbook describing many aspects of clinical trial methodology with particular reference to cancer.

Cohen J. (1977) *Statistical Power Analysis for the Behavioral Sciences.* Academic Press, New York.
    The author provides discussion and tables for determining sample sizes required in many situations and stresses the importance of statistical power analysis.

Donner A. (1984) Approaches to sample size estimation in the design of clinical trials – a review. *Statistics in Medicine*, **3**, 199-214.
    Reviews and gives the corresponding formulae for size estimation in many types of clinical trial. These include clinical trials for comparing two proportions, detecting a particular relative risk, demonstrating equivalence, allowing for stratification of subjects, time to some critical event, patient dropout, length of time required to achieve maximum benefit of treatment, patient accrual by cohorts, group randomization and cross-over designs.

Fleiss J. L. (1981) *Statistical Methods for Rates and Proportions.* 2nd ed. Wiley, New York.
    Gives extensive tables for the comparison of binomial proportions and also random number tables.

Friedman L. M., Furberg C. D. & DeMets D. L. (1982) *Fundamentals of Clinical Trials.* John Wright, Boston.
    Discusses all aspects of clinical trials both practical and theoretical from design to analysis.

Hills M. & Armitage P. (1979) The two-period cross-over clinical trial. *Br. J. Clin. Pharmac.* **8,** 7-20.
    Describes problems associated with and the design and analysis of cross-over trials in which the effects of different treatments are compared on the same subject during different treatment periods, see also Armitage & Hills (1982), Millar (1983) & Pocock (1983).

Johnson F. N. & Johnson S. (1977) *Clinical Trials.* Blackwell Scientific Publications, Oxford.
    Discusses the problems of testing drugs and other therapeutic techniques together with the considerations necessary in planning clinical trials.

Lachin J. M. (1981) Introduction to sample size determination and power analysis for clinical trials. *Controlled Clinical Trials,* **2,** 93-113.
    Gives a comprehensive review and formulae for sample size calculations in clinical trials.

Lee Y. J. (1983) Interim recruitment goals in clinical trials. *J. Chron. Dis.* **36,** 379-385.
    Discusses the establishment of interim recruitment goals for clinical trials both to aid the investigator and the sponsors of a particular trial to decide, in the early stages, if the final recruitment goal can be achieved.

Meydrech E. F. & Kupper L. (1978) Cost considerations and sample size requirements in cohort and case-controlled studies. *Amer. J. Epid.* **107,** 201-205.
    Discusses methods for incorporating cost considerations into determining the sample size requirements of cohort and case-control studies.

Peto R., Pike M. C., Armitage P., Breslow N. E., Cox D. R., Howard S. V., Mantel N., MacPherson K., Peto J. & Smith P. G. (1976). Design and analysis of randomized clinical trials requiring prolonged observation of each patient: I, Introduction and design. *Br. J. Cancer* **34,** 585-612.

Peto R., Pike M. C., Armitage P., Breslow N. E., Cox D. R., Howard S. V., Mantel N., MacPherson K., Peto J. & Smith P. G. (1977). Design and analysis of randomized clinical trials requiring prolonged observation of each patient: II, Analysis and examples. *Br. J. Cancer,* **35,** 1-39.
    These two papers provide the basic references for the design and analysis of randomized clinical trials requiring prolonged

follow-up of each subject. Results of trials in patients with acute leukaemia and myelomatosis are used for illustration.

Pocock S. J. (1983) *Clinical Trials: A Practical Approach.* Wiley. Chichester.

Describes the principles and practice of clinical trials and gives a detailed account of how to conduct them.

Schoenfeld D. A. (1983) Sample-size formula for the proportional-hazards regression model. *Biometrics,* **39,** 499-503.

Gives a formula for determining the number of subjects necessary to test the equality of two survival distributions taking concomitant information into account. This should be useful in designing clinical trials with a heterogeneous patient population.

Schwartz D., Flamant R. & Lellouch J. (1980) *Clinical Trials:* Academic Press, London.

This is a translation of 'L'essai therapeutique chez l'homme' by the same authors. As well as being a practical guide to those engaged in planning, analysis and interpretation of clinical trials, it discusses the distinction between pragmatic and explanatory trials and how this affects the planning and analysis of a trial.

Taulbee J. D. & Symons M. J. (1983) *Sample Size and Duration for Cohort Studies of Survival Time with Covariables.* *Biometics,* **39,** 351-360.

The determination of sample size and duration of cohort studies is considered for an exponential survival model involving two parameters, one for the underlying hazard, a second for a single concomitant variable. The results of George & Desu (1974) are extended to take account of censoring and an example involving more than two subject groups is given.

14.5 REFERENCES

Armitage P. (1975) *Sequential Medical Trials.* 2nd ed. Blackwell Scientific Publications, Oxford.

Armitage P. & Hills M. (1982) The two-period crossover trial. *Statistician,* **31,** 119-131.

George S. L. & Desu M. M. (1974) Planning the size and duration of a clinical trial studying the time to some critical event. *J. Chron. Dis.* **27,** 15-24.

Le C. T. (1983) Sample size requirements. *J. Chron. Dis.* **36,** 663-664.

Lee E. T. (1980) *Statistical Methods for Survival Data Analysis.* Lifetime Learning Publications, Belmont, California.

Lee Y. J., Catane R., Rozencweig M., Bono V. H., Muggia F. M., Simon R., & Staquet M. J. (1979) Analysis and interpretation of response rates for anticancer drugs. *Cancer Treatment Reports,* **63,** 1713-1720.

Lee Y. J., Staquet M., Simon R., Catane R. & Muggia F. (1979) Two-stage plans for patient accrual in Phase II cancer clinical trials. *Cancer Treatment Reports,* **63,** 1721-1726.

Lesser M. L. (1982) Design and implementation of clinical trials. In Miké V. & Stanley K. E. *Statistics in Medical Research: Methods and Issues with Applications in Cancer Research.* Wiley, New York.

Makuch R. W. & Simon R. (1982) Sample size requirements for comparing time-to-failure among *k* treatment groups. *J. Chron. Dis.* **35,** 861-867.

Millar K. (1983) Clinical trial design: the neglected problem of asymmetrical transfer in cross-over trials. *Psychological Med.* **13,** 867-873.

Palta M. (1982) Determining the required accrual rate for fixed-duration clinical trials. *J. Chron. Dis.* **35,** 73-77

Palta M. & McHugh R. (1979) Adjusting for losses to follow-up in sample size determination for cohort studies. *J. Chron. Dis.* **32,** 315-326.

Palta M. & McHugh R. (1980) Planning the size of a cohort study in the presence of both losses to follow-up and non-compliance. *J. Chron. Dis.* **33,** 501-512.

Pasternack B. S. (1972) Sample sizes for clinical trial design for patient accrual cohorts. *J. Chron. Dis.* **25,** 673-681.

Peto R. (1978) Clinical trial methodology. *Biomedicine Special Issue,* **28,** 24-36.

Rubinstein L. V., Gail M. H. & Santner T. J. (1981) Planning the duration of a comparative clinical trial with loss to follow-up and a period of continued observation. *J. Chron. Dis.* **34,** 469-479.

Whitehead J. (1983) *The Design and Analysis of Sequential Clinical Trials.* Wiley, New York.

Author Index

Abbas S.   57
Alling D.W.   37
Altman D.G.   3, 4, 56, 57, 90, 91, 198, 199
Armitage P.   3, 4, 6, 7, 10, 17, 56, 57, 96, 98, 203, 206, 207

Bailey N.T.J.   6, 7
Bennett J.E.   37
Berkson J.   10, 17
Blackwelder W.C.   36, 37
Bono V.H.   205, 207
Bland M.   90, 91
Bradshaw M.   37
Breslow N.E.   96, 98, 206
Bulpitt C.J.   206
Buyse M.E.   206
Byar D.P.   199

Campbell M.J.   14, 17, 57
Casagrande J.T.   14, 17
Catane R.   179, 180, 205, 207
Cate T.R.   37
Chang M.A.   36, 37
Chant A.D.B.   134, 135
Cobbs C.G.   37
Cochran W.G.   14, 17
Cohen J.   14, 17, 79, 80, 83, 89, 90, 91, 205, 206
Couper W.D.   82, 83
Cox D.R.   96, 98, 206
Cox G.M.   14, 17

deMets D.L.   198, 199, 206
Desu M.M.   133, 135, 204, 207
Diegert C.   14, 17
Diegert K.V.   14, 17
Dismukes W.E.   37
Documenta Geigy   56, 57
Doll R.   62, 64
Donnor A.   206
Duma R.J.   37
Dunnett C.W.   36, 37

Elwood P.C.   15, 17, 57

Faithfull N.S.   179, 180
Familiari S.   15, 17
Fields B.   37
Flamant R.   198, 199
Fleiss J.L.   14, 17, 198, 199, 206
Fleming T.R.   178, 180
Fisher R.A.   6, 7, 90, 91, 206
Freedman L.S.   96
Friedman L.M.   198, 199, 206
Fujino Y.   170, 171
Furberg C.D.   198, 199, 206

Gail M.H.   203, 204, 207
Gallis H.   37
Gehan E.A.   169, 170, 171
Gent M.   36, 37
George S.L.   133, 135, 204, 207
Gore S.   3, 4, 56, 57, 198, 199
Green S.B.   199
Guenther W.C.   80, 83

Hartley H.O.   206
Haywood H.   37
Herrern L.   56, 57
Herson J.   170, 172, 179, 180
Hills M.   206, 207
Howard S.V.   96, 98, 206

Johnson F.N.   206
Johnson S.   206

Katongole-Mbidde E.   171, 172
Kleeberg U.R.   169, 172
Kupper L.   206
Kyalwazi S.K.   171, 172

Lachin J.M.   206
Le C.T.   3, 4, 204, 207
Lee E.T.   133, 135, 170, 172, 207
Lee Y.J.   179, 180, 205, 206, 207
Lehman E.L.   80, 83
Lellouch J.   198, 199
Leonard J.   37
Lesser M.L.   80, 83, 205, 207
Lewis J.A.   62, 64
Lewith G.T.   1, 4, 14, 16, 17
Lindley D.Y.   6, 7, 206
Luzza G.   15, 17

Machin D.   1, 4, 14, 16, 17, 134, 135
McPherson K.   96, 98, 206
Mainland D.   56, 57
Makuch R.   3, 4, 36, 37, 203, 204, 205, 207
Mantel N.   14, 17, 96, 98, 206
McGee Z.   37
McHugh R.   3, 4, 204, 207
Medoff G.   37
Meydrech E.F.   206
Millar K.   207
Miller J.C.P.
Moses L.E.   198, 199
Muggia F.M.   179, 180, 205, 207
Mulder J.H.   169, 172
Multicenter Study Group   97, 135

Numerical Algorithm Group   197, 199

Oakford R.V.   198, 199
Olweny C.L.M.   171, 172

Palta M.   204, 207
Pasternack B.S.   204, 207
Pearson E.S.   206
Peto J.   96, 98, 206
Peto R.   96, 98, 203, 206, 207
Pike M.C.   14, 17, 96, 98, 206
Pocock S.J.   3, 4, 6, 7, 10, 14, 17, 198, 199, 207
Postorino S.   15, 17

Rand Corporation   198, 199
Reinhold H.S.   179, 180
Richter J.R.   133, 135
Rozencweig M.   169, 172, 205, 207
Rubinstein L.V.   203, 204, 207
Rumke P.   169, 172

Sande M.A.   37
Santner T.J.   203, 204, 207
Schoenfeld D.A.   133, 135, 207
Schwartz D.   198, 199, 207
Scott W.F.   6, 7, 206
Simon R.   3, 4, 36, 37, 179, 180, 203, 204, 205, 207
Skegg E.C.G.   62, 64
Smith C.R.   14, 16, 17
Smith P.G.   14, 17, 96, 98, 206
Staquet M.J.   179, 180, 205, 206, 207
Sutcliffe M.I.   56, 57
Sweetnam P.M.   15, 17
Sylvester R.J.   206
Symons M.J.   207

Taulbee J.D.   207
Thomas D.   169, 172
Toya T.   171, 172
Turiano S.   15, 17
Turner D.T.L.   134, 135
Tytun A.   14, 17

Upton G.J.G.   10, 17
Ury H.K.   14, 17

van Der Zee J.   179, 180
van Rhoon G.C.   179, 180

Warner J.F.   37
Waters W.E.   57
Whitehead J.   203, 207
Wilson A.B.   62, 64
Wike-Hooley J.L.   179, 180
Woollard M.L.   82, 83

Yates F.   6, 7, 10, 17, 90, 91, 206

Zelen M.   198, 199

Subject Index

Page references in italics refer to tables.

Accrual rate, determination of 202
Acetylcysteine trails, examples 97, 134
Acupuncture/placebo, example 1–2
Adverse reactions, and background incidence 60–1, 62–4, *65–78*
Allocation of subjects, unequal 12
Analgesic trial, example 64
Angina prevalence, example 57
Asprin trial, example 15
Association 89

Background incidence of adverse reactions 60–1
Bed sore treatment trials, example 97
Bias, selection 3
Binomial proportions 10 *et seq*
Blind assessment
single 1
double 1
Blocked randomization 195, 196
Bronchitis, examples 97, 134

Cancer
colon, example 96–7
efficacy of anticancer agents 167
equivalence of treatment example 36–7
gastric, example 63
hyperthermia with, example 177
Cardiac arrhythmias, example 62–3
Censored observations 94, 95, 132
Chi-squared test, with continuity correction 10
Comparative studies, non-randomized 201
Confidence intervals 3, 4
Confidence intervals, sample size for 54–7, *58*
Contingency tables *10*
Continuity correction 10, 11
Control groups 1
Correction factor, for sampling from a finite population 56, 57, *59*
Correlation coefficient 89–91, *92–3*
Critical event 94
Cross-over trials 10

Design
considerations 1–4
statistical tables for 201–3
Diabetics, impotence prevalence, example 57

Distribution
binomial 178
exponential for survival *see under* survival curve comparisons
normal 5–7, *8–9*
Poisson 60, 132
Double blind procedure 1
Duration of study
for exponential survival 131–2, 133–4, *138–66*
from loss to follow-up with continued study 201–2

Effect
large 79, 89
medium 89
small 79, 89
Efficacy, minimum, equation 167
End points defined 1
Entry rate 132–3
Epileptic fit control, example 178
Equivalence, testing for 3, 35–7, *38–53*
Errors, Type I and Type II 2
Ethical considerations 3
Exponential survival *see under* survival curve comparisons

F (variance ratio) test 203
Factorial designs 201
False negative 2
False positive 61
Finite populations, disease prevalence and sampling 55, 56, *59*
Fisher's exact test 10
Fleming's single stage procedure 176–8, *179–94*
Follow-up 133

Gehan's method *see under* phase II trials

Hazard
constant 132
exponential 132
function 132
rate 94
Historical controls 12–13, 201
Hypertension treatment examples 81, 82, 91
Hyperthermia with cancer, example 177
Hypothesis
alternative 2–3
null 2
one-sided 178

Infinite populations, disease prevalence and sampling 54–5
IQ determination, example 81

Levamisole examples 96, 133
Logrank test *see under* survival curve comparisons
Lung disease trial, example 90–1

Mann-Whitney U test 79, 80
Matched data
notation for 10
sample size for 13
Means, comparison of two 79–83, *83–8*
Median survival 95, 132
Meningitis, equivalence of treatment, example 37
McNemar Test 10
Multicompound therapy, and drug toxicity 3
Multiplying factor, for comparing populations 14, 16, *34*
Myocardial infarction trial, example 15

Non-parametric 82
Non-randomized comparative studies 201
Normal distribution 5–7
Number of subjects *see* sample size

Odds ratio 13
One-sided (one-tailed) tests 2–3

Paired data *see* matched data
Pearson's correlation coefficient 89–91
Peptic ulcer treatment, example 15
Phase II trials
Fleming's single stage procedure 176–8, *179–94*
Gehan's method 167–70
probability of successive treatment failures 169–9, *170*
sample size for first stage 167, 168, 169, *171*
sample size for second stage 167–8, 169, *172–5*
response rate 203
two stage patient accrual 203
Phase III trial 178
Placebo 1
Poisson distribution 60, 132

Population
  finite 55
  infinite 54
Post marketing surveillance 60–64
Post-operative pyrexia, example 133
Power of a trial 4
  comparing two treatments 6, 82
  inadequate 3
Precision 169
Pregnancy prolongation, example 82
Pre-menstrual syndrome trial, example
  16
Prevalence of a disease, estimating
  54–7
Probability 2
Probability density function 5
Proportions, binomial
  comparison of 10–17
    multiplying factor for 14, 16, *34*
    sample sizes for 14–16, *18–33*
  confidence limits 54–7, *58, 59*
  testing equivalence of 35
    sample sizes 35–7, *38–53*
p-value 2

Random number tables *198–200*
  use of 196–7
Randomization 3, 195–7, *198–200*
  blocked 198
  stratified 198
Response rate, in phase II trials 203

Sample size 1–2, 4, 6
  and cohort studies 202
  for comparing time to failure
    202–3
  for comparing two proportions 11,
    12–16, *18–33*
  for comparison of exponential

distributions 131, 132–3,
  135–7
  for comparison of survival rates
    94–5, 96–7, *115–29*
  for confidence limits 54–7, *58, 59*
  for correlation coefficients 89–91,
    92–3
  for more than two alternatives 201
  for phase II trials
    single-stage 176, 177–8, *179–94*
    two-stage 167–8, 169, *171–5*
  for post-marketing surveillance
    to detect adverse reactions
      60–1, 62, 63–4, *66–78*
    to produce adverse reactions
      62–3, *65*
  t-test
    one-sample 79, 80, *83–5*
    two-sample 80, 81–3, *86–8*
  tables 201
  for testing the equivalence of
    proportions 35–7, *38–53*
  using analysis of variance 203
  Wilcoxon rank-sum 80, 82–3
Scrub-up trial, example 15
Selection bias 3
Significance 2, 3
  nominal 64
  overall 64
  tests 3
Single-stage procedure, for phase II
  trials 176–8, *179–94*
Size of sample *see* sample size
Standard deviation 5, 79
Standardized Normal distribution 5
Stratified randomization 196–7
Student's *t*-test 79
Survival
  exponential *see under* survival curve
    comparisons
  median 95
    for an exponential distribution 131
  two groups compared 10, 36–7

Survival curve comparisons
  exponential 131–4
    sample size 131, 132–3, *135–7*
    study duration 131–2, 133–4,
    138–66
  logrank 94–8
    critical event table 96, *98–114*
    sample size table 96–7, *115–29*
    survival rate table 96, 97, *130*

t-test
  one-sample 79, 80, *83–5*
  paired 81
  two-sample 80, 81–3, *86–8*
Test
  one-sided and two-sided 2–3
  size 2
TNS trial, example 14, 16
Toxicity, drug, and multicompound
  therapy 3
Treatment, allocation ratio 11
Treatment groups, number of 203
Treatments, choosing 1
Trials, results from previous 12–13,
  201
Two-sided (two-tailed) test 2–3
Type I and Type II errors,
  probability 2

Ultrasound trials, example 97

Variance analysis, for sample size 203
Variance ratio (*F*) test 203

Wilcoxon rank-sum 79, 80
Withdrawals from trials, patient 95,
  132
Wound infection rate, example 15

Yates' correction 10, 16

Thump-Thump

LEARNING ABOUT YOUR HEART

WRITTEN BY PAMELA HILL NETTLETON
ILLUSTRATED BY BECKY SHIPE

Thanks to our advisers for their expertise, research, and advice:
Angela Busch, M.D., All About Children Pediatrics, Minneapolis, Minnesota

Susan Kesselring, M.A., Literacy Educator
Rosemount-Apple Valley-Eagan (Minnesota) School District

PICTURE WINDOW BOOKS
MINNEAPOLIS, MINNESOTA

Managing Editor: Bob Temple
Creative Director: Terri Foley
Editor: Kristin Thoennes Keller
Editorial Adviser: Andrea Cascardi
Copy Editor: Laurie Kahn
Designer: Melissa Voda
Page production: The Design Lab
The illustrations in this book were rendered digitally.

Picture Window Books
5115 Excelsior Boulevard
Suite 232
Minneapolis, MN 55416
1-877-845-8392
www.picturewindowbooks.com

Printed in the United States of America.

Library of Congress Cataloging-in-Publication Data
Nettleton, Pamela Hill.
  Thump-thump: learning about your heart / by Pamela Hill Nettleton ;
illustrated by Becky Shipe.
    p. cm. — (The amazing body)
Includes bibliographical references and index.
Summary: An introduction to the parts of the heart and circulatory system
and how they function.
  ISBN 1-4048-0255-X (lib. bdg.)
1. Heart—Juvenile literature. [1. Heart. 2. Circulatory system.] I. Shipe,
Becky, 1977– ill. II. Title.
  QP111.6 .N48 2004
  612.1'7—dc22   2003018189

Did you know that your heart began beating before you were born? It never stops, even when you sleep. Your heart keeps you alive!

Squeeze your hand into a fist and then relax it. Do it over and over again. That's sort of how your heart works.

Do you know where your heart is located? It's under the hard bone in the center of your chest and a little to your left.

Your heart is a muscle. Each time it tightens, it pumps blood all through your body.

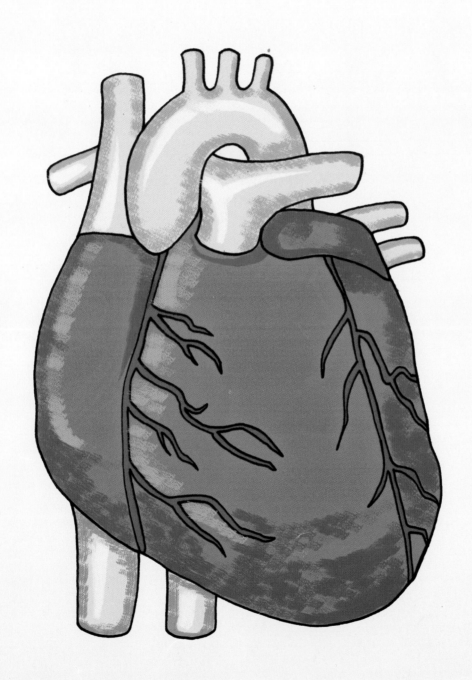

Your heart is part of your circulatory system. This system runs through your body. It includes your heart, your blood, and your blood vessels.

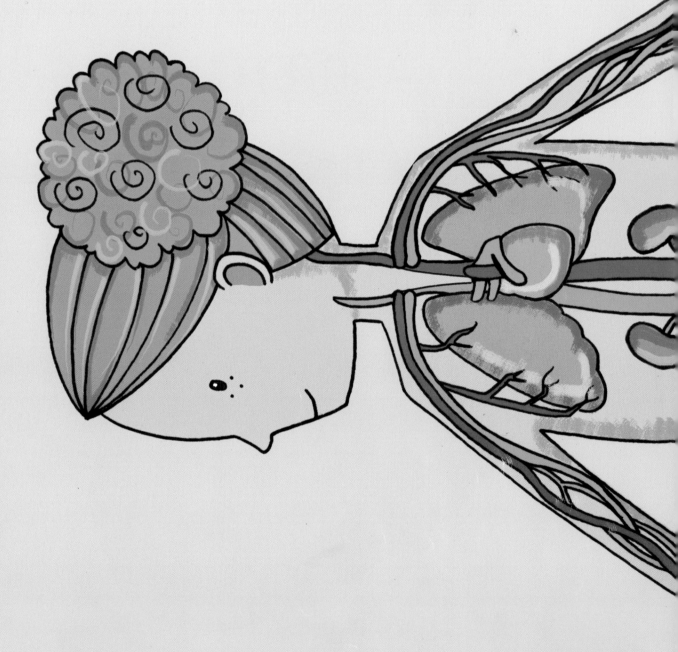

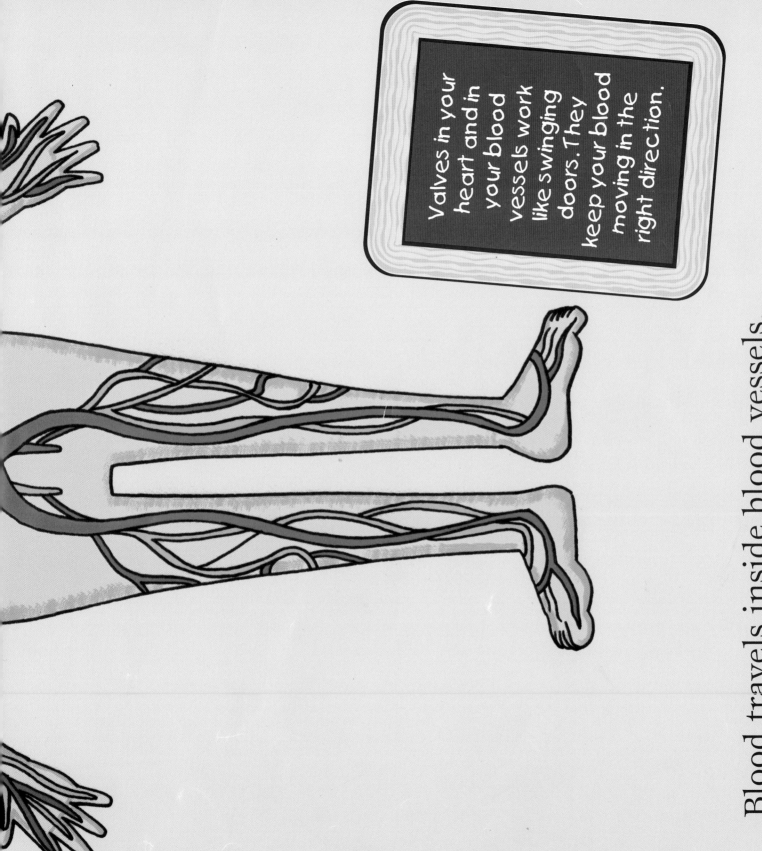

Valves in your heart and in your blood vessels work like swinging doors. They keep your blood moving in the right direction.

Blood travels inside blood vessels. They are like little pipes that bend and move.

Your heart has four parts, called chambers.
Each side of your heart has two chambers,
one on top and one on the bottom.

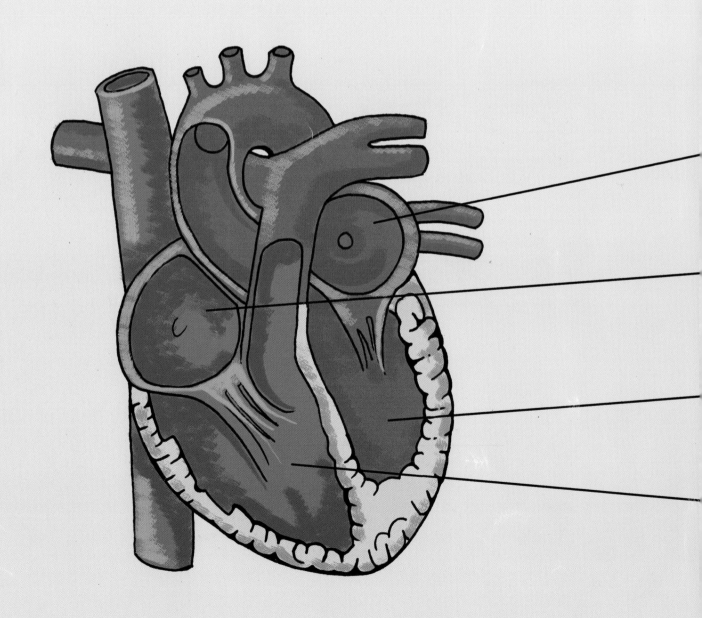

The top chambers, called the left atrium and the right atrium, fill with blood. Then they dump it into the bottom ones. The bottom chambers, called the right ventricle and the left ventricle, squirt the blood out of your heart.

Left Atrium

Right Atrium

Left Ventricle

Right Ventricle

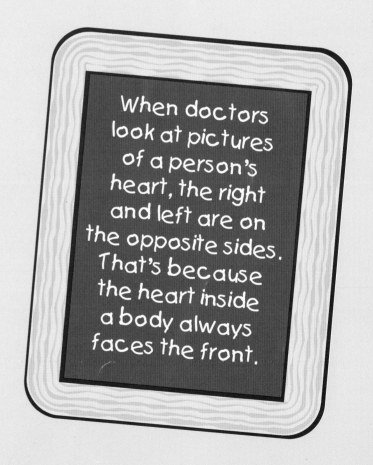

When doctors look at pictures of a person's heart, the right and left are on the opposite sides. That's because the heart inside a body always faces the front.

Your heart pumps blood to your lungs.
There, your blood mixes with oxygen
from the air you breathe and goes back
to the heart.

An adult's heart beats, or pumps, about 70 times each minute. A child's heart beats a little faster.

Your heart pumps this fresh blood to other parts of your body, such as your arms, legs, and brain. These other parts need the oxygen and energy carried in the blood.

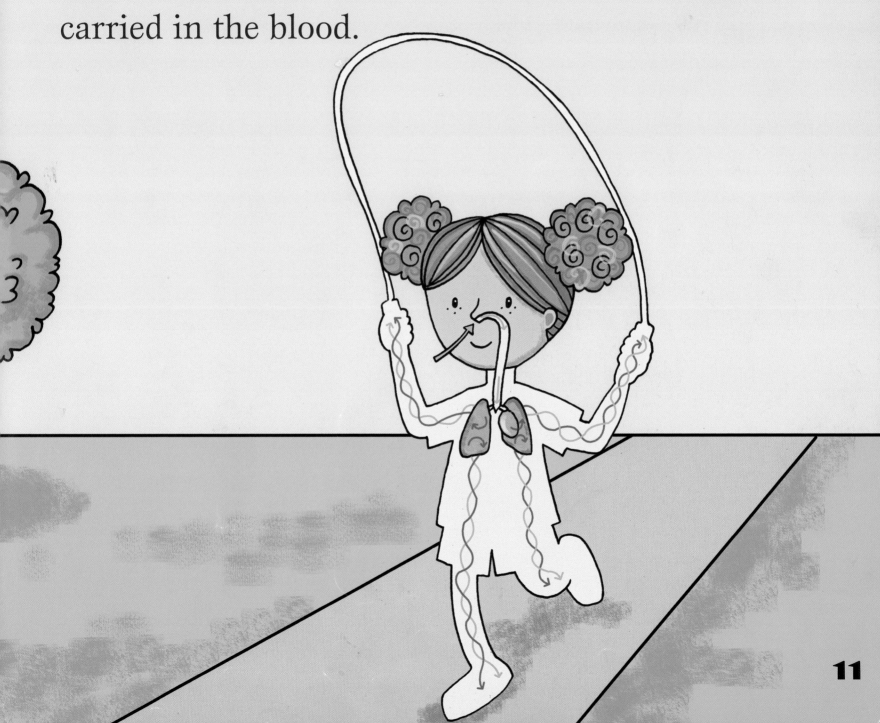

Your blood picks up waste as it moves through the body. It also gets low on oxygen. All the blood in your body makes its way back to the heart. From there, your heart pumps the used blood back to the lungs so the cycle can start again.

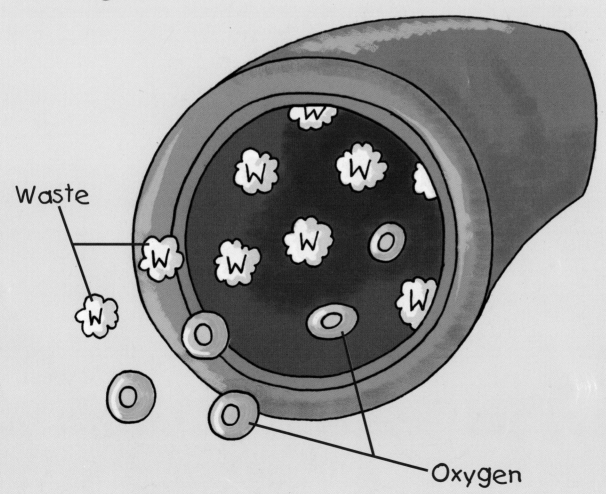

Waste

Oxygen

If you lined them up, all the blood vessels in your body would wrap around the earth more than two times.

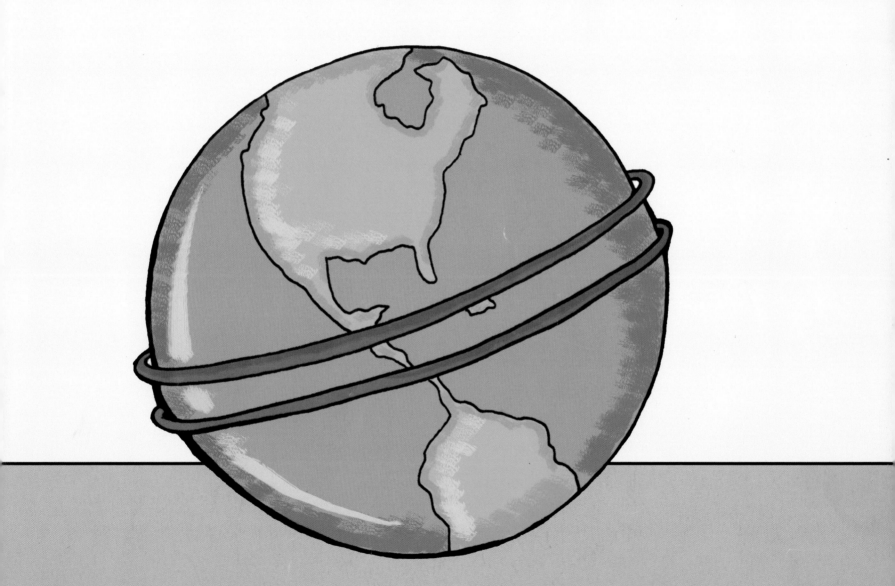

Your heart beats fast when you are afraid.
It also beats fast when you run or jump.
It speeds up because your blood needs
more oxygen at those times.

When you rest, your heart beats slowly.

Hearts can get sick.
When people get older,
their hearts might get
tired. Their hearts might
not work very well.

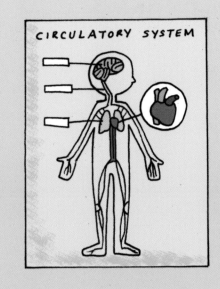

When a heart stops working, it's called a heart attack. Sometimes doctors can get the heart beating again. Sometimes they can't.

Cardiologists are doctors who help hearts.
A doctor can put a little machine called
a pacemaker next to a person's heart.

A pacemaker helps the heart to beat. It helps a person with a sick heart feel better.

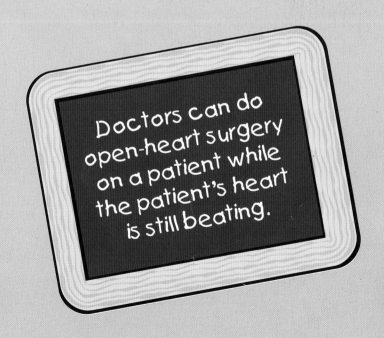

Doctors can do open-heart surgery on a patient while the patient's heart is still beating.

Your heart loves healthy food. Your heart loves exercise. Your heart loves it when you do not smoke.

CHECK YOUR PULSE!

Your pulse is the feeling of blood being pumped through your blood vessels. You can find your pulse. Press your pointer and your middle fingers on the inside of your wrist, below your thumb. Can you feel a small beat? That's your pulse!

Now, watch a clock with a second hand, and count how many beats your heart makes in a minute. It should be around 70.

TOOLS OF THE TRADE

Your doctor uses a stethoscope to listen to your heart. A stethoscope is like a headphone set with a little microphone attached to it. Your doctor sets the little microphone on your chest over your heart. Then he or she listens with the earpieces. Your doctor hears: "Lub-dub! Lub-dub! Lub-dub!" That's what your heart sounds like! It is the sound of your valves closing.

GLOSSARY

atrium (AY-tree-uhm)—either of the two parts of your heart that fill with blood

blood vessels (BLUHD VESS-uhlz)—the narrow tubes that blood flows through

chamber (CHAYM-bur)—one of the four parts of your heart

circulatory system (SUR-kyuh-luh-tor-ee SISS-tuhm)—the system that moves blood throughout your body

lungs (LUHNGZ)—the organs in your chest that help you breathe

muscle (MUHSS-uhl)—a body part that helps your body move

oxygen (OK-suh-juhn)—a colorless gas found in the air

pulse (PUHLSS)—the steady beat of blood moving through your body

pump (PUHMP)—to empty or fill using a pumping motion

valve (VALV)—a moveable part in your blood vessels and in your heart. Valves are like doors that open and close to push blood through in one direction.

ventricle (VEN-truh-kuhl)—either of the two parts of your heart that squirt blood out

TO LEARN MORE

At the Library

Furgang, Kathy. *My Heart.* New York: PowerKids Press, 2001.

Simon, Seymour. *The Heart: Our Circulatory System.* New York: Morrow Junior Books, 1996.

Viegas, Jennifer. *The Heart: Learning How Our Blood Circulates.* New York: Rosen Pub. Group, 2002.

On the Web

Fact Hound offers a safe, fun way to find Web sites related to this book. All of the sites on Fact Hound have been researched by our staff.
http://www.facthound.com

1. Visit the Fact Hound home page.
2. Enter a search word related to this book, or type in this special code: 140480255X.
3. Click the FETCH IT button.

Your trusty Fact Hound will fetch the best sites for you!

INDEX

atrium, 8–9, 22
blood, 5, 6, 7, 9, 10–11, 12, 14, 22, 23
blood vessels, 6–7, 13, 23
cardiologists, 18
chambers, 8–9, 22
chest, 4

circulatory system, 6–7
heart attack, 17
lungs, 10, 12
muscle, 5, 22
oxygen, 10–11, 12, 14
pacemaker, 18–19
pulse, 23

pumps, 5, 10–11, 12
septum, 22
stethoscope, 23
valves, 7, 22, 23
ventricle, 8–9, 22